Lecture Notes in Computer Science 570

Edited by G. Goos and J. Hartmanis

Advisory Board: W. Brauer   D. Gries   J. Stoer

G. Schmidt   R. Berghammer (Eds.)

Graph-Theoretic Concepts in Computer Science

17th International Workshop WG '91
Fischbachau, Germany, June 17-19, 1991
Proceedings

Springer-Verlag
Berlin Heidelberg New York
London Paris Tokyo
Hong Kong Barcelona
Budapest

G. Schmidt   R. Berghammer (Eds.)

Graph-Theoretic Concepts in Computer Science

17th International Workshop, WG '91
Fischbachau, Germany, June 17-19, 1991
Proceedings

Springer-Verlag
Berlin Heidelberg New York
London Paris Tokyo
Hong Kong Barcelona
Budapest

Series Editors

Gerhard Goos
Universität Karlsruhe
Postfach 69 80
Vincenz-Priessnitz-Straße 1
W-7500 Karlsruhe, FRG

Juris Hartmanis
Department of Computer Science
Cornell University
5148 Upson Hall
Ithaca, NY 14853, USA

Volume Editors

Gunther Schmidt
Rudolf Berghammer
Department of Computer Science, Munich Bundeswehr University
Werner-Heisenberg-Weg 39, W-8014 Neubiberg, Germany

CR Subject Classification (1991): B.6.1, B.7.1, E.1, F.1.2-3, F.2, F.4, G.2.2, H.2.1, H.2.4, I.3.5

ISBN 3-540-55121-2 Springer-Verlag Berlin Heidelberg New York
ISBN 0-387-55121-2 Springer-Verlag New York Berlin Heidelberg

Typesetting: Camera ready by author
Printing and binding: Druckhaus Beltz, Hemsbach/Bergstr.
45/3140-543210 - Printed on acid-free paper

Preface

The 17th International Workshop on Graph-Theoretic Concepts in Computer Science (WG 91) was held June 17 – 19, 1991 at The Judges' Home (Richterheim) Fischbachau in Southern Bavaria. It was organized by the Informatics Faculty of the Munich Bundeswehr University.

The WG workshops by now have a successful tradition in Central Europe as is shown by the list of their places and organizers:

Berlin	1975	U. Pape
Göttingen	1976	H. Noltemeier
Linz	1977	J. Mühlbacher
Castle Feuerstein near Erlangen	1978	M. Nagl, H.-J. Schneider
Berlin	1979	U. Pape
Bad Honnef	1980	H. Noltemeier
Linz	1981	J. Mühlbacher
Neunkirchen near Erlangen	1982	H.-J. Schneider, H. Göttler
Haus Ohrbeck near Osnabrück	1983	M. Nagl, J. Perl
Berlin	1984	U. Pape
Castle Schwanberg near Würzburg	1985	H. Noltemeier
Monastery Bernried near München	1986	G. Tinhofer, G. Schmidt
Castle Banz near Bamberg	1987	H. Göttler and H.-J. Schneider
Amsterdam	1988	J. van Leeuwen
Castle Rolduc near Aachen	1989	M. Nagl
Johannesstift Berlin	1990	R. H. Möhring

The program committee of WG 91 consisted of

Giorgio Ausiello	University of Rome "La Sapienza", Italy
Bruno Courcelle	University of Bordeaux, France
Jan van Leeuwen	University of Utrecht, The Netherlands
Ernst Mayr	University of Frankfurt/Main, Germany
Rolf Möhring	Technical University of Berlin, Germany
Manfred Nagl	Technical University of Aachen, Germany
Hartmut Noltemeier	University of Würzburg, Germany
Gunther Schmidt	Munich Bundeswehr University, Germany
Hans-Jürgen Schneider	University of Erlangen-Nürnberg, Germany
Gottfried Tinhofer	Technical University of Munich, Germany

In October 1990 a Call for Papers was sent out where contributions were solicited describing original results in the study and applications of graph-theoretic concepts in various fields of computer science. The 55 submissions that reached us (some nearly two weeks late as a consequence of the gulf crisis and postal strikes) have been carefully refereed with at least three and in nearly all cases four reports.

Finally, 26 papers were selected for personal presentation. The workshop was attended by 64 participants coming from universities and companies of Austria, Belgium, Canada, France, Germany, Greece, Italy, Japan, The Netherlands, Poland, Czekoslovakia, Romania, Switzerland, the United Kingdom, and the United States of America.

The present volume contains the papers that were presented. They all underwent careful revision after the meeting based on the comments and suggestions of the audience and the reports from the referees.

Our special thanks go to the "Staatsrat Hermann Schmitt Foundation", owners of the Richterheim, for their willingness to accept a computer science meeting among their course of judiciary ones. The Richterheim proved to be a perfect choice with regard to lecturing and lodging facilities. We are really grateful to the persons listed below for refereeing and selecting the papers for presentation. Last but not least, the editors would like to express their sincere gratitude to all those who helped organize the workshop: Ludwig Bayer, Hedwig Berghofer, Ingrid Festini, Wolfram Kahl, Peter Kempf, Lothar Schmitz, and Joachim Schreiber.

Munich, October 1991   Gunther Schmidt, Rudolf Berghammer

Referees and Subreferees

Auer E.
Aurenhammer F.
Ausiello G.
Babel L.
Barke E.
Bayer L.
Berghammer R.
Bermond J.-C.
Bertolazzi P.
Bier P.
Blömer J.
Bond J.,
Courcelle B.
d'Amore F.
Dahlhaus E.
di Battista G.
Ebert J.
Felsner S.
Feuerstein E.
Formann M.
Franciosa P.
Gambosi A.
Gomm D.
Gritzner T.

Habib M.
Kahl W.
Kempf P.
Kuchen H.
Lautemann C.
Lengauer T.
Lenzerini M.
Litovsky I.
Lou V.
Marchetti-Spaccamela A.
Mayr E.
Möhring R. H.
Nagl M.
Noltemeier H.
Perl J.
Plewan H.-J.
Proskurowski A.
Raspaud A.
Ripphausen-Lipa H.
Rossmanith P.
Rote G.

Rzehak H.
Schiele W.
Schiermeyer I.
Schmalhofer F.
Schmidt G.
Schmitz L.
Schneider H.-J.
Schreiber J.
Schürr A.
Simon K.
Steffen B.
Thomas W.
Tinhofer G.
van Leeuwen J.
Vogler W.
Wagner D.
Wagner F.
Weihe K.
Westfechtel B.
Zielonca W.
Zierer H.
Zündorf A.

Contents

Complexity Problems and Others

Path-Oriented Algorithms

Applications to VLSI

Disjoint Cycle Problems

Approximating Treewidth, Pathwidth, and Minimum Elimination Tree Height

Hans L. Bodlaender[*]   John R. Gilbert[†]   Hjálmtýr Hafsteinsson[‡]
Ton Kloks[§]

Abstract

We show how the value of various parameters of graphs connected to sparse matrix factorization and other applications can be approximated using an algorithm of Leighton et al. that finds vertex separators of graphs. The approximate values of the parameters, which include minimum front size, treewidth, pathwidth, and minimum elimination tree height, are no more than $O(\log n)$ (minimum front size and treewidth) and $O(\log^2 n)$ (pathwidth and minimum elimination tree height) times the optimal values. In addition we examine the existence of bounded approximation algorithms for the parameters, and show that unless $P = NP$, there are no absolute approximation algorithms for them.

1   Introduction

Many problems in science and engineering require the solving of linear systems of equations. As the problems get larger it becomes increasingly important to exploit the sparsity inherent in many such linear systems. Often each equation only involves a few of the variables. By taking advantage of that fact we are able to solve substantially larger linear systems.

Solving the symmetric positive definite linear system $Ax = b$ via Cholesky factorization involves computing the Cholesky factor L, such that $A = LL^T$, and then solving the triangular systems $Ly = b$ and $L^T x = y$. If A is sparse we usually do some preprocessing on its associated graph, $G(A)$. Various parameters of this graph dictate how fast and efficiently we can solve the system. Among these parameters are treewidth, minimum front size, minimum maximum clique, and minimum elimination tree height. Having small front size is important in the multifrontal method [DR83, Liu90] and an ordering minimizing the elimination tree height minimizes the parallel time required to factor A. All the above parameters depend on the ordering on the rows and columns of A. Unfortunately determining the orderings that give the optimal values of these parameters is NP-complete [ACP87, GJ79, Pot88]. Therefore we have to be content with approximations.

The notion of treewidth has several other applications (see e.g. [Arn85]). It is closely related to the pathwidth, which has among others important applications in the theory of VLSI layout. The pathwidth is equivalent to several other parameters of graphs, including the minimum chromatic number of an interval graph containing the graph as a subgraph and the node search number of a

---

[*]Department of Computer Science, Utrecht University, P.O. Box 80.089, 3508 TB Utrecht, The Netherlands. The research of this author is partially supported by the ESPRIT II Basic Research Actions of the EC under Contract No. 3075 (project ALCOM).

[†]Xerox Palo Alto Research Center, 3333 Coyote Hill Road, Palo Alto, California 94304 USA, and University of Bergen, Norway.

[‡]Department of Computer Science, University of Iceland, 101 Reykjavík, Iceland.

[§]Department of Computer Science, Utrecht University. The research of this author is supported by the Foundation for Computer Science (S.I.O.N.) of the Netherlands Organization for Scientific Research (N.W.O.) and by the ESPRIT II Basic Research Actions of the EC under Contract No. 3075 (project ALCOM).

graph. The pathwidth problem is also equivalent to the gate matrix layout problem. See [Möh90] for an overview.

In this paper we will show how to use a recent result of Leighton et al. (see lemma 4.1 in [KARR90], and also [LR88]) on approximating graph separators to find approximations to the above parameters. These approximations will be no more than $O(\log n)$ or $O(\log^2 n)$ times the optimal values. Some of these results were obtained independently by Klein et al. [KARR90].

We will start with a few definitions. After that we explore the relationship between treewidth, pathwidth, minimum front size, minimum elimination tree height, and other related concepts. Then we present the result of Leighton et al. and our approximation algorithms. Finally we discuss bounded approximations for minimum elimination tree height, treewidth, and pathwidth.

2 Definitions

We assume that the reader is familiar with standard graph theoretic notation (see [Har69]). The subgraph of $G = (V, E)$ *induced* by $W \subseteq V$ is denoted by $G[W]$.

The class of *k-trees* is defined recursively as follows. The complete graph on k vertices is a k-tree. A k-tree with $n + 1$ vertices ($n \geq k$) can be constructed from a k-tree with n vertices by adding a vertex adjacent to all vertices of one of its k-vertex complete subgraphs, and only to these vertices. A *partial k-tree* is a graph that contains all the vertices and a subset of the edges of a k-tree.

A *tree-decomposition* of a graph $G = (V, E)$ is a pair $(\{X_i \mid i \in I\}, T = (I, F))$ with $\{X_i \mid i \in I\}$ a collection of subsets of V, and T a tree, such that

- $\bigcup_{i \in I} X_i = V$.

- for all $(v, w) \in E$, there exists an $i \in I$ with $v, w \in X_i$.

- For all $i, j, k \in I$: if j is on the path from i to k in T, then $X_i \cap X_k \subseteq X_j$.

The third condition can be replaced by the equivalent condition: for all $v \in V$, $\{i \in I | v \in X_i\}$ forms a connected subtree of T. The *treewidth* of a tree-decomposition $(\{X_i \mid i \in I\}, T = (I, F))$ is $\max_{i \in I} |X_i| - 1$. The treewidth of a graph is the minimum treewidth over all possible tree-decompositions of that graph. It can be shown that G has treewidth at most k, if and only if G is a partial k-tree (see e.g. [vL90]).

The problem of finding the treewidth of a given graph G is NP-complete [ACP87]. However, many NP-complete graph problems can be solved in polynomial and even linear time if restricted to graphs with constant treewidth (see e.g. [ALS91, Bod90].) For constant k, determining whether the treewidth of G is at most k, and finding a corresponding tree-decomposition can be done in $O(n \log^2 n)$ time (see [BK91]). The first step of such an algorithm is to find a tree-decomposition of G which has not optimal, but still constant bounded treewidth [RS86b, Lag90].

A *path-decomposition* of a graph $G = (V, E)$ is a tree-decomposition $(\{X_i \mid i \in I\}, T = (I, F))$, such that T is a path. The *pathwidth* of a path-decomposition $(\{X_i \mid i \in I\}, T = (I, F))$ is $\max_{i \in I} |X_i| - 1$. The pathwidth of a graph is the minimum pathwidth over all possible path-decompositions of that graph. The notion of pathwidth has several important applications, e.g., in VLSI-layout theory (see [Möh90]).

The *elimination game* on a graph G repeats the following step until there are no more vertices. Pick a vertex v, delete it from the graph, and add edges between the neighbours of v that are not already adjacent. These added edges are called *fill edges*. The *filled graph* G_π^* is obtained by adding to G all the fill edges that occur when the elimination game is played using the order π on the vertices of G.

Let $C_\pi(v)$ be the set of uneliminated neighbours of vertex v when playing the elimination game with order π on the graph G. The treewidth of G can alternately be defined as the minimum over all orderings π of $\max_{v \in V} |C_\pi(v)|$. (See e.g. [Arn85].)

The *elimination tree* T is defined as follows. Vertex j is the parent of vertex i in T (with $j > i$) iff j is the lowest numbered among the higher numbered neighbours of i in the filled graph G^*.

The elimination tree T of a graph $G(A)$ describes the dependencies between the columns of the matrix A during column-oriented Cholesky factorization. If vertex i is the child of vertex j in T, then column i has to be computed before column j in A's Cholesky factorization. Consequently, we can compute all the columns on the same level in the tree simultaneously. Thus, the height of the elimination tree is a reasonable measure of the parallel time required to factor A. A completely dense matrix has a tree that is just one long chain, since each column depends on all the previous ones. In the case of sparse matrices the shape of the elimination tree can vary. By reordering the rows and columns of A (equivalent to renumbering the vertices of $G(A)$) we can, to some degree, restructure the elimination tree. Thus the first step in parallel solution of sparse linear systems is to reorder the rows and columns of A in order to reduce the height of the elimination tree.

An α *vertex separator (α edge separator)* of a graph $G = (V, E)$ is a set of vertices $S \subseteq V$ (set of edges $S \subseteq E$) such that every connected component of the graph $G[V - S]$ (the graph $(V, E - S)$) obtained by removing S from G has at most $\alpha \cdot |V|$ vertices.

For $W \subseteq V$, an α *vertex separator of W (α edge separator of W)* in $G = (V, E)$ is a set of vertices $S \subseteq V$ (set of edges $S \subseteq E$) such that every connected component of the graph $G[V - S]$ (the graph $(V, E - S)$) has at most $\alpha \cdot |W|$ vertices of W.

3 Relationships

Below we will show the relationships among the various parameters using a series of lemmas. Many of these results are not new, but we present them all here in order to demonstrate how closely linked these parameters are. In addition this will make it easier to see how the result of Leighton et al. can be used to find approximations to the different parameters. The following lemma appears also in [RTL76].

Lemma 3.1 *A graph G has treewidth k iff the smallest maximum clique over all filled graphs G^*_π of G is $k + 1$.*

Proof: In playing the elimination game the set $C_\pi(v) \cup \{v\}$ of v and its uneliminated neighbours becomes a clique in G^*_π. Thus if G has treewidth k (i.e. $\min_\pi \max_{v \in V} |C_\pi(v)| = k$) then the smallest maximum clique over all filled graphs G^*_π of G is at least $k + 1$. If we have a minimum maximum clique of size c in G^*_π then when its first vertex v is eliminated in the elimination game the rest of the vertices will become $C_\pi(v)$, and its size is $c - 1$. Thus the size of the minimum maximum clique is no more than $k + 1$ if the treewidth is k. $\square$

The multifrontal method [DR83, Liu90] organizes the factorization of a sparse matrix into a sequence of partial factorizations of small dense matrices, the goal being to make better use of hierarchical storage, vector floating-point hardware, or sometimes parallelism. Figure 1 shows one elimination step of the method: Here $\mathbf{v}$ is only the nonzeros below the diagonal in the column being eliminated, and the *frontal matrix F* contains only the rows and columns corresponding to nonzeros in the column being eliminated. The *update matrix $B - \mathbf{v}\mathbf{v}^T/d$* is dense, and is saved for use in later elimination steps. Many such matrices may be saved at the same time, but only enough main memory for one frontal matrix is needed. The *front size* of A is the dimension of the largest update matrix, or one less than the dimension of the largest frontal matrix.

The front size of A can also be characterized as the largest number of nonzeros below the diagonal in any column of its Cholesky factor, or the maximum number of neighbors of any vertex when it is eliminated in the elimination game.

Lemma 3.2 *The minimum front size of a graph $G(A)$ with treewidth k over all orderings is k.*

$$F = \begin{bmatrix} d & \mathbf{v}^T \\ \mathbf{v} & B \end{bmatrix} = \begin{bmatrix} \sqrt{d} & \mathbf{0} \\ \mathbf{v}/\sqrt{d} & I \end{bmatrix} \begin{bmatrix} 1 & \mathbf{0} \\ \mathbf{0} & B - \mathbf{v}\mathbf{v}^T/d \end{bmatrix} \begin{bmatrix} \sqrt{d} & \mathbf{v}^T/\sqrt{d} \\ \mathbf{0} & I \end{bmatrix}$$

Figure 1: A step in the multifrontal method

Proof: If $G(A)$ has treewidth k, then there is some ordering π, such that the largest set $C_\pi(v)$ of higher numbered neighbours of any vertex v has size k. If we order the columns and rows of A according to π then at each step the nonzero elements of the vector $\mathbf{v}$ in Figure 1 correspond to the uneliminated neighbours of v in the elimination game. The frontal matrix has as many columns and rows as there are non-zeros in $\mathbf{v}$. Thus the largest frontal matrix has the same size as the largest $C_\pi(v)$, and the front size of $G(A)$ is the same as the treewidth. $\square$

The following lemma, which can be traced back to C. Jordan (see [Kön36]) will be useful when proving things about the elimination tree.

Lemma 3.3 *Given a tree T we can, in time $O(n)$, find a vertex v such that $T \setminus \{v\}$ has no component of size greater than $\frac{n}{2}$.*

In other words, the algorithm finds an $\frac{1}{2}$ vertex separator of T of size 1. This result can easily be generalized as follows:

Lemma 3.4 *Given a tree T and a subset of the vertices W, there is a vertex v in T such that every component of $T \setminus \{v\}$ contains at most $\frac{1}{2}|W|$ vertices of W.*

Proof: Let $\delta = \frac{1}{2}|W|$. Consider the following algorithm. Start at any vertex c. If every component of $T \setminus \{c\}$ contains at most δ vertices of W we are done. Otherwise, let c' be the neighbour of c, in the component of $T \setminus \{c\}$ that contains more than δ vertices of W. Replace c by c' and repeat the process. Notice that the component of $T \setminus \{c'\}$ containing c has less than δ vertices of W. So the algorithm terminates. $\square$

The following theorem, (see also e.g. [RS86a]), given here with a short proof, plays an important role in our approximation algorithm.

Theorem 3.5 *Let $G = (V, E)$ be a graph with treewidth $\leq k$. Let $W \subseteq V$. Then there exists a $\frac{1}{2}$ vertex separator of W in G of size at most $k + 1$.*

Proof: Consider an elimination tree T_π such that G_π^* has treewidth k. Pick a vertex v as indicated in the previous lemma. Let S be the set of ancestors of v adjacent in G_π^* to a vertex in the subtree of T_π rooted at v (i.e. $S = C_\pi(v)$). This set is a separator of size less than or equal to k. This follows from the fact that $S \cup \{v\}$ forms a clique in G_π^*, and since the largest clique in G_π^* has size $k + 1$, S cannot contain more than k vertices. So we can take $S \cup \{v\}$ as the required separator. $\square$

The next pair of lemmas illustrates the relationship between elimination trees and vertex separators of graphs.

Lemma 3.6 *If G and its subgraphs have α vertex separators of size s, then there is an elimination tree of height $O(s \log n)$.*

Proof: If we apply the nested dissection ordering (see [Geo73]), i.e., order the vertices of the first separator last, then the vertices of the next level of separators, and so on, then the height of the resulting elimination tree is at most $s \log_{1/\alpha} n$. Since α is a constant, the height is $O(s \log n)$. $\square$

Lemma 3.7 *If G has an elimination tree of height h then it and its subgraphs have $\frac{1}{2}$ vertex separators of size h.*

Proof: If we have an elimination tree T_π of height h, then we can use Lemma 3.3 to find a vertex v such that no component of $T_\pi \setminus \{v\}$ contains more than $\frac{n}{2}$ vertices. The set of ancestors of v adjacent in G_π^* to a vertex of the subtree rooted at v including v itself, is a separator. Its size obviously cannot be more than the height of the elimination tree. This argument can be applied recursively to the remaining components, since the heights of their elimination trees cannot be more than h. □

Next we compare the minimum elimination tree height to the minimum size of the maximum clique in G_π^*.

Lemma 3.8 *If the minimum maximum clique of G_π^* has size k then the minimum height elimination tree of G is lower than $k \log n$.*

Proof: Assume that the minimum maximum clique of G_π^* has size k and it occurs when we use ordering π on the vertices. Consider the elimination tree T_π, which is the elimination tree of G using ordering π. We can pick a vertex v in T_π, such that no component of $T_\pi \setminus \{v\}$ contains more than $\frac{n}{2}$ vertices, using Lemma 3.3. let S be the set of ancestors of v adjacent in G_π^* to a vertex in subtree of T_π rooted at v. Then $S \cup \{v\}$ is a separator of size less than or equal to k, since this set forms a clique in G_π^* and the largest clique in G_π^* has size k. We continue recursively finding separators of size at most k in the components, none of which is larger than $\frac{n}{2}$. We can then use nested dissection to order the vertices, ordering the vertices of the first separator last, and so on. This will give us an elimination tree of height at most $k \log n$. (The same construction was used in [Gil88].) □

Now we give a relationship between pathwidth and the other parameters. As path-decompositions are a special case of tree-decompositions, the treewidth of a graph is never larger than the pathwidth. We also have the following, interesting relationship.

Lemma 3.9 *If G has an elimination tree with height k, then the pathwidth of G is at most k.*

Proof: Number the leaves of the elimination tree $v_1, \ldots, v_r$, from left to right. Let X_j, $(1 \le j \le r)$ consist of v_j and all ancestors of v_j in the elimination tree. Now $(\{X_i \mid 1 \le i \le r\}, T = (\{1, 2, \cdots, r\}, \{(i, i+1) \mid 1 \le i < r\})$ is a path-decomposition of the filled graph G_π^* and hence of G with pathwidth k. □

As a direct consequence we have that the treewidth of G is no larger than the height of an elimination tree and the minimum maximum clique of G^* is no larger than the height plus one.

Let us now summarize these relationships. Minimum front size is equal to the treewidth and minimum largest clique of G_π^* is one more than the treewidth of G. The minimum elimination tree height is no less than those three parameters, and at most $\log n$ times them. The pathwidth of a graph is "between" the treewidth of a graph, and its minimum elimination tree height. We can summarize these relationships as follows:

- treewidth = min front size = min max clique -1

- treewidth $\le$ pathwidth $\le$ min height $\le$ treewidth $\cdot \log n$.

4 Approximations of vertex separators

In [LR88] Leighton and Rao present approximation algorithms for various separator problems, including the problem of finding minimum size balanced edge separators. Recently, Leighton et al. [Le90] obtained similar results for vertex separators, as reported in [KARR90]. We will use the following result.

Theorem 4.1 (Le90) *There exists a polynomial algorithm that, given a graph $G = (V, E)$ and a set $W \subseteq V$, finds a $\frac{2}{3}$ vertex separator $S \subseteq V$ of W in G of size $O(w \cdot \log n)$, where w is the minimum size of a $\frac{1}{2}$ vertex separator of W in G.*

When we now apply theorem 3.5, we get the following result, which is the fundamental step in our approximation algorithm.

Theorem 4.2 *There exist a constant $\beta \geq 1$ and a polynomial time algorithm that, given a graph $G = (V, E)$ and a set $W \subseteq V$, finds a $\frac{2}{3}$ vertex separator of W in G of size $\beta \cdot \log n \cdot k$, where $n = |V|$, and k is the treewidth of G.*

Proof: Theorem 3.5 tells us that there exists a $\frac{1}{2}$ vertex separator of W in G of size $k + 1$. The result hence follows by using the algorithm of theorem 4.1 and taking β to be constant hidden in the O of theorem 4.1 times a small factor to account for the factor $\frac{k+1}{k}$. $\square$

In the remainder of the paper, β is assumed to be the constant implied in this theorem.

5 Approximation algorithms

In this section we give a polynomial time approximation algorithm for the treewidth problem that is at most a factor of $O(\log n)$ off optimal. Clearly, from the analysis in Section 3 this directly implies polynomial time approximations for minimum maximum cliques and minimum front size that are a factor of $O(\log n)$ off optimal, and for minimum elimination trees that is a factor of $O(\log^2 n)$ off optimal. Readers familiar with the approximation algorithms for constant treewidth of Lagergren [Lag90] and of Robertson and Seymour [RS86b] may note some similarities. Our algorithm also has some similarities to Lipton, Rose, and Tarjan's version of nested dissection [LRT79].

Our approximation algorithm consists of calling $makedec(V, \emptyset)$, where $makedec$ is the following recursive procedure:

proc $makedec(Z, W)$;
    (*comment: Z and W are disjoint sets of vertices.*)
    if $3 \cdot |Z| \leq |W|$ **then**
        return a tree-decomposition with one single node, containing $Z \cup W$.
    else *perform the following steps:*
        Find a 2/3 vertex separator S of W in $G[Z \cup W]$ with the algorithm of theorem 4.2.
        Find a 2/3 vertex separator S' of $Z \cup W$ in $G[Z \cup W]$, with the algorithm of
        theorem 4.2.
        Compute the connected components of $G[Z \cup W - (S \cup S')]$.
        Suppose these have vertices $Z_i \cup W_i$ with $Z_i \subseteq Z$ and $W_i \subseteq W$, for all $1 \leq i \leq t$.
        For $i := 1$ **to** t **do**
            call $makedec(Z_i, W_i \cup S \cup S')$.
        end for
        Now return the following tree-decomposition:
        take a root-node $r_{Z,W}$, containing $W \cup S \cup S'$ (i.e. $X_{r_{Z,W}} = W \cup S \cup S'$).
        Then add all tree-decompositions returned by the calls of $makedec(Z_i, W_i \cup S \cup S')$
        with an edge from the root of each to $r_{Z,W}$.
    end if
end proc

Claim 5.1 Makedec(Z, W) *returns a tree-decomposition of $G[Z \cup W]$ such that the root-node of the tree-decomposition contains all vertices in W. If $|W| \leq 6\beta k \log n$, where k is the treewidth of G and $n = |V|$, then the treewidth of this tree-decomposition is at most $8\beta k \log n$.*

Proof: We prove this by induction on the recursive structure of the *makedec* procedure. Clearly the claim is true in case $3 \cdot |Z| \leq |W|$. We first show that for all edges $(v, w) \in E$, $v, w \in G[Z \cup W]$, there exists a node i in the returned tree-decomposition with $v, w \in X_i$. Suppose this is not the case. First consider an edge $(v, w) \in E$ with $v, w \in G[Z \cup W]$. If v and w both are in the set W, then $v, w \in X_{r_{Z,W}}$. Otherwise, v and w both belong to a set $Z_i \cup W_i \cup S \cup S'$. By induction, there is a set X_j in the tree-decomposition returned by $makedec(Z_i, W_i \cup S \cup S')$ with $v, w \in X_j$.

We now show that $\{i \in I \mid v \in X_i\}$ forms a connected subtree in the decomposition-tree for all $v \in Z \cup W$. If $v \notin X_{r_{Z,W}}$, then this holds by induction, as v then belongs to exactly one set Z_i. Otherwise, for each of the subtrees under $r_{Z,W}$, either v does not appear in any of the nodes in this subtree, or the nodes containing v form, by induction, a connected subtree of this subtree, and include the root of this subtree, i.e., the child of $r_{Z,W}$ that is in this subtree. The result now follows. Therefore the procedure indeed outputs a tree-decomposition of $G[Z \cup W]$.

We now have to show that the treewidth of the tree-decomposition is at most $8\beta k \log n$. By induction, it is sufficient to show that $|X_{r_{Z,W}}| \leq 8\beta k \log n$, and that $|W_i \cup S \cup S'| \leq 6\beta k \log n$. Clearly the first holds (use the assumption on the size of W, and use theorem 4.2 to bound the size of S and S'.) The second holds, as $S \cup S'$ is an $\frac{2}{3}$ separator of W in $G[Z \cup V]$, and hence each W_i is of size at most $2/3|W| \leq 4\beta k \log n$, whence $|W_i \cup S \cup S'| \leq 6\beta k \log n$. $\square$

Thus, we have obtained the following result:

Theorem 5.2 *There exists a polynomial time algorithm that, given a graph $G = (V, E)$ with $|V| = n$, finds a tree-decomposition of G with treewidth at most $O(k \log n)$, where k is the treewidth of G.*

This result implies approximation algorithms for the other parameters discussed in this paper. Clearly, by lemmas 3.1 and 3.2 we have also a polynomial time algorithm that, given a graph G, solves the minimum maximum clique problem and the minimum front size problem within $O(\log n)$ times optimal. We also have:

Theorem 5.3 *There exists a polynomial time algorithm that, given a graph $G = (V, E)$ with $|V| = n$, finds an elimination tree of G with height at most $O(h \log^2 n)$, where h is the minimum height of an elimination tree of G.*

Proof: Find the tree-decomposition of G with the algorithm of theorem 5.2. Then use lemma 3.7. We obtain an elimination tree of G with height at most $\log n \cdot O(\log n) \cdot k$, where k is the treewidth of G. Observe that k is smaller than or equal to the pathwidth of G, and hence, by lemma 3.8, k is at most h. $\square$

Similarly we can obtain:

Theorem 5.4 *There exists a polynomial time algorithm that, given a graph $G = (V, E)$ with $|V| = n$, finds a path-decomposition of G with pathwidth at most $O(k \log^2 n)$, where k is the pathwidth of G.*

6 Absolute approximations

In this section we consider absolute approximations, i.e. polynomial time algorithms that give solutions within an additive constant of the optimal solution. We show that when $P \neq NP$, then no such algorithms exist for the minimum height elimination tree problem, for treewidth (and hence for minimum front size and minmax clique), or for pathwidth.

Given an approximation algorithm $\mathcal{A}$ for a minimization problem we can distinguish between three kinds of performance guarantees on it. In *absolute approximations* the approximate solution

$\mathcal{A}(I)$ is within a constant off the optimal solution $\mathcal{OPT}(I)$, i.e. $\mathcal{A}(I) - \mathcal{OPT}(I) \leq K$. Second, the approximate solution can be a constant factor off the optimal one, i.e., $\mathcal{A}(I) \leq C\,\mathcal{OPT}(I)$, with $C \geq 1$. Finally the difference between the optimal and approximate solutions can depend on the size of the problem, i.e. $\mathcal{A}(I) \leq f(n)\,\mathcal{OPT}(I)$. The algorithms we have presented above all have performance guarantees of the last kind with $f(n) = O(\log n)$ or $f(n) = O(\log^2 n)$. The hardest of these bounds to achieve is the absolute bound and very few NP-complete problems have absolute approximation algorithms. We will now prove that the minimum height elimination tree problem has no absolute approximation algorithms unless $P = NP$.

Theorem 6.1 *If $P \neq NP$ then no polynomial time approximation algorithm $\mathcal{A}$ for the minimum height elimination tree problem can guarantee $\mathcal{A}(I) - \mathcal{OPT}(I) \leq K$ for a fixed constant K.*

Proof: Assume we have a polynomial time absolute approximation algorithm $\mathcal{A}$, so that $\mathcal{A}$ always gives an elimination tree with height at most K more than the optimal. We will show that then we can solve the mutual independent set problem (MUS) in polynomial time. The *MUS problem* is the following: *Given a bipartite graph $B = (P, Q, E)$, are there sets V_1 ($V_1 \subseteq P$) and V_2 ($V_2 \subseteq Q$), with $|V_1| = |V_2| = k$, such that V_1 and V_2 are mutually independent?* That is: no edge joins a vertex in V_1 to a vertex in V_2. The MUS problem has been shown to be NP-complete [Pot88]. The MUS problem is equivalent to the blockfolding problem (see [Möh90]), which is shown to be NP-complete in [EL84].

Let $B = (P, Q, E)$ be a bipartite graph. Its corresponding *biclique* $C = (P, Q, E \cup P^2 \cup Q^2)$ is the graph that contains enough extra edges to make each of P and Q into cliques. A bipartite graph is a *chain graph* if the adjacency sets of vertices in P form a chain, i.e., the vertices of P can be ordered such that

$$\mathrm{Adj}(v_1) \supseteq \mathrm{Adj}(v_2) \supseteq \cdots \supseteq \mathrm{Adj}(v_p).$$

Yannakakis [Yan81] has shown that if we add edges to the bipartite graph B to make it a chain graph B' then adding the same edges to B's corresponding biclique C makes it a chordal graph C'. The graph C' is called a *chordal completion of the biclique* C. Pothen [Pot88] has proved that B has mutually independent sets of size k iff there exists a chordal completion C' with elimination tree of height $n - k - 1$.

Suppose we had a polynomial time algorithm $\mathcal{A}$ for the minimum elimination tree problem, such that $\mathcal{A}(I) - \mathcal{OPT}(I) \leq K$. We solve MUS by making a new bipartite graph $\hat{B} = (P_1 \cup \cdots \cup P_{K+1}, Q_1 \cup \cdots \cup Q_{K+1}, \hat{E})$ that contains $K+1$ copies of B and additional edges between the copies. If there is an edge between vertices v and w in B ($v \in P$, $w \in Q$), then $\hat{B}$ has an edge between v_i and w_j ($v_i \in P_i$ and $w_j \in Q_j$), for $i, j = 1, \ldots, K+1$. The new graph $\hat{B}$ has $(K+1)n$ vertices and $(K+1)^2 m$ edges. In Figure 2 we show $\hat{B}$ when $K = 1$.

Using the lemma from [Pot88] mentioned above we see that $\hat{B}$ has mutually independent sets of size $(K+1)k$ iff there exists a chordal completion $\hat{C}$ with an elimination tree of height $(K+1)(n-k) - 1$. Assuming that we have an approximation algorithm $\mathcal{A}$ that gives us an elimination tree with height no more than K off the minimum we apply it to $\hat{C}$. If the resulting elimination tree has height between $(K+1)(n-k) - 1$ and $(K+1)(n-k) + K - 1$ then we can extract the $K+1$ elimination trees corresponding to the copies of the bipartite graph B that $\hat{B}$ was made up of. That gives us an elimination tree for C' of height $(n-k) - 1$ (or lower). If the elimination tree computed by algorithm $\mathcal{A}$ is higher than $(K+1)(n-k) + K - 1$ then the elimination tree corresponding to the original bipartite graph B (actually its chordal completion C') cannot have height $(n-k) - 1$. Otherwise we could join those together and obtain an elimination tree for $\hat{C}$ with height $(K+1)(n-k) - K - 1$, which is at least $2K$ lower than the solution given by $\mathcal{A}$. Thus we could use the polynomial time algorithm $\mathcal{A}$ to solve an NP-complete problem. $\square$

Similar results can be proven for the treewidth problem. We need the following lemma.

Lemma 6.2 *Let $(\{X_i \mid i \in I\}, T = (I, F))$ be a tree-decomposition of $G = (V, E)$. Let $W \subseteq V$ be a clique in G. Then there exists an $i \in I$ with $W \subseteq X_i$.*

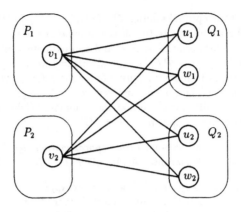

Figure 2: The graph $\hat{B}$ when $K = 1$

See [BM90] for a short proof of this lemma.

Theorem 6.3 *If $P \neq NP$ then no polynomial time approximation algorithm $\mathcal{A}$ for the treewidth problem (and hence for minimum front size, and minimum maximum clique) can guarantee $\mathcal{A}(G) - \mathcal{OPT}(G) \leq K$ for a fixed constant K.*

Proof: Assume we have a polynomial time algorithm $\mathcal{A}$, that given a graph $G = (V, E)$, finds a tree-decomposition of G with treewidth at most K larger than the treewidth of G. Let a graph $G = (V, E)$ be given. Let $G' = (V', E')$ be the graph obtained by replacing every vertex of G by a clique of $K + 1$ vertices, and adding edges between every pair of adjacent vertices in G, i.e. $V' = \{v_i \mid v \in V, 1 \leq i \leq K + 1\}$, $E' = \{(v_i, w_j) \mid (v = w \wedge i \neq j) \vee (v, w) \in E\}$. We examine the relationship between the treewidth of G and the treewidth of G'.

Suppose we have a tree-decomposition of G, $(\{X_i \mid i \in I\}, T = (I, F))$ with treewidth L. One easily checks that $(\{Y_i \mid i \in I\}, T = (I, F))$ with $Y_i = \{v_j \mid v \in X_i, 1 \leq j \leq K + 1\}$ is a tree-decomposition of G' with treewidth $(L + 1)(K + 1) - 1$. It follows that treewidth$(G') \leq$ (treewidth$(G) + 1) \cdot (K + 1) - 1$.

Next suppose we have a tree-decomposition $(\{Y_i \mid i \in I\}, T = (I, F))$ of G' with treewidth M. Let $X_i = \{v \in V \mid \{v_1, v_2, \cdots, v_{K+1}\} \subseteq Y_i\}$. We claim that $(\{X_i \mid i \in I\}, T = (I, F))$ is a tree-decomposition of G with treewidth $(M + 1)/(K + 1) - 1$. Let $(v, w) \in E$. Note that $v_1, v_2, \cdots, v_{K+1}, w_1, w_2, \cdots, w_{K+1}$ form a clique in G'. Hence, by lemma 6.2 there exists an $i \in I$ with $\{v_1, \cdots, v_{K+1}, w_1, \cdots w_{K+1}\} \subseteq Y_i$, and thus $v, w \in X_i$. Let $j \in I$ be on the path in T from $i \in I$ to $k \in I$. If $v \in X_i \cap X_k$, then $\{v_1, \cdots, v_{K+1}\} \subseteq Y_i \cap Y_k$, and hence by definition of tree-decomposition $\{v_1, \cdots, v_{K+1}\} \subseteq Y_j$, so $v \in X_j$. Clearly, $\max_{i \in I} |X_i| \cdot (K + 1) \leq \max_{i \in I} |Y_i|$. This finishes the proof of our claim. It follows that treewidth$(G') \geq$ (treewidth$(G) + 1) \cdot (K + 1) - 1$, and hence that treewidth$(G') = $ (treewidth$(G) + 1) \cdot (K + 1) - 1$.

Now we are able to describe the polynomial time algorithm for the treewidth problem: let G be the input graph. Make G', and apply algorithm $\mathcal{A}$ to G'. Apply the construction described above to make a tree-decomposition of G. This must be a tree-decomposition with minimum treewidth: if the treewidth of G is k, then the treewidth of G' is $(k + 1)(K + 1) - 1$, hence $\mathcal{A}$ outputs a tree-decomposition of G' with treewidth at most $(k + 1)(K + 1) - 1 + K$, hence the algorithm described above outputs a tree-decomposition of G with treewidth at most $\lfloor ((k + 1)(K + 1) + K)/(K + 1) - 1 \rfloor = k$. Thus we would have a polynomial time algorithm for treewidth. $\square$

In the same way we can prove the following theorem. With a different terminology this result was also proved by Deo et al. [DKL87].

Theorem 6.4 *If $P \neq NP$ then no polynomial time approximation algorithm $\mathcal{A}$ for the pathwidth problem can guarantee $\mathcal{A}(G) - \mathcal{OPT}(G) \leq K$ for a fixed constant K.*

7 Conclusion

We have presented a way to find bounded approximations to various parameters of graphs. To be precise: for treewidth, minimum front size and minimum maximum clique we obtain approximations that are never more than $O(\log n)$ times optimal, and for pathwidth and minimum height elimination tree we obtain approximations that are never more than $O(\log^2 n)$ times optimal.

An open problem is to find algorithms that give solutions that are only a constant times optimal for any of the parameters discussed in this paper. In this paper we have shown that there are no absolute approximations to the considered problems, but it is not known if we can find solutions that are only a constant factor off the optimal.

Two related problems are how to permute A so that its Cholesky factor has the minimum fill or the minimum operation count. Klein et al. [KARR90] use a nested dissection algorithm somewhat similar to ours to give approximation algorithms for these measures getting within $O(\log^4 n)$ and $O(\log^6 n)$ times optimal (respectively) for graphs of bounded degree.

Very recently, Seymour and Thomas have obtained a polynomial algorithm for the related notion of branchwidth, when retricted to planar graphs [ST90]. As the branchwidth and treewidth of a graph differ by at most a factor of 1.5 [RS91], this gives a polynomial time approximation algorithm for treewidth of planar graphs with performance ratio 1.5, and approximation algorithms for pathwidth and minimum height elimination tree of planar graphs with performance ratio $\log n$. An interesting question is whether there exists a polynomial time algorithm that solves treewidth exactly on planar graphs.

References

[ACP87] S. Arnborg, D. G. Corneil, A. Proskurowski. Complexity of finding embeddings in a k-tree. *SIAM Journal of Algebraic and Discrete Methods*, 8:277–284, 1987.

[ALS91] S. Arnborg, J. Lagergren, and D. Seese. Easy problems for tree-decomposable graphs, *J. of Algorithms*, 12:308–340,1991.

[Arn85] S. Arnborg. Efficient algorithms for combinatorial problems on graphs with bounded decomposability – A survey. *BIT*, 25:2–23, 1985.

[Bod90] H. L. Bodlaender. Polynomial algorithms for Graph Isomorphism and Chromatic Index on partial k-trees. *J. Algorithms*, 11:631–644, 1990.

[BK91] H. L. Bodlaender and T. Kloks. Better algorithms for the pathwidth and treewidth of graphs. *Proc. 18th ICALP*, pages 544–555. Springer Verlag, Lect. Notes in Comp. Sc. 510, 1991.

[BM90] H. L. Bodlaender and R. H. Möhring. The pathwidth and treewidth of cographs. In *Proc. 2nd Scandinavian Workshop on Algorithm Theory*, pages 301–309. Springer Verlag Lect. Notes in Computer Science vol. 447, 1990.

[DKL87] N. Deo, M. S. Krishnamoorty and M. A. Langston. Exact and approximate solutions for the gate matrix layout problem. *IEEE Transactions on Computer-Aided Design*, 6:79–84, 1987.

[EL84]   J.R. Egan and C.L. Liu. Bipartite folding and partitioning of a PLA, *IEEE Trans. CAD*, 3:191–199, 1984.

[DR83]   I. S. Duff and J. K. Reid. The multifrontal solution of indefinite sparse symmetric linear equations. *ACM Transactions on Mathematical Software*, 9:302–325, 1983.

[GJ79]   M. R. Garey and D. S. Johnson. *Computers and Intractability: A Guide to the Theory of NP-Completeness*, W.H. Freeman and Company, 1979.

[Geo73] J. A. George. Nested dissection of a regular finite element mesh. *SIAM Journal of Numerical Analysis*, 10:345-363, 1973.

[Gil88] J. R. Gilbert. Some nested dissection order is nearly optimal. *Information Processing Letters*, 6:325-328, 1987/88.

[Har69] F. Harary. *Graph Theory*, Addison-Wesley, 1969.

[KARR90] P. Klein, A. Agrawal, R. Ravi, and S. Rao. Approximation through multicommodity flow. In *Proceedings of the 31th Annual Symposium on Foundations of Computer Science*, IEEE, pages 726–737, 1990.

[Kön36] D. König. *Theorie der Graphen*, Reprinted by Chelsea Publishing Company, New York, 1950.

[Lag90] J. Lagergren. Efficient parallel algorithms for tree-decomposition and related problems. In *Proceedings of the 31th Annual Symposium on Foundations of Computer Science*, IEEE, pages 173–182, 1990.

[vL90]   J. van Leeuwen. Graph Algorithms. In *Handbook of Theoretical Computer Science. A: Algorithms and Complexity Theory*, North Holland, Amsterdam, 1990.

[LR88]   F. T. Leighton and S. Rao. An approximate max-flow min-cut theorem for uniform multi-commodity flow problems with applications to approximate algorithms. In *Proceedings of the 29th Annual Symposium on Foundations of Computer Science*, IEEE pages 422–431, 1988.

[Le90]   F. T. Leighton. Personal communication, 1990.

[LL88]   C. E. Leiserson and J. G. Lewis. Orderings for parallel sparse symmetric factorization. An unpublished manuscript, 1988.

[LRT79] R. J. Lipton, D. J. Rose and R. E. Tarjan. Generalized nested dissection. *SIAM Journal of Numerical Analysis* 16:346-358, 1979.

[Liu90] J. W. H. Liu. The multifrontal method for sparse matrix solution: theory and practice. Tech. Report CS-90-04, York University, North York, Ontario, Canada. To appear.

[Möh90] R. H. Möhring. Graph problems related to gate matrix layout and PLA folding. In *Computing Supplementum* 7 (G. Tinhofer et al, eds), page 17–51. Springer-Verlag, Vienna, 1990.

[Pot88] A. Pothen. The complexity of optimal elimination trees. Tech. Report CS-88-16, Pennsylvania State University, 1988.

[RS86a] N. Robertson and P. D. Seymour. Graph minors. II. Algorithmic aspects of tree-width. *J. Algorithms*, 7:309–322, 1986.

[RS86b] N. Robertson and P. D. Seymour. Graph minors. XIII. The disjoint paths problem. Manuscript, 1986.

[RS91] N. Robertson and P. D. Seymour. Graph minors. X. Obstructions to tree-decompositions. To appear in J. Combinatorial Theory, Ser. B., 1991.

[RTL76] D.J. Rose, R.E. Tarjan and G.S.Lueker. Algorithmic aspects of vertex elimination on graphs. *SIAM J. Comput.* Vol. 5, No. 2, 266–283, 1976.

[ST90] P.D. Seymour and R. Thomas. Call Routing and the Ratcatcher. Manuscript, 1991.

[Yan81] M. Yannakakis. Computing the minimum fill-in is NP-complete. *SIAM Journal of Algebraic and Discrete Methods*, 2:77–79, 1981.

Monadic Second-Order Evaluations
on Tree-Decomposable Graphs[*]

B. COURCELLE   M. MOSBAH[†]

Bordeaux I university, Laboratoire d'Informatique,[‡]
351, cours de la Libération, 33405 Talence, France

Introduction

Many $\mathcal{NP}$-complete problems become polynomial when restricted to particular sets of graphs. A number of such cases are discussed in [Jo85]. More informative than isolated results are *metaresults* exhibiting classes of sets of graphs, classes of problems having polynomial algorithms on the sets of graphs of the corresponding classes, and uniform descriptions of these algorithms. Such an approach is that of [TNS], [BLW], [ALS] and [HKV89].

The present paper follows this line of research where the notion of a problem is extended into that of an *evaluation*. An evaluation is a function that associates with every graph a value in some set S, say $\mathbb{N}$, $\mathbb{R}$, $\mathcal{P}(\mathbb{N})$. Hence, a problem is an evaluation where $S = \{\textbf{true}, \textbf{false}\}$. However, the value of an evaluation can also be a set of vertices, or a set of edges of the given graph. It follows that we can consider the problem of evaluating *optimal subgraphs* of the given graphs, and we can consider the problems raised in [BLW].

We deal with directed or undirected graphs, with possible loops and multiple edges, possibly labeled, weighted, and ordered. We consider sets of such graphs that are of uniformly bounded tree-width, or, alternatively, that are definable by *hyperedge replacement* grammars. The sets of series-parallel graphs, of Halin graphs, of outerplanar graphs are examples of such sets. So are, for each k, the set of partial k-trees, and the set of graphs of bandwidth at most k.

Tree-decompositions or derivation trees relative to grammars are essential in that they permit to describe graphs as *"tree gluings"* of elementary graphs of bounded size. Graph evaluations of the appropriate type, (we shall say following Habel[H89], that they are *compatible* with the grammar), can be computed by means of one bottom-up traversal of the derivation tree or of the tree-decompostion, (like in attribute grammars when there are only synthesized attributes.)

The central result (Theorem 2.2) of this paper is the description, in a uniform way, of a "large" class of compatible evaluations. This will be done in a formalism that associates logic and algebra, or more precisely, monadic second-order logic and semiring homomorphisms.

Let us present briefly and informally the basic ideas.
Let φ be a monadic second-order formula with set W of free variables. For every grammar-generated graph G, (we consider it as a logical structure), we denote by $\boldsymbol{sat}(\varphi)(G)$ the set of W-assignments in G that satisfy the formula φ. The fundamental result of Courcelle[Co90a] says that $\boldsymbol{sat}(\varphi)(G)$ can be evaluated bottom-up on any derivation tree of G.

Some evaluations v can be expressed by $v(G) = h(\boldsymbol{sat}(\varphi)(G))$. This expression does not give immediately an efficient way of computing $v(G)$, because $\boldsymbol{sat}(\varphi)(G)$ is frequently a very large, although finite, set. If h is a homomorphism, in an appropriate sense, the mapping $v(G)$

---

[*]This work has been supported by the "Programme de Recherches Coordonneés: Mathématiques et Informatique" and the ESPRIT Basic Research Action 3299 "Computing by Graph Transformations" .

[†]Electronic mail: courcell@geocub.greco-prog.fr and mosbah@geocub.greco-prog.fr

[‡]Laboratoire associé au CNRS

can be evaluated *directly* bottom-up on any derivation tree of G, without needing the costly computation of $sat(\varphi)(G)$.

We obtain in this way a new proof of some results established by Arnborg et al.[ALS88], and linear algorithms in some cases not covered by the *extended monadic second order logic* introduced in their paper. Also, we obtain a syntactic expression for a large class of *compatible* evaluations, including the special cases considered in [H89,HKV89]. We also obtain linear or polynomial algorithms for the sample optimization problems considered in [BLW] because h can be a choice function that selects some set of maximal or minimal weight. We actually apply the method introduced in this paper to a large class of problems, described syntactically in a uniform way, and we answer the questions raised in its conclusion.

1 Notations and definitions:

We let S be a finite set of sorts, we let F be an S-signature, and we let $M = \langle (M_s)_{s \in S}, (f_M)_{f \in F} \rangle$ be an F-algebra. In particular, if $f \in F$ is of profile $s_1 \times s_2 \times \cdots \times s_n \longrightarrow s$, then f_M is a total mapping $M_{s_1} \times M_{s_2} \times \cdots \times M_{s_n} \longrightarrow M_s$. The functions f_M are called the *operations* of M. A *derived operation* of M is a function $M_{s_1} \times M_{s_2} \times \cdots \times M_{s_n} \longrightarrow M_s$ defined by a term t of sort s built with the operation symbols from F, variables $x_1, \cdots, x_n$ of respective sorts $s_1, \cdots, s_n$ and additional constants denoting fixed elements of M.

We shall use algebras where $S \subseteq \mathbb{N}$, and M_s is the set of graphs of tree-width at most some fixed k, having s distinguished vertices called *sources*.

Typical operations are *gluings*, that take two graphs and glue them by their sources into larger graphs. For more details, the reader is refered to [Co90a,Co90b,BC87].

The set of subsets of a set D is denoted by $\mathcal{P}(D)$. Its set of finite subsets is denoted by $\mathcal{P}_f(D)$.

Inductively computable evaluations

(1.1) DEFINITION : *Inductive families of evaluations.*

Let D be any set. A *family of evaluations* on M is a set $\mathcal{E}$ of unary mappings such that each mapping v in $\mathcal{E}$ maps $M_{\alpha(v)}$ into D. The object $\alpha(v)$ belongs to S and is called the type of v.

A family of evaluations $\mathcal{E}$ is F-*inductive* if for every f in F of profile $s_1 \times s_2 \times \cdots \times s_n \longrightarrow s$, for every v in $\mathcal{E}$ of type s, there exists an operator $\theta_{v,f}$ on D and a sequence of length $(m_1 + m_2 + \cdots + m_n)$ of elements of $\mathcal{E}$, $(v_{1,1}, \ldots, v_{1,m_1}, v_{2,1}, \ldots, v_{2,m_2}, \ldots, v_{n,m_n})$ such that :

1. $\alpha(v_{i,j}) = s_i$ for all $j = 1, \ldots, m_i$.

2. For all $d_1 \in M_{s_1}, \ldots, d_n \in M_{s_n}$:

$$v(f(d_1, \ldots, d_n)) = \theta_{v,f}(v_{1,1}(d_1), \ldots, v_{1,m_1}(d_1), v_{2,1}(d_2), \ldots, v_{2,m_2}(d_2), \ldots, v_{n,m_n}(d_n)) \ .$$

The sequence $(\theta_{v,f}, v_{1,1}, \ldots, v_{n,m_n})$ is called a *decomposition* of v w.r.t f, and $\theta_{v,f}$ its *decomposition operator*.

In words, the existence of such a decomposition means that the value of v for any object of the form $f_M(d_1, \ldots, d_n)$ can be determined and computed from the values of finitely many mappings of $\mathcal{E}$ at $d_1, \ldots, d_n$.

Lemma 1.1 *If $\mathcal{E}$ is F-inductive, then it is G-inductive where G is any set of derived operations constructed over F.*

(1.2) DEFINITION : *Inductively computable evaluations*

Let $s_0 \in S$ be a sort of interest. A mapping $v : M_{s_0} \longrightarrow D$ is F-*inductively computable*, if there exists a family of evaluations $\mathcal{E}$ such that:

1. $v \in \mathcal{E}$.

2. $\mathcal{E}$ has finitely many functions of each type.

3. $\mathcal{E}$ is F-inductive, with effectively given decompositions.

We shall use these definitions in the case where M is the H_A-algebra of finite graphs defined in [BC87] and the evaluations are functions on graphs defined in *monadic second-order logic* as explained below.

Graphs of type k :

We consider finite, edge labeled, directed or undirected graphs, equipped with a sequence of distinguished k vertices called the *sources*. $src_G(i)$ denotes the ith element of this sequence. A is the set of edge labels.

We denote by $G_k(A)$ be the set of all finite graphs of type k over A.

Moreover, we consider that two *isomorphic* graphs are equal.

Graph operations:

In this abstract, we do not detail graph operations. We use, the basic operations (disjoint sum, source fusion, source redefinition). By combinig these operations in derived operations, one obtains more powerful operations like series and parallel compositions.

We quote from [BC87], [ACPS], and [Co90b] the following results:

Fact: *1. A set $L \subseteq G_n(A)$ is expressible by finitely many of the operations of H_A iff it is of bounded tree-width.*

2. This is the case of partial k-trees, of sets of graphs and hypergraphs generated by hyperedge replacement grammars, and of k-terminal recursive families of graphs in the sense of Wimer [W].

2 Monadic second-order evaluations on graphs:

(2.1) DEFINITION. *Graphs as logical structures.*

In order to express propreties of graphs in $G_k(A)$, we define the symbols:

v: the *vertex* sort,

e: the *edge* sort,

s_i, a constant of sort **v**, for each i, $1 \leq i \leq k$,

$\mathbf{edg}_a$, a predicate symbol of arity **evv**, for each a, $a \in A$.

With $G \in G_k(A)$, we associate the logical structure $|G| = \langle V_G, E_G, (s_{iG})_{i \in [k]}, (\mathbf{edg}_{aG})_{a \in A} \rangle$, where V_G is the domain of sort **v**, E_G is the domain of sort **e**, s_{iG} is the ith source of G, and $\mathbf{edg}_{aG}(e, v_1, v_2) = true$ iff $lab_G(e) = a$ and e links v_1 to v_2.

(2.2) DEFINITION. *Monadic second-order logic.*

Let $\mathcal{W}$ be a finite sorted set of variables $\{u, u', \ldots, U, U', \ldots\}$, each of them having a sort $\sigma(u), \sigma(u'), \ldots, \sigma(U), \sigma(U'), \ldots$ in $\{\mathbf{v}, \mathbf{e}\}$. We denote by $\mathcal{W}_s$ the set $\mathcal{W} \cup \{s_1, \ldots, s_k\}$. Uppercase letters denote set variables and lowercase letters denote object variables or constants.

The set of atomic formulas consists of:

$u = u'$ with $u, u' \in \mathcal{W}_s$, $\sigma(u) = \sigma(u')$,

$u \in U$ with $u, U \in \mathcal{W}_s$, $\sigma(u) = \sigma(U)$,

$\mathbf{edg}_a(u, u'_1, u'_2)$ with $u, u'_1, , u'_2 \in \mathcal{W}_s$, $\sigma(u) = \mathbf{e}, \sigma(u'_1) = \sigma(u'_2) = \mathbf{v}$.

The language of *Monadic second-order logic* (MS) is the set of formulas formed with the above atomic formulas together with the Boolean connectives and quantifications over object and set variables.

In the following, we consider MS-formulas with set variables only. This is not a loss of generality because each MS-formula can be translated into an equivalent MS-formula using only set variables (see [Co90a] for details). We denote by $\Phi^h_{A,k}(\mathcal{W})$ the set of MS-formulas of height at most h and variables in $\mathcal{W}$. (The height of a formula is the depth of nested quantifications.) To shorten our writing, we will fix h and A and refer to the previous set by Φ_k. In fact, without loss of generality, these parameters can be fixed in the following definitions. However, we shall work with several types of graphs at the same time, hence, k will have to vary. For every graph G, D_G will denote the set $E_G \cup V_G$. (We take always $E_G \cap V_G = \emptyset$.). Furthermore, we shall let D be a countably infinite set (say the set of integers) such that $D_G \subseteq D$ for all graphs G.

A $\mathcal{W}$-assignment in G is a mapping ν associating with every variable X in $\mathcal{W}$ a subset of D_G such that $\nu(X) \subseteq E_G$ if X is of sort e and $\nu(X) \subseteq V_G$ if it is of sort v. Such an assignment will be written $\nu = (\nu_1, \ldots, \nu_n)$ with $\nu_i = \nu(X_i)$ in the usual case where $\mathcal{W}$ is $\{X_1, \ldots, X_n\}$.

For each k-graph G, we let $sat(G) : \Phi_k \longrightarrow \mathcal{P}(\mathcal{P}(D_G)^n)$ be the mapping such that for every $\varphi(X_1, \ldots, X_n)$ in Φ_k, $sat(G)(\varphi)$, also denoted by $sat(G, \varphi)$, is the set of assignments $\nu = (\nu_1, \ldots, \nu_n) \in \mathcal{P}(D_G)^n$ such that $(G, \nu) \models \varphi$. If no assignment satisfies φ in G, then $sat(G, \varphi) = \emptyset$.

Theorem 2.1 For every k, for every φ in Φ_k, the mapping $sat(\varphi) : \mathbf{G}_k(A) \longrightarrow \mathcal{P}_f(\mathcal{P}_f(D)^n)$ such that $sat(\varphi)(G) = sat(G, \varphi)$ is H_A-inductively computable.

Proof: We will use the three basic lemmas of [Co90a], nevertheless we will detail the proof for only one lemma since the proofs of the two others are similar. Moreover, we prefer to deal with a general graph operation as we shall see below. We let $\mathcal{W} = \{X_1, \cdots, X_n\}$. If $\nu' = (\nu'_1, \ldots, \nu'_n)$ and $\nu'' = (\nu''_1, \ldots, \nu''_n)$ are two assignments in G' and G'' respectively, then the assignment $\nu := \nu' \cup \nu''$ in $G' \oplus G''$ is defined as $(\nu'_1 \cup \nu''_1, \ldots, \nu'_n \cup \nu''_n)$ ($\nu'_i \cup \nu''_i$ is a shorthand writing of $\nu'(X_i) \cup \nu''(X_i)$ where X_i is in $\mathcal{W}$.)
We shall keep in mind that the sets we handle in the following are sets of n-tuples of sets. Let us define some operations on sets of n-tuples.
Two sets A and B are called *separated* if for all $i = 1, \ldots, n$

$$\left(\bigcup\{\alpha_i \, / \, (\alpha_1, \ldots, \alpha_n) \in A\}\right) \cap \left(\bigcup\{\beta_i \, / \, (\beta_1, \ldots, \beta_n) \in B\}\right) = \emptyset$$

If A and B are two sets of n-tuples of sets, then we define an "*extended*" union $\cup$ by $A \cup B = \{\alpha \cup \beta / \alpha \in A, \beta \in B\}$. If, in addition, A and B are *separated*, then we shall write $A \uplus B$ instead of $A \cup B$.
It is clear that $\emptyset = \{(\emptyset, \emptyset, \ldots, \emptyset)\}$ is the neutral element of $\cup$ (and of $\uplus$), i.e, for each A, $A \cup \emptyset = A$. We observe that the empty set $\emptyset$ is the *annihilator* element of $\cup$. That is, for each A, $A \cup \emptyset = \emptyset$. (Note that $\emptyset$ is an element of $\mathcal{P}_f(D)^n$ whereas $\emptyset$ denotes the usual empty set).
The disjoint set union will be written $\uplus$. That is, if A and B are two sets such that $A \cap B = \emptyset$ then $A \uplus B$ is nothing but $A \cup B$. We assume that $A \uplus B$ is undefined if $A \cap B \neq \emptyset$. It is evident that the neutral element of $\uplus$ is $\emptyset$.
Similarly, $A \uplus B$ is not defined if A and B are not separated. Note in particular that $A \uplus B$ is defined if $A = \emptyset$, and that $A \uplus B$ is defined if $A = \emptyset$ or $A = \emptyset$.

Let us consider the case where a graph is obtained as the disjoint sum of two other graphs, then:

Lemma 2.1 Let $k = k' + k''$.
Given φ in Φ_k, one can construct a finite sequence of formulas $\psi'_1, \psi'_2, \ldots, \psi'_m$ in $\Phi_{k'}$, a finite sequence of formulas $\psi''_1, \psi''_2, \ldots, \psi''_m$ in $\Phi_{k''}$ such that for every k'-graph G', for every k''-graph G'':

$$sat(G' \oplus G'', \varphi) = \biguplus_{1 \leq j \leq m} sat(G', \psi'_j) \uplus sat(G'', \psi''_j).$$

Proof: It was proved by Courcelle[Co90a] that given a formula φ in Φ_k, one can construct a finite sequence of formulas $\varphi'_1, \varphi'_2, \ldots, \varphi'_{n'}$ in $\Phi_{k'}$, a finite sequence of formulas $\varphi''_1, \varphi''_2, \ldots, \varphi''_{n''}$

in $\Phi_{k''}$ and an $(n' + n'')$-place Boolean expression B such that, for every k'-graph G', for every k''-graph G'', for every assignment ν' in G', for every assignment ν'' in G'': if $\nu = \nu' \cup \nu''$ and $G = G' \oplus G''$,then

$$\varphi_G(\nu) = B[\varphi'_{1G'}(\nu'), \ldots, \varphi'_{n'G'}(\nu'), \varphi''_{1G''}(\nu''), \ldots, \varphi''_{n''G''}(\nu'')]^1$$

This Boolean expression can be put in a disjunctive form, that is, $\bigvee\limits_{1 \leq i \leq r} \lambda'_{iG'}(\nu') \wedge \lambda''_{iG''}(\nu'')$

where λ'_i and λ''_i are, in turn, some Boolean combinations of the φ'_j's and the φ''_j's respectively. The formulas λ'_i and λ''_i are,respectively, in $\Phi_{k'}$ and $\Phi_{k''}$, because each set Φ_k is closed under Boolean operations. We have then the following identity:

$$sat(G, \varphi) = \bigcup\limits_{1 \leq i \leq r} sat(G', \lambda'_i) \boxtimes sat(G'', \lambda''_i). \tag{1}$$

It is easy to come up to this statement by proving a double inclusion, i.e, that an assignment of any handside of (1) belongs to the other. (Since G' and G'' are disjoint, $sat(G', \lambda'_i)$ and $sat(G'', \lambda''_i)$ are always separated.)

Now, we shall transform this union into a disjoint one. For each nonempty subset I of $[r]$, we let $\lambda'_I = \bigwedge\limits_{i \in I} \lambda'_i \wedge \bigwedge\limits_{i \notin I} \neg\lambda'_i$ and $\lambda''_I = \bigwedge\limits_{i \in I} \lambda''_i \wedge \bigwedge\limits_{i \notin I} \neg\lambda''_i$. We assert that,

$$sat(G, \varphi) = \biguplus\limits_{I,J \subseteq [r], I \cap J \neq \emptyset} sat(G', \lambda'_I) \boxtimes sat(G'', \lambda''_J). \tag{2}$$

Corollary 2.1 *Let F be a derived signature of H_A. For every k and φ in Φ_k, the mapping $sat(\varphi)$ is F-inductively computable. The corresponding decomposition operators are expressible in terms of $(\uplus, \boxtimes)$.*

Now, we shall define a class of evaluations on graphs that can be computed inductively, in some sense "directly".

(2.4) DEFINITION: *MS-evaluation*
An *evaluation structure* is a structure $\mathcal{R} = < S, \sqcup, \sqcup, \perp, \varepsilon >$. In many cases, it will be a semiring, but we do not require this in general.
The notion of a homomorhpism $h : < \mathcal{P}_f(\mathcal{P}_f(D)^n), \cup, \uplus, \emptyset, \emptyset > \longrightarrow \mathcal{R}$ is standard. (Note the correspondence of $\cup$ with $\sqcup$, of $\uplus$ with $\sqcup$, of $\emptyset$ with $\perp$ and of $\emptyset$ with ε.)
The following weaker notions will be useful. A $(\uplus, \boxtimes)$-*homomorphism* $h : < \mathcal{P}_f(\mathcal{P}_f(D)^n), \uplus, \boxtimes, \emptyset, \emptyset >$ into $\mathcal{R}$ is a mapping such that:

$$h(\emptyset) = \perp,$$
$$h(\emptyset) = \varepsilon,$$
$$h(A \uplus B) = h(A) \sqcup h(B) \quad \text{if } A \cap B = \emptyset,$$
$$h(A \boxtimes B) = h(A) \sqcup h(B) \quad \text{if } A \text{ and } B \text{ are separated}$$

Similarly, we define $(\cup, \boxtimes)$-homomorphisms and $(\uplus, \cup)$-homomorphisms.
An *MS-evaluation* is a mapping $v : G_k(A) \longrightarrow S$ where $\mathcal{R} = < S, \sqcup, \sqcup , \perp, \varepsilon >$ is an evaluation structure, that is of the form $v = h \circ sat(\varphi)$ where φ is a MS-formula, and h is a $(\uplus, \boxtimes)$-homomorphism : $< \mathcal{P}_f(\mathcal{P}_f(D)^n), \cup, \uplus, \emptyset, \emptyset > \longrightarrow \mathcal{R}$.
We must remember that, as defined in Section 2, $sat(\varphi)$ is a mapping $G_k(A) \longrightarrow \mathcal{P}_f(\mathcal{P}_f(D)^n)$. Hence, $h \circ sat(\varphi)$ is a mapping $G_k(A) \longrightarrow S$.
In words, an MS-evaluation is defined by a homomorphism which takes as input the tuples computed by the mapping sat. Thus, its value for a graph G can be actually determined from those for the subgraphs composing G. Moreover, its values for the basic graphs ($\mathbf{0}$, $\mathbf{1}$ and edges) must be known since these graphs are the leaves of the parse tree of any graph.

---

[1] $\varphi_G(\nu)$ denotes the Boolean value *true* iff $(G, \nu) \models \varphi$ and false otherwise. Hence, $\varphi_G(\nu) = true$ iff $sat(G, \varphi) \neq \emptyset$.

Evaluation	Definition	Description
$v_1(G)$	$sat(\varphi)(G)$	The set of all simple paths from $src_G(1)$ to $src_G(2)$.
$v_2(G)$	$Card(sat(\varphi)(G))$	The number of simple paths from $src_G(1)$ to $src_G(2)$.
$v_3(G)$	$NonEmpty(sat(\varphi)(G))$	This evaluation has the value **true** if $sat(\varphi)(G) \neq \emptyset$ and **false** otherwise. It indicates the existence of a simple path from $src_G(1)$ and $src_G(2)$.
$v_4(G)$	$Max\{\ Card(X)\ /X \in sat(\varphi)(G)\}$	The maximal length of a simple path from $src_G(1)$ to $src_G(2)$. (If $v_1(G) = \emptyset$ then $v_4(G) = -\infty$).
$v_5(G)$	$Min\{\ Card(X)\ /X \in sat(\varphi)(G)\}$	The length of the shortest path from $src_G(1)$ to $src_G(2)$, ($= +\infty$ if $v_1(G) = \emptyset$).
$v_6(G)$	$\sum\{Card(X)\ /\ X \in sat(\varphi)(G)\}$	The sum of the lengths of all simple paths joining $src_G(1)$ and $src_G(2)$.
$v_7(G)$	$Average(sat(\varphi)(G))$	The average length of a simple path from $src_G(1)$ to $src_G(2)$, ($= v_6(G)/v_2(G)$).
$v_8(G)$	$Dif(sat(\varphi)(G))$	The difference between the maximum and the minimum lengths of simple paths from $src_G(1)$ to $src_G(2)$. That is, $v_4(G) - v_5(G)$. ($= -\infty$ if $v_1(G) = \emptyset$).

Figure 1: Some evaluations

Theorem 2.2 *Every MS-evaluation is H_A-inductively computable.*

Proof: This is a consequence of Theorem 2.1. In fact, it suffices to compose h and sat in Lemma 2.1 and to use the fact that h is a homomorphism.

(2.3) Examples

We let $k = 2$ and $\varphi(X)$ be the MS-formula saying that X is the set of edges of a simple path linking the first source to the second one. Let $G \in G_2(A)$. We shall consider the following evaluations expressed in terms of $sat(\varphi)(G)$ described in the table of Figure 1.

The evaluations v_2, v_3, v_4, v_5 are MS-evaluations. The corresponding evaluation structures are listed in Figure 2. The others v_6, v_7, v_8 are not, but they are computable in terms of auxiliary MS-evaluations so that they are H_A-inductively computable. Let us check that v_2 is really an MS-evaluation. If A and B are two elements of $\mathcal{P}_f(\mathcal{P}_f(D))$ (two finite sets of finite subsets of D), then

$$Card(A \uplus B) = Card(A) + Card(B)$$

We also have

$$
\begin{aligned}
Card(A \uplus B) &= Card(\{\alpha \cup \beta / \alpha \in A, \beta \in B\}) \\
&= Card(A) \times Card(B)
\end{aligned}
$$

because A and B are separated, i.e, are such that $(\bigcup\{\alpha_i\ /\ (\alpha_1, \ldots, \alpha_n) \in A\}) \cap (\bigcup\{\beta_i\ /\ (\beta_1, \ldots, \beta_n) \in B\}) = \emptyset$ for all $i = 1, \ldots, n$. Hence $Card$ is a $(\uplus, \uplus)$-homomorphism of $< \mathcal{P}_f(\mathcal{P}_f(D)), \cup, \uplus, \emptyset, \emptyset >$ into $< \mathbb{N}, +, \times, 0, 1 >$ and v_2 is an MS-evaluation.

3 Building algorithms

Let F be a finite derived signature of H_A (its set of sorts is a finite subset $\mathcal{S}(F)$ of $\mathbb{N}$). Let $k \in \mathcal{S}(F)$ and v be an evaluation : $G_k(A) \longrightarrow S$, for some set S.

MS-evaluation	S	⊔	⊎	⊥	ε
v_2	$\mathbb{N}$	$+$	$\times$	0	1
v_3	$\{\text{true}, \text{false}\}$	$\vee$	$\wedge$	false	true
v_4	$\mathbb{N} \cup \{-\infty\}$	Max	$+$	$-\infty$	0
v_5	$\mathbb{N} \cup \{+\infty\}$	Min	$+$	$+\infty$	0

Figure 2: Evaluation structures

A *(v,k,F)-algorithm* is an algorithm that takes as input a term t in $M(F)_k$ and produces the value $v(\mathbf{val}(t))$, where $\mathbf{val}(t)$ is the value of t in $G_k(A)$, i.e, the k-graph denoted by this term.

Letting the **size** of a graph G be $\text{size}(G) = \mathbf{Card}(V_G) + \mathbf{Card}(E_G)$, we have $\text{size}(G) \leq m.|t|$ for some constant m, where $t \in M(F)$ and $G = \mathbf{val}(t)$. Conversely, $|t| \leq m'.\text{size}(G) + 1$ for some constant m'. It follows that if an algorithm decides a graph property in time $O(|t|^k)$ where $\mathbf{val}(t) = G$, then one can also say that its time complexity is $O(\text{size}(G)^k)$.

3.1 Proposition

If v is F-inductively computable, then there exists a (v,k,F)-algorithm with time complexity $O(|t|.\eta)$, where η is an upper bound on the complexity of the computation of each right handside of an equation as in Definition (1.1).

Proof: We first present a basic algorithm and we shall describe later how to improve it.

Let $(\mathcal{E}_s)_{s \in S(F)}$ be the finite set of evalutions that we have by the definition of an inductively computable evaluation.

Let $t \in M(F)_k$ be considered as usual as a finite tree. Each node u of t has a label f in F. The sort of f will be called the *sort of u*, and is actually the type of the graph defined by the term t/u, namely the subtree issued from node u of t.

With each node u of t, we associate *attribute* occurrences $w(u)$, for each $w \in \mathcal{E}_s$, where s is the sort of u. The intended value of an attribute occurrence $w(u)$ is $w(\mathbf{val}(t/u))$. We now explain how it can be computed from the values of other attributes at the successor nodes of u.

Let u be a node, let $w(u)$ be an attribute at u, and f be the symbol of F that labels u. Then, by Definition 1.1 :

$$w(u) = \theta_{w,f}(w_{1,1}(u_1), \ldots, w_{1,m_1}(u_1), w_{2,1}(u_2), \ldots, w_{2,m_2}(u_2), \ldots, w_{n,m_n}(u_n)). \tag{1}$$

where $(\theta_{w,f}, w_{1,1}, \ldots, w_{n,m_n})$ is the decomposition of w relative to f, and $u_1, \ldots, u_n$ is the sequence of successors of u.(This assumes that the rank of f is n; if $n = 0$ then $\theta_{w,f}$ is a constant value).

It is thus clear that one can compute bottom-up on the tree t all the attributes associated with all its nodes. We are actually in the case of a purely synthesized attribute grammar. See [DJL] for a survey of attribute grammars and their evaluation algorithms. Among the attributes of the root, one finds $v(\varepsilon) = v(\mathbf{val}(t))$, namely the value to be returned as output of the algorithm.

The time complexity can be evaluated as

$$\sum_{u \in \text{Node}(t)} \sum_{w \in \mathcal{E}_{\sigma(u)}} \eta(w, u) \tag{2}$$

where $\eta(w, u)$ is the time complexity for the evaluation of $w(u)$ by equation (1). An upper bound can be given as follows

$$|t|.\mathbf{Max}\{\mathbf{Card}(\mathcal{E}_s) \ / \ s \in \mathcal{S}(F)\}.\eta \tag{3}$$

where $\eta(w, u) \leq \eta$ for all w and u. $\square$

MS-evaluation	S	$\sqcup$	$\sqcup\!\!\!\sqcup$	$\perp$	ε	Complexity bound
Boolean values	$\{\mathbf{true}, \mathbf{false}\}$	$\vee$	$\wedge$	$\mathbf{false}$	$\mathbf{true}$	$O(|t|)$
Cardinality	$\mathbb{N}$	$+$	$\times$	0	1	$O(|t|)$
MaxWeight	$\mathbb{R} \cup \{-\infty\}$	Max	$+$	$-\infty$	0	$O(|t|)$
MinWeight	$\mathbb{R} \cup \{+\infty\}$	Min	$+$	$+\infty$	0	$O(|t|)$
SetCard	$\mathcal{P}_f(\mathbb{N})$	$\cup$	$+$	$\emptyset$	$\{0\}$	$O(|t|^3)$
Universal	$\mathcal{P}_f(\mathcal{P}_f(D)^n)$	$\cup$	$\uplus$	$\emptyset$	$\emptyset$	$O(2^{|t|})$

Figure 3: A catalogue of evaluation structures

3.2 Issues for the construction of efficient (v, k, F)-algorithms

The usability of this technique depends on the following facts:

1. For each w, f, one needs to define a subroutine implementing $\theta_{w,f}$. Clearly, good data structures·for storing and computing the values of attributes must be designed.

2. It is clear that using as few sorts as possible, and, as small sets $\mathcal{E}_s$ as possible, improves time and space complexity.

Concerning point 2, we use use derived operators in order to decrease the number of sorts, of operations, and of operators to be used. This tends also to minimize the number of auxilary formulas. Theorems 2.1 and 2.2 are stated in terms of logical formulas. However, it seems intractable to use them in the way they are established, because the proofs involve very large sets of auxiliary formulas. In concrete cases, one should rather work in terms of graph properties, knowing what they mean, and forgetting the logical formulas.

4 A catalogue of evaluation structures

We review the evaluation structures already presented in Example (2.3), that we sum up in the table of Figure 3. They are presented by order of increasing complexity. All these structures are detailed in [CM90].

- MaxWeight and MinWeight are evaluations defined on weighted graphs. More precisely, if $A \in \mathcal{P}_f(\mathcal{P}_f(D))$ and w is a function : $D \longrightarrow \mathbb{R}$, then

$$\mathbf{MaxWeight}(A) = \mathrm{Max}\{w(X) \ / \ X \in A\}$$

$$\mathbf{MinWeight}(A) = \mathrm{Min}\{w(X) \ / \ X \in A\}$$

where, for $X \subseteq D$, we have $w(X) = \sum\{w(x)/x \in D\}$.
Note that maximum and minimum cardinalities (as in the Example 2.3) are particular cases of these evaluations.

- SetCard is an evaluation such that $\mathbf{Setcard}(A)= \{\mathbf{Card}(\alpha) \ / \ \alpha \in A\}$ and $N+M = \{n + m \ / \ n \in N, m \in M\}$ for $N, M \subseteq \mathbb{N}$.

- Universal is the evaluation that computes the set $sat(G, \varphi)$ as a whole.

- Let us mention the evaluation *"sum of cardinalities"* defined by $\sum \mathbf{Card}(A) = \sum\{\mathbf{Card}(\alpha)/\alpha \in A\}$, so that $\sum\mathbf{Card}$ maps $\mathcal{P}_f(\mathcal{P}_f(D))$ into $\mathbb{N}$. It is clear that

$$\sum\mathbf{Card}(A \uplus B) = \sum\mathbf{Card}(A) + \sum\mathbf{Card}(B)$$

The definition of $\sum \mathbf{Card}(A \uplus B)$ needs the auxiliary use of $\mathbf{Card}(A)$ and $\mathbf{Card}(B)$:

$$\sum \mathbf{Card}(A \uplus B) = \mathbf{Card}(B) \cdot \sum \mathbf{Card}(A) + \mathbf{Card}(A) \cdot \sum \mathbf{Card}(B)$$

(The verification is easy).

It follows that an evaluation of the form $\sum \mathbf{Card}(sat(G, \varphi))$ is not an MS-evaluation, but is nevertheless H_A-inductively computable, with help of the auxiliary evaluations $\mathbf{Card}(sat(G, \varphi))$. The corresponding algorithm is linear with uniform cost measure.

4.1 Evaluation formulas

We now assume that, instead of one weighted function $w : D \longrightarrow \mathbb{R}$, we have n of them, $w_i : D \longrightarrow \mathbb{R}$, not necessarily all different. We let $w_i(X) = \sum \{w_i(x) \ / \ x \in X\}$ for X finite, $X \subseteq D$.

Let $\varphi(X_1, \ldots, X_n)$ be a formula. Let G be a graph. For every $(X_1, \ldots, X_n) \in \mathcal{P}_f(D_G)^n$, we let :

$$\theta(X_1, \ldots, X_n) = c_1 w_1(X_1) + \ldots + c_n w_n(X_n)$$

where $c_1, \ldots, c_n$ are real numbers. Hence θ is a linear evaluation term in the sense of [ALS].

For every set of tuples A, we let :

$$\Theta(A) = \{\theta(X_1, \ldots, X_n) \ / \ (X_1, \ldots, X_n) \in A\}$$

and

$$
\begin{aligned}
\Theta(G, \varphi) &= \{\theta(X_1, \ldots, X_n) \ / \ (X_1, \ldots, X_n) \in sat(G, \varphi)\} \\
&= \Theta(sat(G, \varphi)).
\end{aligned}
$$

We let:

$$\mathbf{Max}\Theta(A) = \mathbf{Max}\{\theta(\alpha) \ / \ \alpha \in A\}$$

$$\mathbf{Min}\Theta(A) = \mathbf{Min}\{\theta(\alpha) \ / \ \alpha \in A\}.$$

Then, we have

$$\mathbf{Max}\Theta(A \cup B) = \mathbf{Max}(\mathbf{Max}\Theta(A), \mathbf{Max}\Theta(B))$$

$$\mathbf{Min}\Theta(A \cup B) = \mathbf{Min}(\mathbf{Min}\Theta(A), \mathbf{Min}\Theta(B))$$

and

$$\mathbf{Max}\Theta(A \uplus B) = \mathbf{Max}\Theta(A) + \mathbf{Max}\Theta(B)$$

$$\mathbf{Min}\Theta(A \uplus B) = \mathbf{Min}\Theta(A) + \mathbf{Min}\Theta(B)$$

from which it follows that:

$\mathbf{Max}\Theta(G, \varphi)$ and $\mathbf{Min}\Theta(G, \varphi)$ are computable in time $O(|t|)$, if $t \in M(F)$ defines G (where F is fixed and finite). The computation of $\mathbf{Max}\Theta(G, \varphi)$ or $\mathbf{Min}\Theta(G, \varphi)$ is called a *linear EMS extremum* problem in [ALS]. Hence, we obtain another proof of the result of [ALS] stating that a linear EMS extremum problem can be solved in linear time (with uniform cost measure) for graphs of tree-width at most some fixed k (given by their tree-decompostions of width at most k).

4.2 Ordered graphs: a choice function

We now assume that D is linearly ordered by $\leq_D$. (If D is a set of memory locations, it is linearly ordered in a natural way).

Given an MS-formula $\varphi(X)$ and a graph G (with $D_G \subseteq D$), one may consider the problem of computing the *lexicographically first maximal* set X such that $\varphi(X)$ holds in G, i.e. the unique set $X \subseteq D_G$ such that

1. $G \models \varphi(X)$

2. if $Y \supset X$ (and $Y \neq X$), then $G \models \neg\varphi(Y)$ (maximality of X)

3. If X' also satisfies (1) and (2) then $X \leq_{lex} X'$ (since X and X' are subsets of a totally ordered set, one can order them by increasing order and compare them by the lexicographic order associated with $\leq_D$)

Such a set X will be denoted by $\mathbf{LFM}(G, \varphi)$.

The complexity of finding lexicographically first maximal sets satisfying certain properties has been investigated by Miyano[Mi].

It is shown in [CM90] that, for every monadic second order formula $\varphi(X)$, the evaluation LFM is inductively computable. Each attribute has a value $\subseteq D_G$ (or $\bot$) which can be represented by a Boolean vector of length $\mathbf{Card}(D_G)$. The basic operations can be performed in time $O(\mathbf{Card}(D_G))$. Hence, one obtains a global time complexity $O(|t^2|)$.

Another deterministic choice function can be constructed with minimality for inclusion instead of maximality. One obtains, in this way, methods for constructing optimal subgraphs as in [BLW] in linear time. Details will be given in the full paper.

5 Hyperedge replacement graph grammars and compatible functions

In the present section, we compare our results with those of [HKV89].

Let Γ be a hyperedge replacement grammar, and f be a function from graphs to values. This function is Γ-*compatible*[HKV89] if there exists a finite set of functions $f_0 = f, f_1, \ldots, f_n$, such that for every i, for every G, if G is obtained by a derivation sequence

$$u \longrightarrow_\Gamma H \overset{*}{\longrightarrow}_\Gamma G$$

where H has nonterminals $u_1, \ldots, u_m$ then,

$$f_i = h(f_{j_1}(G_{j'_1}), \ldots, f_{j_m}(G_{j'_m}))$$

where $u_i \overset{*}{\longrightarrow}_\Gamma G_i$ and $G = H[G_1/u_1, \ldots, G_m/u_m]$ (and h is a fixed mapping associated with H).

In our words, this means that f is inductively computable with respect to the graph operations associated with the right handsides of the grammar. (In the above definition, the graph operation associated with H maps $(G_1, \ldots, G_m)$ to $H[G_1/u_1, \ldots, G_m/u_m]$, the result of the simultaneous substitution of G_i for each u_i in H. See [BC87,Co90b] for more details).

It follows that every inductively computable evaluation and in particular every MS-evaluation is Γ-compatible for every hyperedge replacement graph grammar Γ.

If f is an evaluation, from graphs to integers, then one can consider the following decision problems, where $n \in \mathbb{N}$ and Γ is an hyperedge replacement grammar, both given as inputs:

1. Does there exist $G \in L(\Gamma)$ such that $f(G) \leq n$? (or $\geq n$, or $= n$) ? More difficult seems to be:

2. Does there exist m such that $f(G) \leq m$ for all $G \in L(\Gamma)$?

It is proved in [HKV89] that problems of the form 1 are decidable for evaluations such that all the decomposition operators can be written with $+, \times, \mathbf{max}, \mathbf{min}$, and that problems of the form 3 are decidable too for those not using $\mathbf{min}$.

It follows that problems of the forms 1 and 2 are decidable for evaluations of the forms $\mathbf{Card}(sat(\cdot, \varphi))$, $\mathbf{MaxCard}(sat(\cdot, \varphi))$ and $\sum \mathbf{Card}(sat(\cdot, \varphi))$, and that problems of the form 1 are also solvable for those of the form $\mathrm{MinCard}(sat(\cdot, \varphi))$ where, of course, φ is a MS-formula.

Acknowledgements: We thank H.J. Kreowski, S. Miyano, and T. Nishizeki for helpful remarks and suggestions on the topics of this paper.

References

[ACP] S. ARNBORG, D.G. CORNEIL, AND A. PROSKUROWSKI. Complexity of finding embeddings in a k-tree, *SIAM J. Alg. Disc. Meth.*, 8(1987), 277-287.

[ACPS] S. ARNBORG, B. COURCELLE, A. PROSKUROWSKI AND D. SEESE. An algebraic theory of graph reduction, Report 90-02, Bordeaux-1 University (1990).

[ALS] S. ARNBORG, J. LAGERGREN, AND D. SEESE. Problems easy for tree decomposable graphs.(extended abstract), *Proc. 15th ICALP.* Springer Verlag, Lect. Notes in Comp. Sc. 317(1988) 38-51.

[AP] S. ARNBORG, AND A. PROSKUROWSKI. Characterization and recognition of partial 3-trees, *SIAM J. Alg. Disc. Meth.*, 7(1986), 305-314.

[BC87] M. BAUDERON AND B. COURCELLE. Graph expressions and graph rewritings, *Math. systems Theory* 20(1987), 83-127.

[Bo89] H.L. BODLAENDER. Complexity of path forming games, Report RUU-CS-89-29, Utrecht University (1989).

[Bo88] H.L. BODLAENDER. Polynomial algorithms for chromatic index and graph isomorphism on partial k-trees, in *Proc. 1st Scandinavian Workshop on Algorithmic Theory.* Springer Verlag, Lect. Notes in Comp. Sc. 318(1988) 223-232.

[BLW] J.A. BERN, E. LAWLER AND A. WONG. Linear time computation of optimal subgraphs of decomposable graphs, *J. Algorithms* 8(1987), 216-235.

[Co90a] B. COURCELLE. The monadic second-order logic of graphs I: Recognizable sets of finite graphs, *Information and Computation* 85(1990), 12-75.

[Co90b] B. COURCELLE. Graph rewriting: an algebraic and logic approach, in *Handbook of Computer Science, Volume B*, J. Van Leeuwen Ed., Elsevier (1990), 193-242.

[Co91] B. COURCELLE. The monadic second-order logic of graphs V: On closing the gap between definability and recognizibility, *Theoret. Compt. Sci.* 80(1991) 153-202.

[CM90] B. COURCELLE, M. MOSBAH. Monadic second-order evaluations on tree-decomposable graphs, Report 90-110, Bordeaux-1 University (1990).

[DJL] P. DERANSART, M. JOURDAN, B. LORHO. Attribute grammars, Springer Verlag, Lect. Notes in Comp. Sc. 323(1988).

[GJ] M.R. GAREY AND D.S. JOHNSON. *Computers and Intractability*, W.H. Freeman and Company, San Francisco (1979).

[H89] A. HABEL. *Hyperedge Replacement Grammars and Languages*, PhD. thesis, Bremen University (1989).

[HKV89] A. HABEL, H.J. KREOWSKI AND W. VOGLER. Decidable boundness problems for hyperedge replacement graph grammars, Springer Verlag, Lect. Notes in Comp. Sc. 351(1989) 275-289.

[Jo85]   D.S. JOHNSON. The NP-completeness column: An ongoing guide (16th), *J. algorithms* 6(1985), 434-451.

[Mi]   S. MIYANO. The lexicographically first maximal subgraph problems: P-completeness and NC algorithms, *Math. systems Theory* 22(1989) 47-73.

[VLT]   J. VALDES, E. LAWLER AND R. TARJAN. The recognition of series-parallel digraphs, *SIAM J. Comput.* 11(1982) 289-313.

[TNS]   K. TAKAMIZAWA, T. NISHIZEKI AND N. SAITO. Linear-time computability of combinatorial problems on series-parallel graphs, *J. Assoc. Comput. Mach* 29(1982) 623-641.

[W]   T. WIMER. Linear algorithms on *k*-terminal graphs, PhD. thesis, Clemson University(1987).

Optimal Embedding of Complete Binary Trees into Lines and Grids

R. Heckmann, R. Klasing*, B. Monien*, W. Unger

Department of Mathematics and Computer Science
University of Paderborn, West Germany

e-mail : Ralf.Heckmann@uni-paderborn.de, Ralf.Klasing@uni-paderborn.de,

Burkhard.Monien@uni-paderborn.de, Walter.Unger@uni-paderborn.de

Abstract

We consider several graph embedding problems which have applications in parallel and distributed computing and which have been unsolved so far. Our major result is that the complete binary tree can be embedded into the square grid of the same size with almost optimal dilation (up to a very small factor). To achieve this, we first state an embedding of the complete binary tree into the line with optimal dilation.

1 Introduction

Graph embedding problems have gained importance in different areas of computer science (for a survey, cf. [MS90], [Ro88]). E.g. in the field of interconnection networks for parallel computer architectures, graph embeddings can generally be used to model the simulation of network and algorithm structures on a different network.

Several types of networks have been considered, like hypercubes, shuffle-exchange networks, X-trees, binary trees, grids and lines. We will take a look at the latter ones. The importance of the tree-like structures arises from their use as data and algorithm structures. Furthermore, interconnection networks are also arranged as trees and meshes. Finally, embeddings of graphs into meshes are of special interest in VLSI layout (see e.g. [Ull84]).

The quality of an embedding can be measured by certain cost criteria. One of these criteria which is considered very often is the dilation. The dilation of an embedding is the maximum distance between images of adjacent nodes. It is a measure for the communication time needed when simulating one network on another.

---

*The work of these authors was supported by grant Mo 285/4-1 from the German Research Association (DFG).

A lot of work has been done in the area of VLSI layout. The general problem of laying out trees (i.e. the embedding of trees into 2-dimensional meshes) is discussed in [FP80] and [RS81]. Layouts of complete binary trees are investigated in several papers. The well-known H-tree construction from [Ull84] and the improved embeddings in [Go87] and [SY88] all have $O(\sqrt{n})$ dilation and constant expansion (i.e. the number of processors only increases by a constant factor), whereas the modified H-tree construction from [PRS81], [Ull84] has $O(\sqrt{n}/\log n)$ dilation at a constant expansion matching the lower bound of $\Omega(\sqrt{n}/\log n)$ from [Ull84]. An embedding of the complete binary tree into its optimal square mesh, i.e. a square mesh of the same size, with dilation $O(\sqrt{n})$ [and edge congestion 2] is presented in [Zi90].

As our main result, we will present a near optimal embedding of the complete binary tree into its optimal square grid. In section 4 we will prove the following theorem:

Theorem 2:

The complete binary tree $T(k)$ can be embedded into its optimal square mesh with dilation at most

$$\left\lceil \frac{2^{(k+1)/2}-2}{k-1} \right\rceil + 2 \quad \text{if} \quad k \text{ is odd},$$

$$\left\lceil \frac{2^{(k+1)/2}-2}{k-2} \right\rceil + 5 \quad \text{if} \quad k \text{ is even}.$$

This theorem yields a near optimal embedding, because the dilation differs only by a factor of $(1 + \epsilon)$ from the trivial lower bound

$$\left\lceil \frac{\text{diameter of the mesh}}{\text{diameter of the tree}} \right\rceil = \left\lceil \frac{2 \cdot (\lceil \sqrt{2^{k+1}-1} \rceil - 1)}{2k} \right\rceil \geq \left\lceil \frac{2^{(k+1)/2}-2}{k} \right\rceil .$$

In terms of the number n of the nodes, we are able to state for the first time that the dilation for embedding the complete binary tree into its optimal square mesh is in $\Theta(\sqrt{n}/\log n)$.

Our mapping does not even exploit the whole mesh structure. It only needs a substructure shaped like a "twofold comb". The details of this construction are investigated in section 4.

For the proof of Theorem 2, we will use the embedding of a complete binary tree into the line. In section 3, we will construct an algorithm, called tree_into_line, which computes such an embedding:

Theorem 1:

The algorithm tree_into_line computes an embedding of any complete binary tree into the line with optimum dilation $l = \lceil (2^k - 1)/k \rceil$.

Thus, the algorithm matches the trivial lower bound

$$\left\lceil \frac{(\text{number of nodes}) - 1}{\text{diameter}} \right\rceil = \left\lceil \frac{2^{k+1}-2}{2k} \right\rceil = \left\lceil \frac{2^k - 1}{k} \right\rceil$$

for the dilation of any embedding. To our knowledge, this is the first time that an optimal upper bound is reported. According to [CCDG82], only the trivial upper bound of (no. of nodes) $- 1 = 2^{k+1} - 2$ has been proven so far. In [Ch88], an optimal upper bound of $\lceil (2^k - 1)/k \rceil$ is stated.. But the construction given there is incomplete. In this paper, we extend the construction of [Ch88] and correct it in this way. The idea of our algorithm laying out the complete binary tree in the line is surprisingly simple. Basically, it proceeds by laying out the nodes of the tree roughly from left to right into "blocks" of a certain size. For later use — the embedding of the complete binary tree into the mesh — the algorithm distributes the leaves almost evenly into the line. The technical details and an analysis of the algorithm are given in section 3.

The embedding of graphs into lines is a well-known and important problem. This is the so-called Bandwidth Minimization Problem (BMP). Also, embeddings of graphs into lines are motivated·by applications in sparse matrices.

There is a wealth of results about the BMP (for a survey, cf. [CCDG82]). It is known that the BMP is NP-complete for trees with maximum vertex degree 3, [GJ79], and even for a very restricted class of trees, the so-called caterpillars with hair length at most 3, [Mo86]. Also, an $O(\log n)$ times optimal algorithm for caterpillar graphs is stated in [HMM90]. Not many results are known for trees in general (see e.g. [CCDG82, chap. 4.2]).

In section 5 we present some more applications and results (about X-trees and k-ary trees) for our technique devised in section 3.

2   Definitions

(For the definitions, cf. [MS90], [Ro88].) The *complete binary tree* of height k, denoted by $T(k)$, is the graph whose nodes are all binary strings of length at most k and whose edges connect each string x of length i $(0 \leq i < k)$ with the strings xa, a in $\{0, 1\}$, of length $i + 1$. The node ϵ, where ϵ is the empty string, is the *root* of $T(k)$ and a node x is at *level* i, $i \geq 0$, in $T(k)$ if x is a string of length i. The nodes at level k are the *leaves* of the tree. For a node x at level i, $0 \leq i < k$, the node $x0$ is called the *left son* of x, $x1$ the *right son*. x is called the *parent* of $x0$ and $x1$. For any node x, the nodes xu, $u \in \{0, 1\}^*$, are called *descendants* of x, and x is called an *ancestor* of xu. A *binary tree* is a connected subgraph of $T(k)$, for some $k \geq 0$. Ternary trees and general trees can be defined in a similar way.

The *d-dimensional mesh* of dimensions $a_1, a_2, \ldots, a_d$, denoted by $[a_1 \times a_2 \times \ldots \times a_d]$, is the graph whose nodes are all d-tuples of positive integers $(z_1, z_2, \ldots, z_d)$, where $1 \leq z_i \leq a_i$, for all i $(0 \leq i \leq d)$, and whose edges connect d-tuples which differ in exactly one coordinate by one.

Let G and H be finite undirected graphs. An *embedding* of G into H is a one-to-one mapping f from the nodes of G to the nodes of H. G is called the *guest* graph and H is called the *host* graph of the embedding f. The *dilation* of the embedding f is the maximum distance in the host between the images of adjacent guest nodes. Its *expansion* is the ratio of the number of nodes in the host graph to the number of nodes in the guest, i.e. $|\text{nodes}(H)|/|\text{nodes}(G)|$. When hosts are chosen from a collection C and no graph K in C satisfies $|\text{nodes}(G)| \leq |\text{nodes}(K)| \leq |\text{nodes}(H)|$, then H is called an *optimal host* in C for G. The *edge congestion* of the embedding f is the maximum number of edges that are routed through a single edge of H. (A *routing* is a mapping r of G's edges to paths in H, $r(v_1, v_2) = $ a path from $f(v_1)$ to $f(v_2)$ in H.) When embedding graphs into simple paths, dilation is customarily called *bandwidth*. The bandwidth of a graph is the minimum bandwidth of all such embeddings.

3 Bandwidth of Complete Binary Trees

In this section we will describe the embedding of a complete binary tree into the line with optimal dilation. Let T be a complete binary tree of height k with root r. We have to describe an embedding of T into the line with dilation $l = \lceil (2^k - 1)/k \rceil$.

To be more precise: Let r' (r'') be the left (right) son of r and T' (T'') the complete subtree with r' (r'') as root. We will describe the embedding of T' into k blocks of length l such that r' is located in the rightmost block. By embedding T'' as a mirror image of the embedding of T' we get an embedding of T by placing r in the middle between the two embeddings of T' and T''. So we just have to describe the embedding of T'.

Before presenting the precise algorithm we will give an informal and incomplete description. Let $B_1, B_2, ...B_k$ be k blocks of length l, i.e. $B_i = \{(i-1) \cdot l, ..., i \cdot l - 1\}$ for $1 \leq i \leq k$. These blocks will be filled step by step. In the first step block B_1 is filled with the l leftmost leaves of T'. In the i-th step the block B_i is filled with the parents of nodes located within B_{i-1}. The remaining free positions in B_i are filled with the leftmost unplaced leaves of T'. The algorithm described so far matches the algorithm from Chung [Ch88]. We will see in the following that this algorithm is incorrect.

To get a better view of this filling up technique and to develop a full description we take a closer look at the case $k = 2^{k'}$ for some integer k'. In this case $l = 2^{2^{k'}-k'}$ holds, thus l is a power of 2. Block B_1 contains l leaves (nodes of level k). In the block B_2 there are $l/2$ nodes of level $k-1$ and $l/2$ leaves located and block B_3 contains $l/4$ nodes of level $k-2$, $l/4$ nodes of level $k-1$ and $l/2$ leaves. This may be continued until we reach the situation where there is only one node of one level within one block. This situation, reached in block $B_{\log(l)+1}$ is shown in Figure 1.

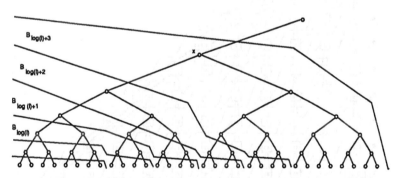

Figure 1: Situation within the algorithm

Also block $B_{\log(l)+2}$ may take all parents of nodes of block $B_{\log(l)+1}$, but after block $B_{\log(l)+2}$ we have to consider also the descendants of node x as well. In our algorithm we will insert also descendants of node x into block $B_{\log(l)+3}$ (see Figure 1). With this modification we are now able to state our full algorithm.

Algorithm tree_into_line

1:    Let T be the complete binary tree of height k with root r and T' (T'') the left (right) subtree of r. Let $l = \lceil (2^k - 1)/k \rceil$ and $B_1, B_2, ..., B_k$ be k blocks of length l.

2:    Place the l leftmost leaves of T' into B_1.

3:    For all i ranging from 2 to k fill block B_i as follows:

     3.1:   Place all parents of nodes in B_{i-1} — which are not already placed — within block B_i. Visit the nodes of B_{i-1} from left to right and place the nodes in block B_i also from left to right.

     3.2:   Place all descendants — which are not already laid out — of nodes located in block B_i into block B_i.

     3.3:   Fill the remaining positions of block B_i with the leftmost leaves of T' which are not placed already.

4:    Place r as the last node in block B_k.

5:    Embed T'' as a mirror image of T' to the right of r.

In Figure 2 we give a full example of our algorithm for the case $k = 12$. In this Figure we represent x nodes of the same level by a square marked with the number x. The dashed lines are edges between nodes within the same block.

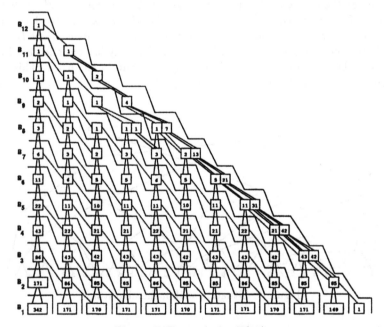

Figure 2: Example for $T(12)$

Theorem 1:

The algorithm `tree_into_line` computes an embedding of any complete binary tree into the line with optimum dilation $l = \lceil (2^k - 1)/k \rceil$.

Proof:

The proof of correctness for this algorithm uses some complex estimations, which we will not present here. Just the basic ideas are given.

When l is not a power of 2, then there may be in any block B_i a node and some of its descendants (see Figure 2). But these descendants are placed by step 3.1 into the block, they are parents of nodes within block B_{i-1}.

As long as there are no nodes placed into a block by step 3.2 the correctness of the algorithm follows from:

- Step 3.1 does not exceed the claimed dilation, because of the left to right motion within step 3.1.

- The number of parents of a block B_i is less than l, thus they fit into block B_{i+1}.

The two conditions from above are easy to verify. Using some not very difficult estimations we conclude: Step 3.2 inserts only nodes into the blocks B_j for $\lceil k - \log(k) \rceil < j \leq k$. Thus algorithm `tree_into_line` works correctly for the first $\lceil k - \log(k) \rceil$ blocks.

When step 3.2 is involved we have to estimate the number of nodes inserted into a block B_i by step 3.2. These estimations will be presented in the final paper. □

In the next section this embedding is used to embed the complete binary tree into a mesh. For that embedding we need some very precise information about the positions on the line to which the leaves of T are mapped.

Lemma 1:

Let T be a complete binary tree with height k and $l = \lceil (2^k - 1)/k \rceil$. Consider the embedding of T determined by the algorithm `tree_into_line`. Let $x_1, x_2, ... x_{2^k}$ with $x_i < x_{i+1}$ for $1 \leq i < 2^k$ be the positions to which the leaves of T are mapped. Then $2 \cdot i - l \leq x_i \leq 2 \cdot i + l$ holds for all $1 \leq i \leq 2^k$.

4 Embedding Complete Binary Trees into Square Meshes

In this section, we will show how to use the optimum bandwidth embedding from Section 3 in order to embed the complete binary tree $T(k)$ into its optimal square mesh with low dilation. To be more precise, we prove the following theorem:

Theorem 2:

The complete binary tree $T(k)$ can be embedded into its optimal square mesh with dilation at most

$$\left\lceil \frac{2^{(k+1)/2}-2}{k-1} \right\rceil + 2 \quad \text{if} \quad k \text{ is odd,}$$

$$\left\lceil \frac{2^{(k+1)/2}-2}{k-2} \right\rceil + 5 \quad \text{if} \quad k \text{ is even.}$$

Proof: We distinguish two cases:

A) k odd

If k is odd, then $T(k)$ has to be embedded into an $(m \times m)$-mesh M where $m = 2^{(k+1)/2}$. We partition $T(k)$ into subtrees $T_0, T_1, \ldots, T_m$ as shown in Figure 3.

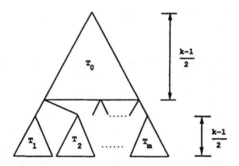

Figure 3: Partitioning of $T(k)$, k odd

Let $\tilde{f}$ be the minimum bandwidth embedding of $T((k-1)/2)$ (into the path of length $m-1$) as specified in section 3. We will now describe the embedding f of $T(k)$ into M. The overall embedding scheme of f is displayed in Figure 4.

1. First, the tree T_0 is embedded into the $(m/2)$-th row of M by using $\tilde{f}$ and leaving the last node in that row empty.

2. Then, for each i, $1 \leq i \leq m$, the i-th node of level $(k+1)/2$ of $T(k)$, i.e. the root of T_i, is mapped to the i-th node of the $(m/2+1)$-st row of M.

3. Finally, for each i, $1 \leq i \leq m$, the tree T_i is embedded into the i-th column of the mesh M by using the embedding $\tilde{f}$ and leaving the $(m/2)$-th row of M empty.

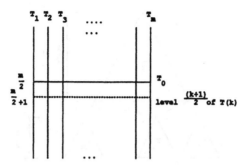

Figure 4: Embedding of $T(k)$, k odd

As mentioned before, we do not even need the whole mesh structure, but only a substructure shaped like a "twofold comb". For the correctness and low dilation of our embedding, we use the following properties of the embedding $\tilde{f}$ as constructed in Section 3:

- $\tilde{f}$ always maps the root of a tree to the middle node of the line. Therefore, the root of T_i, $1 \leq i \leq m$, is mapped to the same position by the second and the third step of the embedding f.

- $\tilde{f}$ distributes the leaves of T_0 nearly evenly on the $(m/2)$-th row of M according to Lemma 1 of Section 3. As the sons of the leaves are also distributed evenly on the $(m/2+1)$-th row of M, they are within a small distance from the parent, at most (bandwidth of T_0) $+ 2$.

- $\tilde{f}$ has optimal bandwidth. Therefore, if u, v are adjacent nodes in $T(k)$ within the same T_i, $0 \leq i \leq m$, then the maximum distance between their image nodes in M is at most (bandwidth of T_i).

Overall, the dilation of f is at most

$$\text{(bandwidth of } T((k-1)/2)) + 2 \;=\; \left\lceil \frac{2^{(k-1)/2} - 1}{(k-1)/2} \right\rceil + 2 \;=\; \left\lceil \frac{2^{(k+1)/2} - 2}{k-1} \right\rceil + 2 \;.$$

B) k even

If k is even, then $T(k)$ is first embedded into an almost square $(m \times 2m)$-mesh M where $m = 2^{k/2}$ by using the same technique as above. This time, we partition $T(k)$ into subtrees $T_0, T_1, \ldots, T_m$ as shown in Figure 5.

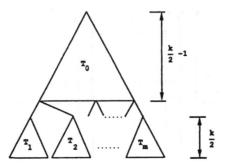

Figure 5: Partitioning of $T(k)$, k even

By applying the same embedding scheme as above and by making use of the optimum bandwidth embeddings of the subtrees (as derived in section 3), we achieve the situation displayed in Fig. 6.

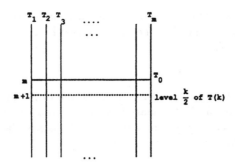

Figure 6: Embedding of $T(k)$, k even

After this is done, the grid M is transformed into a square grid M' by

- first, stretching the distances in the rows by a factor of $\sqrt{2}$,

- and then, squeezing the columns by a factor of $\sqrt{2}$.

In the first step, all the positions in the horizontal direction are multiplied by a factor of $\sqrt{2}$ and possibly rounded to the next highest integer.

In the second step, the columns containing the nodes of the tree are squeezed together to form a snakelike line. For each column, this can be done in such a way that the previous distances are reduced by a factor of $\sqrt{2}$ plus a small additive constant of c. In the same way, this can be achieved simultaneously for all columns containing tree nodes.

Our technique to square up grids should be contrasted to *squeezing* and *folding* techniques used in [AR82] and [Ell88]. Our approach works a lot better, because we only have to square up a structure shaped like a "twofold comb". We do not have to consider all the edges of the mesh M. To complete the proof, we state the dilation of our final embedding:

If u, v are adjacent nodes in $T(k)$ within the same T_i, $1 \leq i \leq m$, then the maximum distance between their image nodes in M' is at most

$$\left\lceil \frac{1}{\sqrt{2}} \cdot (\text{bandwidth of } T_i) \right\rceil + 4 = \left\lceil \frac{1}{\sqrt{2}} \cdot (\text{bandwidth of } T(k/2)) \right\rceil + 4$$

$$= \left\lceil \frac{1}{\sqrt{2}} \cdot \left\lceil \frac{2^{k/2} - 1}{k/2} \right\rceil \right\rceil + 4 \leq \left\lceil \frac{2^{(k+1)/2} - 2}{k} \right\rceil + 5 .$$

If u, v are adjacent nodes in $T(k)$ within T_0 or with u at level $k/2 - 1$ and v at level $k/2$, then by applying Lemma 1 of Section 3 we also have a maximum distance of

$$\left\lceil \sqrt{2} \cdot (\text{bandwidth of } T_0) \right\rceil + 3 = \left\lceil \sqrt{2} \cdot (\text{bandwidth of } T((k-2)/2)) \right\rceil + 3$$

$$= \left\lceil \sqrt{2} \cdot \left\lceil \frac{2^{(k-2)/2} - 1}{(k-2)/2} \right\rceil \right\rceil + 3 \leq \left\lceil \frac{2^{(k+1)/2} - 2}{k - 2} \right\rceil + 5$$

between their image nodes in M'. Hence, we achieve the desired dilation.

5 Conclusions

In this paper, we have devised a technique to determine the exact bandwidth of the complete binary tree and presented an embedding into the grid which is nearly optimal. But there are some more results achievable by our technique. It can also be applied for X-trees, d-dimensional meshes and k-ary trees. The corresponding results which can be achieved are as follows:

By applying a recursive embedding technique, we can determine an upper bound for the bandwidth of the complete ternary tree which is at most (optimal bandwidth) $* (1 + \epsilon)$, where ϵ is a very small constant, $\epsilon > 0$.

For the bandwidth of k-ary trees we have also proved near optimal embeddings.

In addition, our technique also embeds the X-tree with bandwidth $O(n/\log n)$. With a slight modification of the technique, we get an embedding of the X-tree into the square mesh with dilation $O(\sqrt{n}/\log n)$. The factors involved may be improved considerably. Combining this with the fact that every binary tree can be embedded with $O(1)$ dilation into an X-tree of the same size, [Mo91], we have an efficient embedding of any arbitrary binary tree into the mesh.

The techniques described above are also applicable for embeddings into d-dimensional meshes. In this context, they also lead to near optimal results.

As far as embeddings of general trees into meshes are concerned, we will have to do further research on whether and how the bandwidth results can be used for any mesh embedding. Maybe, known results on embeddings between different kinds of meshes could be helpful here, [AK88, AR82, Ell88]. Also, it should be interesting to find out whether the embeddings can still be improved if the whole mesh structure is taken into account.

References

[AK88]   M.J. Atallah, S.R. Kosaraju, "Optimal simulations between mesh-connected arrays of processors", *Journal of the ACM*, vol. 35 (1988), pp. 635-650.

[AR82]   R. Aleliunas, A.L. Rosenberg, "On Embedding Rectangular Grids in Square Grids", *IEEE Transactions on Computers,* vol. C-31 (1982), pp. 907-913.

[CCDG82] P.Z. Chinn, J. Chvátalová, A.K. Dewedney, and N.E. Gibbs. The bandwidth problem for graphs and matrices - a survey. *Journal of Graph Theory,* 6:223 – 254, 1982.

[Ch88]   F.R.K. Chung, "Labelings of Graphs", in *Selected Topics in Graph Theory III*, edited by L.W. Beineke and R.J. Wilson, Academic Press, (1988) pp. 151-168.

[Ell88]   J.A. Ellis, "Embedding Rectangular Grids into Square Grids", *Proceedings of the 3rd Aegean Workshop on Computing (AWOC): VLSI Algorithms and Architectures* (1988), LNCS 319, pp. 181-190.

[FP80]   M.J. Fischer, M.S. Paterson, "Optimal Tree Layout", *Proceedings of the 12th ACM Symposium on the Theory of Computing* (1980), pp. 177-189.

[GJ79]   M.R. Garey, D.S. Johnson, "Computers and Intractability", *W.H. Freeman,* New York, 1979.

[Go87]   D. Gordon, "Efficient Embeddings of Binary Trees in VLSI Arrays", *IEEE Transactions on Computers,* vol. C-36 (1987), pp. 1009-1018.

[HMM90]   J. Haralambides, F. Makedon, B. Monien, "Approximation Algorithms for the Bandwidth Minimization Problem for Caterpillar Graphs", *Proceedings of the 2nd IEEE Symposium on Parallel and Distributed Processing,* (1990), pp. 301-307.

[Mo86]   B. Monien, "The bandwidth minimization problem for caterpillars with hair length 3 is NP-complete", *SIAM J. Alg. and Discrete Methods,* 7 (1986), pp. 505-512.

[Mo91]   B. Monien, "Simulating Binary Trees on X-trees", *Proceedings of the 3rd ACM Symposium on Parallel Algorithms and Architectures* (1991).

[MS90]   B. Monien, I.H. Sudborough, "Embedding one Interconnection Network in Another", *Computing Suppl.* 7 (1990), pp. 257-282.

[PRS81]   M.S. Paterson, W.L. Ruzzo, L. Snyder, "Bounds on minimax edge length for complete binary trees", *Proceedings of the 13th ACM Symposium on the Theory of Computing* (1981), pp. 293-299.

[Ro88]   A.L. Rosenberg, "Graph embeddings 1988: Recent breakthroughs, new directions", *Proceedings of the 3rd Aegean Workshop on Computing (AWOC): VLSI Algorithms and Architectures* (1988), LNCS 319, pp. 160-169.

[RS81]   W.L. Ruzzo, L. Snyder, "Minimum Edge Length Planar Embeddings of Trees", in Kung, Sproull, Steele: *VLSI Systems and Computations,* Computer Science Press (1981), pp. 119-123.

[SY88]   A.D. Singh, H.Y. Youn, "Near Optimal Embedding of Binary Tree Architectures in VLSI", *Proceedings of the 8th International Conference on Distributed Computing Systems* (1988), pp. 86-93.

[Ull84]   J.D. Ullman, "Computational Aspects of VLSI", *Computer Science Press,* 1984.

[Zi90]   P. Zienicke, "Embedding of Treelike Graphs into 2-dimensional Meshes", *Proceedings of the 16th International Workshop on Graph-Theoretic Concepts in Computer Science (WG 90).*

Graph Rewriting Systems and their Application to Network Reliability Analysis

Yasuyoshi Okada, Masahiro Hayashi

NTT Basic Research Lab., NTT Telecommunication Networks Research Lab.
3-9-11, Midori-cho, Musashino, Tokyo 180 Japan
email: okada@nuesun.ntt.jp

.Abstract
We propose a new kind of Graph Rewriting Systems (GRS) that provide a theoretical foundation for using the reduction methods to analyze network reliability, and give the critical pair lemma in this paper.

1 Introduction

Many graph rewriting systems(GRS) and graph grammars have been proposed and have found uses in such diverse fields as the category theory, the semantics of recursively defined functions, record handling in data base systems, compiler techniques, and analyzing biological development and evolution (see [4], [5], [6], [7], [8], [9], [10], [17], [18], [20]). In these systems, rewriting has been primarily identified as a production rule or a derivation rule: from a simpler graph to a more complicated one. Recently, a graph rewriting was introduced as a reduction rule [1] for monadic second order logic. Our goal is to provide a theoretical foundation for network reliability analysis, so we define rewriting as the simplification of graphs. It is therefore natural to define GRS in the framework of "Abstract Reduction Systems"(ARS). That is, GRS are defined as ARS for graphs, just as term rewriting systems (TRS) are ARS for terms. They differ from TRS, however, in that with GRS it seems difficult to define rewriting together with redex occurrence in a graph and it is defficult to well define the convergence of a critical pair. To cope with these difficulties, we introduce a boundary graph (B-Graph), all of whose vertices are labelled and interior vertices are distinguished from boundary vertices. Accordingly, the conventional basic notions in TRS (superposition, critical pair, convergence of critical pair) can be applied to B-graphs. For application to network reliability analysis, we will construct a complete system (strongly normalizing and Church Rosser) from conventional incomplete rewriting rules. This is because a complete system needs no strategy concerning sites and order of rule applications. This new desired system is constructed by using our main theorem, the critical pair Lemma for B-graphs, to eliminate all critical pairs (Knuth-Bendix Algorithm).

2 Graph Rewriting Systems

2.1 Basic Notions

Let us first review the concept and properties of an *Abstract Reduction System (ARS)*, which is a structure $A = \langle A, \longrightarrow \rangle$ consisting of a set A and a binary relation $\longrightarrow$ on A [14]. If we have $(a, b) \in \longrightarrow$ for $a, b \in A$, we write $a \longrightarrow b$ and call b a one-step *reduct* of a. A transitive-reflexive closure of $\longrightarrow$ is denoted by $\overset{*}{\longrightarrow}$. So $a \overset{*}{\longrightarrow} b$ if there is a possibly empty, finite sequence of 'reduction steps' $a \equiv a_0 \longrightarrow a_1 \longrightarrow ... \longrightarrow a_n \equiv b$. Here $\equiv$ denotes the identity of the elements of A. We say that $a \in A$ is a *normal form* if there is no $b \in A$ such that $a \longrightarrow b$. And we say that $b \in A$ *has a normal form* if there is a normal form a such that $b \overset{*}{\longrightarrow} a$. The structure A is *strongly normalizing (SN)* if there are no infinite reduction sequences $a_0 \longrightarrow a_1 \longrightarrow a_2....$ The reduction relation of $\longrightarrow$

is *weakly Church − Rosser*(WCR) if for all $a, b, c \in A$ with $a \longrightarrow b$ and $a \longrightarrow c$, we can find $d \in A$ such that $b \xrightarrow{*} d$ and $c \xrightarrow{*} d$. This d is called the *common reduct* of b and c. The structure A is *Church − Rosser*(CR) if for all $a, b, c \in A$ with $a \xrightarrow{*} b$ and $a \xrightarrow{*} c$, we can find $d \in A$ such that $b \xrightarrow{*} d$ and $c \xrightarrow{*} d$. An A with the SN and CR properties is called *complete*. The following proposition is fundamental to ARS.

Proposition 1 *Let A be an ARS.*
i) If A is WCR and SN, then A is CR.
ii) If A is CR, then A has a unique normal form.

Here i) is Newman's lemma [14], [21]. The concept of the CR property originated in lambda calculus [3], [13] and it appears in many systems - for examples, see [14], [15], [16],[31] for TRS and see [5], [6], [9] for graph grammars. Note that if A is complete, then A has a unique normal form.

We use the concept of an ARS to propose a new type of *Graph Rewriting Systems* (GRS); they follow in the framework of ARS (in the original sketch see [22].) We do by this introducing a boundary graph (B-Graph); here interior vertices are distinguished from boundary vertices and all vertices are labeled. Accordingly, the conventional basic notions in TRS (superposition, critical pairs, and convergence of critical pairs) can be defined for B-Graphs.

Definition 1 Boundary Graph and Subgraph : A *boundary graph* G is a quadruplet : $\langle V_G, E_G, lab_G, typ_G \rangle$, where
V_G : a finite set of elements called *vertices.*
E_G : a finite set of elements called *edges* and that are associated with an unordered pair of vertices $\{u, v\}$ for $u, v \in V_G$.
$lab_G : V_G \longrightarrow \mathbf{N}$ is an injective mapping, where $\mathbf{N}$ is the set of natural numbers. $lab_G(v)$ is called the *label* of v for each of $v \in V_G$.
$typ_G : V_G \longrightarrow T$ is a mapping, where T is a set of two elements $\{interior, boundary\}$. $typ_G(v)$ is called the *type* of v.
A family consisting of B-graphs is denoted by $\mathcal{G}$.

Put $V_G^o = \{v \in V_G | \ typ_G(v) \ is \ interior\}$ and $\partial V_G = \{v \in V_G | \ typ_G(v) \ is \ boundary\}$. Elements of V_G^o are called *interior* vertices of G and elements of ∂V_G are called *boundary* vertices of G. Then $V_G = V_G^o \amalg \partial V_G$ is the disjoint union of two sets. In this paper, interior vertices will be represented by open circles; boundary vertices, by filled circles (see, e.g. Fig. 1.). For every two B-graphs s and t, it is assumed that $\{lab_s(v)|v \in V_s^o\} \cap \{lab_t(v)|v \in V_t^o\} = \emptyset$.

A *boundary subgraph* (B−subgraph) g of G, denoted by $g \subset_G G$ ($g \subset G$) is a quadruple $g = \langle V_g, E_g, lab_g, typ_g \rangle$ that is a B-graph. Here $V_g \subset V_G$, $E_g \subset E_G$, $lab_g = lab_G|_{V_g}$, and $typ_g = typ_G|_{V_g}$ - where | denotes a restriction of a mapping.
Edge, degree, path, connectedness, and *cycle* here are the conventional notions in graph theory (see [11]).

Definition 2 Isomorphism and B-isomorphism between B-Graphs :
i) For $H, K \in \mathcal{G}$, an isomorphism $f : H \longrightarrow K$ is a pair of bijective mappings (f_V, f_E) such that $f_V : V_H \longrightarrow V_K$ and $f_E : E_H \longrightarrow E_K$, where for $e \in E_H$ associated with $\{u, v\} \in V_H$, $f_E(e)$ is associated with an unorderd pair of $\{f_V(u), f_V(v)\}$.
ii) For $H, K \in \mathcal{G}$, $f : H \longrightarrow K$ is a B-isomorphism if f is an isomorphism such that $f_V(\partial V_H) = \partial V_K$ and $(lab_H)(v) = (lab_K)(f(v))$ for all $v \in \partial V_H$.

There is no B-isomorphism in Fig. 1. (b), because the second condition of Definition 2 is not satisfied.

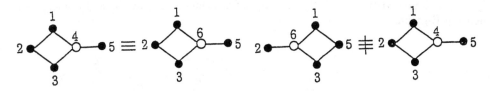

Fig. 1. (a) B-isomorphic   (b) non-B-isomorphic

Definition 3 **Difference B-Graph :** Given $G \in \mathcal{G}$ and $g \subset G$, $v \in V_g$ is a *common vertex* in V_G if there exists $e \in E_G$ and $e \notin E_g$. The set $\{v|v : \text{common vertices}\}$ is denoted by V_{gCG}^{co}. The *difference B-graph* between g and G, denoted by $G\backslash g$, (see Fig. 2) is defined as follows: $G\backslash g = \langle V_{G\backslash g}, E_{G\backslash g}, lab_{G\backslash g}, typ_{G\backslash g} \rangle$ where $V_{G\backslash g} = (V_G - V_g) \cup V_{gCG}^{co}$, $E_{G\backslash g} = (E_G - E_g)$, $lab_{G\backslash g} = lab_G|_{V_{G\backslash g}}$, and $typ_{G\backslash g} = typ_G|_{V_{G\backslash g}}$.

Let us now the define the union of two mappings. For sets A and B, if two partial mappings $f : A \longrightarrow B$ and $f' : A' \longrightarrow B$ coincide on $Dom(f) \cap Dom(f')$, we denote by $f \cup f'$ their common extension into a partial mapping : $A \cup A' \longrightarrow B$, with domain $Dom(f) \cup Dom(f')$.

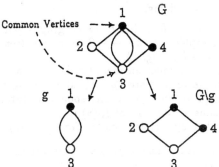

Fig. 2. g, $G\backslash g$, and common vertices.

For different B-graphs with common vertices we can obtaine the convenient property of B-isomorphism.

Proposition 2 *Let G and G' be B-Graphs with $g \subset G$ and $g' \subset G'$. If there exist two B-isomorphisms, $f : G\backslash g \longrightarrow G'\backslash g'$ and $h : g \longrightarrow g'$ such that $f_V|_{v_{gCG}^{co}} = h_V|_{v_{gCG}^{co}}$, then there exists a B-isomorphism $K : G \longrightarrow G'$.*

Proof. By their definition, V_G and E_G can be written as $V_G = V_g \cup V_{G\backslash g}$ and $E_G = E_g \cup E_{G\backslash g}$. Then by $f_V|_{v_{gCG}^{co}} = h_V|_{v_{gCG}^{co}}$, B-isomorphism $K : G \longrightarrow G'$ is a pair (K_V, K_E) defined as $K_V = (f_V \cup h_V)$ and $K_E = (f_E \cup h_E)$.

Definition 4 **Reduction Rule :** A *reduction rule* (or *rewriting rule*) is a pair of B-graphs (s, t) with $s, t \in \mathcal{G}$ subject to the following constraints.
   i) s is connected and $E_s \neq \emptyset$.
   ii) there is a bijective mapping $f : \partial V_s \longrightarrow \partial V_t$ such that $lab_t(f(v)) = lab_s(v)$ for all $v \in \partial V_s$.

An example of a reduction rule is shown in Fig. 3.

Definition 5   Graph Rewriting System :   A *Graph Rewriting System* is a pair $(\mathcal{G}, \mathcal{R})$, where $\mathcal{R}$ is a set of rewriting rules, (s, t). (usually we write $r : s \longrightarrow t$.) Henceforth $(\mathcal{G}, \mathcal{R})$ is simply written as $\mathcal{R}$. The identity of elements $\equiv$ is given by a B-isomorphism between B-graphs.

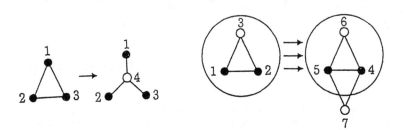

Fig. 3. Reduction Rule.   Fig. 4. Embedding.

Definition 6   Embedding :   Given $s, G \in \mathcal{G}$, a mapping $\sigma : s \longrightarrow s_\sigma \subset G$ is an embedding if σ is an isomorphism such that $\sigma_V(v) \in V_G^o$ with $deg(v) = deg(\sigma_V(v))$ in G for all $v \in V_s^o$. Then s_σ is an *occurrence* of σ in G. We denote this as $G \equiv G[s_\sigma]$. In two embeddings $\sigma_1 : s \longrightarrow M$ and $\sigma_2 : M \longrightarrow G$, $\sigma_2 \circ \sigma_1$ is a pair $\langle (\sigma_2)_V \circ (\sigma_1)_V, (\sigma_2)_E \circ (\sigma_1)_E \rangle$.

Figure 4 shows an example of embedding.

Definition 7   Composition :   Let H and K be two B-graphs, with $h \subset H$ and $k \subset K$, such that there exists an isomorphism $f : h \longrightarrow k$. The *composition* of H and K by f is a new B-graph defined as follows (see Fig. 5):

(1) for all $v \in V_h$ and $e \in E_h$, under an identification of v and $f_V(v)$ to $\tilde{v}$, and an identification of e and $f_E(e)$, the vertices are $V_H \cup V_K$ and the edges are $E_H \cup E_K$.

(2) the type of $\tilde{v}$ is defined according to the correspondence table (Table 1).

(3) the label of $\tilde{v}$ is the same as the label of $f_V(v)$.

The composition of H and K by f is denoted by $Comp_f(H, K)$.

$Comp_f(H, K)$ induces two *natural isomorphisms*. They are $id^H : H \longrightarrow \hat{H} \subset M$ and $id^K : K \longrightarrow \hat{K} \subset M$, where $M \equiv Comp_f(H, K)$.

$v \setminus f_V(v)$	$\circ$	$\bullet$
$\circ$	$\circ$	$\circ$
$\bullet$	$\circ$	$\bullet$

Table 1.

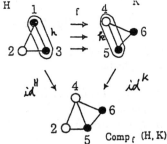

Fig. 5. Composition.

Definition 8   One-step Reduction :   Let $r : s \longrightarrow t$ be a rewriting rule with a bijective mapping $f : \partial V_s \longrightarrow \partial V_t$, and let $\sigma : s \longrightarrow s_\sigma \subset G$ be an embedding; in the rewriting rule r, the labels of every pair of nodes in t and G should be kept distinct by renaming the labels of nodes.

The one-step reduction $G \xrightarrow{r}_\mathcal{R} G'$ is the construction of G' from G according to the following two steps.

step 1) Take $G \setminus s_\sigma$.

step 2) Because $\sigma \circ f^{-1} : \partial V_t \longrightarrow V_{s_\sigma \subset G}^{co}$ is an isomorphism between two subgraphs,

$G' \equiv Comp_{\sigma \circ f^{-1}}(t, G \setminus s_\sigma)$. We then write $Comp_{\sigma \circ f^{-1}}(t, G \setminus s_\sigma)$ by $G[t_\sigma/s_\sigma]$. Then $G' \equiv G[t_\sigma/s_\sigma]$ is a one-step reduct of G. Figure 6 shows an example of one-step reduction.

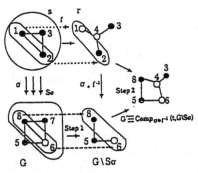

Fig. 6. One-step Reduction.

This s_σ is called a *redex occurrence* in G. In the reductions, an embedding in our *GRS* corresponds to a *match* using a *substitution*.

Definition 9 Trace and Induced Embedding : Let $\rho : \tilde{G} \longrightarrow G$ be an embedding for $\tilde{G}, G \in \mathcal{G}$ and let $r : s \longrightarrow t$ be a rewriting rule with an embedding $\sigma : s \longrightarrow \tilde{G}$.

(i) A one-step reduction $\tilde{G}[s_\sigma] \xrightarrow{r}_\mathcal{R} \tilde{G}[t_\sigma/s_\sigma]$ gives rise to another one-step reduction: $G[s_{\rho\sigma}] \xrightarrow{r}_\mathcal{R} G[t_{\rho\sigma}/s_{\rho\sigma}]$. This is called a *trace* of one (see Fig. 7.).

(ii) The one-step reduction and the trace give two identity maps: for $\tilde{G} \setminus s_\sigma \subset \tilde{G}$ and $\tilde{G} \setminus s_\sigma \subset \tilde{G}[t_\sigma/s_\sigma]$, $I_{s_\sigma}^{\tilde{G}} : \tilde{G} \setminus s_\sigma$ (in $\tilde{G}$) $\longrightarrow \tilde{G} \setminus s_\sigma \subset \tilde{G}[t_\sigma/s_\sigma]$ and for $G \setminus s_{\rho\sigma} \subset G$ and $G \setminus s_{\rho\sigma} \subset G[t_{\rho\sigma}/s_{\rho\sigma}]$, $I_{s_{\rho\sigma}}^{G} : G \setminus s_{\rho\sigma}$ (in G) $\longrightarrow G \setminus s_{\rho\sigma} \subset G[t_{\rho\sigma}/s_{\rho\sigma}]$. They also give two natural isomorphisms: $I^\sigma : t \longrightarrow \hat{t}_\sigma$ and $I^{\rho\sigma} : t \longrightarrow \hat{t}_{\rho\sigma}$. Now put $\alpha = I^{\rho\sigma} \circ (I^\sigma)^{-1} : \hat{t}_\sigma \longrightarrow \hat{t}_{\rho\sigma}$ and $\beta = I_{s_{\rho\sigma}}^{G} \circ \rho \circ (I_{s_\sigma}^{\tilde{G}})^{-1} : \tilde{G} \setminus s_\sigma \longrightarrow G \setminus s_{\rho\sigma}$.

Moreover, $\eta = (\alpha_E \cup \beta_E, \alpha_V \cup \beta_V) : \tilde{G}[t_\sigma/s_\sigma] \longrightarrow G[t_{\rho\sigma}/s_{\rho\sigma}]$ can be obviously an embedding. This η is called an *induced embedding* (see Fig. 7.).

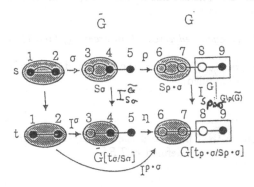

Fig. 7. Trace and Induced Embedding

The next proposition will be used to prove the critical pair Lemma for B-Graphs.

Proposition 3 (Difference Graph Invariance of Induced Embedding)
For $\tilde{G}$ and $G \in \mathcal{G}$, let $\rho : \tilde{G} \longrightarrow G$ be an embedding and let $r : s \longrightarrow t$ be a rewriting rule with an embedding $\sigma : s \longrightarrow \tilde{G}$.

Let $\tilde{G} \equiv \tilde{G}[s_\sigma] \xrightarrow{r}_{\mathcal{R}} \tilde{G}[t_\sigma/s_\sigma]$ be one-step reduction. Then

(a) A trace $G \equiv G[s_{\rho\sigma}] \xrightarrow{r}_{\mathcal{R}} G[t_{\rho\sigma}/s_{\rho\sigma}]$ exists.

(b) An embedding $\rho' : \tilde{G}[t_\sigma/s_\sigma] \longrightarrow G[t_{\rho\sigma}/s_{\rho\sigma}]$ such that $G \setminus \rho(\tilde{G}) \equiv G[t_{\rho\sigma}/s_{\rho\sigma}] \setminus \rho'(\tilde{G}[t_\sigma/s_\sigma])$ also exists.

Proof. (a) is obvious.

(b) It is obvious that there is an induced embedding ρ' that satisfies the condition in (b). In this embedding, we can also get the natural embedding $I^G_{s_\sigma} : G \setminus s_\sigma \longrightarrow G \setminus s_\sigma \equiv G[t_\sigma/s_\sigma]$. Here, $I^G_{s_\sigma}|_{G \setminus \rho(\tilde{G})} : G \setminus \rho(\tilde{G}) \longrightarrow G \setminus \rho'(\tilde{G}[t_\sigma/s_\sigma])$ satisfies the given condition. (See Fig. 7.)

Proposition 4 (WCR Heredity via Embedding)
Let $\rho : \tilde{G} \longrightarrow G$ be an embedding for $\tilde{G}, G \in \mathcal{G}$, let $r_1 : s^1 \longrightarrow t^1$ and $r_2 : s^2 \longrightarrow t^2$ be two rewriting rules, and let $\sigma_1 : s^1 \longrightarrow \tilde{G}$ and $\sigma_2 : s^2 \longrightarrow \tilde{G}$ be two embeddings. Moreover, r_1 and r_2 give two one-step reductions: $\tilde{G}[s_{\sigma_1}] \xrightarrow{r_1}_{\mathcal{R}} \tilde{G}[t_{\sigma_1}/s_{\sigma_1}]$ and $\tilde{G}[s_{\sigma_2}] \xrightarrow{r_2}_{\mathcal{R}} \tilde{G}[t_{\sigma_2}/s_{\sigma_2}]$, and they give two traces of one-step reductions: $G[s_{\rho\sigma_1}] \xrightarrow{r_1}_{\mathcal{R}} G[t_{\rho\sigma_1}/s_{\rho\sigma_1}]$ and $G[s_{\rho\sigma_2}] \xrightarrow{r_2}_{\mathcal{R}} G[t_{\rho\sigma_2}/s_{\rho\sigma_2}]$.

If there exist $\tilde{G}_a$ and $\tilde{G}_b$ such that $\tilde{G}[t^1_{\sigma_1}/s^1_{\sigma_1}] \xrightarrow{*}_{\mathcal{R}} \tilde{G}_a$, $\tilde{G}[t^2_{\sigma_2}/s^2_{\sigma_2}] \xrightarrow{*}_{\mathcal{R}} \tilde{G}_b$ and $\tilde{G}_a \equiv \tilde{G}_b$,
then there exist G_a and G_b such that $G[t^1_{\rho\sigma_1}/s^1_{\rho\sigma_1}] \xrightarrow{*}_{\mathcal{R}} G_a$, $G[t^2_{\rho\sigma_2}/s^2_{\rho\sigma_2}] \xrightarrow{*}_{\mathcal{R}} G_b$ and $G_a \equiv G_b$.

Proof. In each reduction step let $\rho_x : \tilde{G}_x \longrightarrow G_x$ be each induced embedding. Then, because from Proposition 3 the difference graph is invariant in each reduction step, $G \setminus \rho(\tilde{G}) \equiv G_x \setminus \rho_x(\tilde{G}_x)$.

If $\tilde{G}_a \equiv \tilde{G}_b$, then $\rho_a(\tilde{G}_a) \equiv \rho_b(\tilde{G}_b)$. Because labels of boundary nodes are invariant. Therefore, $G_a \equiv G_b$ is concluded from Proposition 2.

Definition 10 Cycle+Edge Reduction Ordering : For $s \in \mathcal{G}$, $s^\sharp$ is the sum of all the cycles and all the edges in s. Then $\mathcal{R}$ has a *cycle + edge reduction ordering* if $s^\sharp > t^\sharp$ for any one-step reduction : $s \longrightarrow t$.

Theorem 1 *If a GRS $\mathcal{R}$ has cycle+edge reduction ordering, then $\mathcal{R}$ is SN.*

Proof. The proof is obvious because the sum of the cycles and edges is finite and decreases monotonically as the rules are applied.

2.2 Main Properties

Overlapping, superposition, critical pairs and the convergence of critical pairs can be defined here in the same way they are defined in *TRS*.

Definition 11   Overlapping :   Let $r_1 : \alpha \longrightarrow \beta$ and $r_2 : \gamma \longrightarrow \delta$ be two rewriting rules and let $\sigma_1 : \alpha \longrightarrow G$ and $\sigma_2 : \gamma \longrightarrow G$ be two embeddings. Two redex occurrences α_{σ_1} and γ_{σ_2} in G are *overlapping* if $E_{\alpha_{\sigma_1}} \cap E_{\gamma_{\sigma_1}} \neq \emptyset$. They are *disjoint* otherwise.

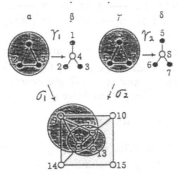

Fig. 8. Overlapping.

Now we define a *superposition* s of two rules $r_1 : \alpha \longrightarrow \beta$ and $r_2 : \gamma \longrightarrow \delta$.

It is assumed that by renaming the labels of vertices in V_α and V_γ, $\{lab_\alpha(v)|v \in \partial V_\alpha\} \cap \{lab_\gamma(v)|v \in \partial V_\gamma\} = \emptyset$.

Definition 12   Superposition :   Take $g \subset \alpha$ (g is not necessarily connected) with an isomorphism $f : g \longrightarrow f(g) \subset \gamma$, where $E_g \neq \emptyset$. The superposition S of r_1 and r_2 is the composition $Comp_f(\alpha, \gamma)$ such that its two natural isomorphisms $\alpha \longrightarrow \alpha'$ and $\gamma \longrightarrow \gamma'$ are embeddings. An example of the superposition of two rules is shown in Fig. 9.

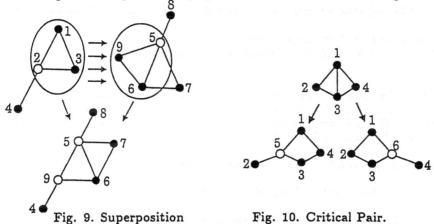

Fig. 9. Superposition   Fig. 10. Critical Pair.

According to two embeddings $\sigma_1 : \alpha \longrightarrow S$ and $\sigma_2 : \gamma \longrightarrow S$, then the superposition S can be reduced in two possible ways: $S[\alpha_{\sigma_1}] \longrightarrow S[\beta_{\sigma_1}/\alpha_{\sigma_1}]$ and $S[\gamma_{\sigma_2}] \longrightarrow S[\delta_{\sigma_2}/\gamma_{\sigma_2}]$.

Definition 13   Critical Pairs in GRS :   The pair of reducts $\langle S[\beta_{\sigma_1}/\alpha_{\sigma_1}], S[\delta_{\sigma_2}/\gamma_{\sigma_2}]\rangle$ is called the *critical pair* obtained by the *superposition* of $\alpha \longrightarrow \beta$ and $\gamma \longrightarrow \delta$.

That is, a critical pair in $\mathcal{R}$ is the result of two non-B-isomorphic reductions from the same B-graph (see Fig. 10).

Definition 14   Convergence of Critical Pairs :   A critical pair $< s, t >$ is *convergent* if $\mathcal{G}$ has elemensts u and $v \in \mathcal{G}$ such that $s \xrightarrow{*} u, t \xrightarrow{*} v$ and $u \equiv v$ (i.e. u is B-isomorphic to v).

For two redex occurrences, the following proposition gives a relationship between over-lapping and superposition.

Proposition 5 *For $G \in \mathcal{G}$, let $r_1 : \alpha \longrightarrow \beta$ and $r_2 : \gamma \longrightarrow \delta$ be two rewriting rules with two embeddings $\eta_1 : \alpha \longrightarrow G$ and $\eta_2 : \gamma \longrightarrow G$, where $\{lab_\alpha(v)|v \in \partial V_\alpha\} \cap \{lab_\gamma(v)|v \in \partial V_\gamma\} = \emptyset$.*

i) If two redex occurrences α_{η_1} and γ_{η_1} in G are disjoint and, if $G \xrightarrow{r_1}_{\mathcal{R}} G_1$ and $G \xrightarrow{r_2}_{\mathcal{R}} G_2$, then there exists G_3 such that $G_1 \xrightarrow{*}_{\mathcal{R}} G_3$ and $G_2 \xrightarrow{*}_{\mathcal{R}} G_3$.

ii) If two redex occurrences α_{η_1} and γ_{η_2} in G are overlapping and k is the smallest B-subgraph containing both α_{η_1} and γ_{η_2}, then there exists

(a) a superposition S of two rules with embedding $\rho : S \longrightarrow \rho(S) \equiv k \subset G$

(b) a critical pair $\langle P, Q \rangle$ such that $G[\alpha_{\eta_1}] \xrightarrow{r_1} G[\beta_{\eta_1}/\alpha_{\eta_1}]$ and $G[\gamma_{\eta_2}] \xrightarrow{r_2} G[\delta_{\eta_2}/\gamma_{\sigma_2}]$ are traces of $S[\alpha_{\sigma_1}] \xrightarrow{r_1} S[\beta_{\sigma_1}/\alpha_{\sigma_1}] \equiv P$ and $S[\gamma_{\sigma_2}] \xrightarrow{r_2} S[\delta_{\sigma_2}/\gamma_{\sigma_2}] \equiv Q$, and

(c) two induced embeddings $\eta_1 : P \longrightarrow G[\beta_{\eta_1}/\alpha_{\eta_1}]$ and $\eta_2 : Q \longrightarrow G[\delta_{\eta_2}/\gamma_{\eta_2}]$.

Proof. i) Obvious. ii) Put $M = \alpha_{\eta_1} \cap \gamma_{\eta_2}$. $\eta_1^{-1}(M) = \alpha_M$ and $\eta_2^{-1}(M) = \gamma_M$. For these two subgraphs $\alpha_M \subset \alpha$ and $\gamma_M \subset \gamma$, $\eta_2^{-1} \circ \eta_1 : \alpha_M \longrightarrow \gamma_M$ is obviously an isomorphism. This isomorphism gives a superposition $S \equiv Comp_{\eta_2^{-1} \circ \eta_1}(\alpha_M, \gamma_M)$ by way of the composition of α and γ.

Put $\rho = \eta_1 \cup \eta_2$. Then ρ satisfies (a). Also let $\sigma_1 : \alpha \longrightarrow S$ and $\sigma_2 : \gamma \longrightarrow S$ be two natural isomorphisms (embeddings) into superposition S. These ρ, σ_1, and σ_2 produce a critical pair $\langle P, Q \rangle$ and two induced embeddings η_1 and η_2 such that they satify (b) and (c). ∎

The following lemma corresponds to the well-known critical pair lemma in TRS.

Lemma 1 Critical Pair Lemma for GRS.
A GRS $\mathcal{R}$ is WCR iff all its critical pairs are convergent.

Proof. The "only if" part is obvious. We we have only to show " if" part. Redex occurrences must be either nonoverlapping or overlapping. If they are nonoverlapping, the proof is obtained from Proposition 5 i). If they are overlapping, the WCR property follows from Propositions 3, 4, and 5 ii). ∎

Theorem 2 *If $\mathcal{R}$ is SN, then $\mathcal{R}$ is CR iff all critical pairs of $\mathcal{R}$ are convergent.*

Proof. Straightforward from Proposition 1 and Lemma 1. ∎

Figure 11 shows an example of a complete set.

3 Application to Network Reliability Analysis

The purpose of network reliability analysis is to evaluate the probabilities (see P_{123} in Fig. 12.) that specific vertices can be connected (or communicated). The network model used in this paper is a family $\mathcal{G}$ of B-graphs (graphs, for short) such that every edge has a probability of connected states called a reliability, and every vertex has a reliability 1.

One of the question in network reliability analysis is how to use the reliabililty of every edge to determine the probability that all vertices are connected.

For general networks, this problem is known to be NP-hard [2], [27].

In conventional network reliability analysis, reliability-preserving reduction methods [28] are known as efficient algorithms [24],[25], [26], [30]. Figure 13 shows the set of three fundamental reduction rules (denoted $\mathcal{R}_0$) [30], [19]. From top to bottom, these rules are *series reduction*, *parallel reduction*, and $\Delta - Y$ *reduction*. Here P_{ij} is the reliability of the edge between vertices i and j and Reliability P_{ij}^k is the reliability of the k-th edge between vertices i and j. $\bar{P}_{ij}^k = 1 - P_{ij}^k$, where $x = P_{12} + P_{23}P_{31} - P_{12}P_{23}P_{31}$, $y = P_{23} + P_{31}P_{12} - P_{12}P_{23}P_{31}$, and $z = P_{31} + P_{12}P_{23} - P_{12}P_{23}P_{31}$.

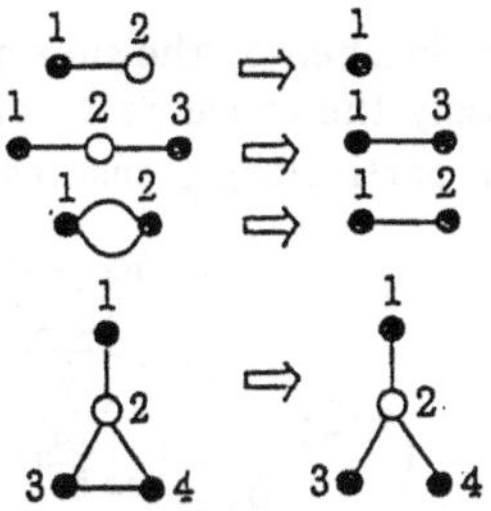

Fig. 11. A Complete Rule Set

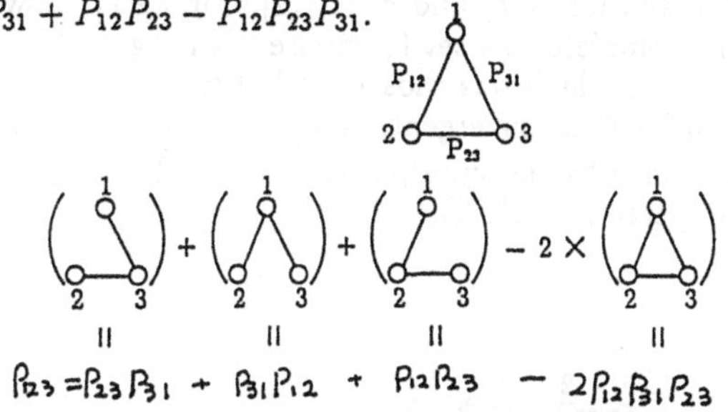

Fig. 12. Reliability

A number of graphs are reducible by $\mathcal{R}_0$ (see Fig. 14.). But when we try to reduce a graph to a simpler one by using these reduction rules, a complicated form often remains (see Fig. 15), where the shaded vertices indicate where $\mathcal{R}_0$ is to be applied. This is called a divergence problem.

We can obtain the reliability of the original graph if only a single edge remains. A much larger graph than a single edge and several methods (e.g. a factoring method [29]) are used simultaneously to derive the connection probability from graphs with more than one edge. These methods, though, are not efficient algorithms. There are many effective reduction methods [23], [24], [25], [26] for a special class of graphs, but they require a complicated reduction strategy.

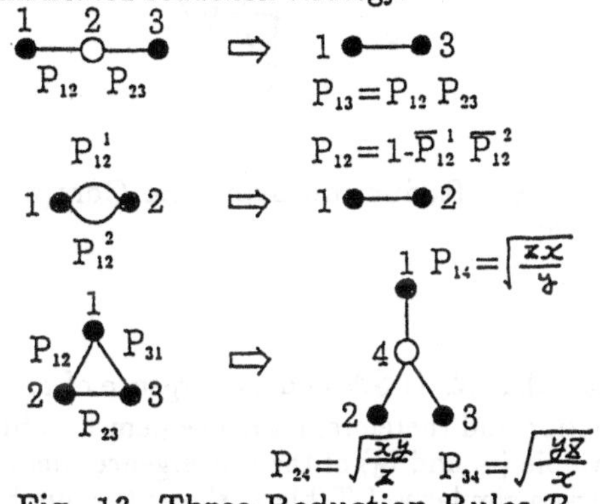

Fig. 13. Three Reduction Rules $\mathcal{R}_0$.

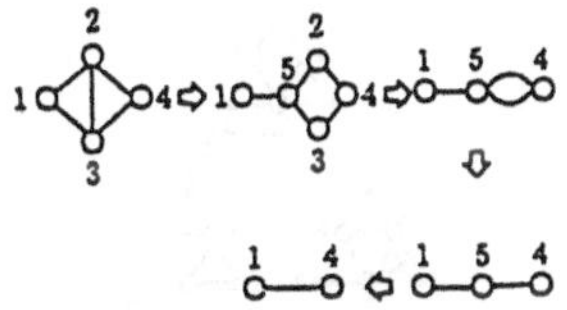

Fig. 14. Graph reduction.

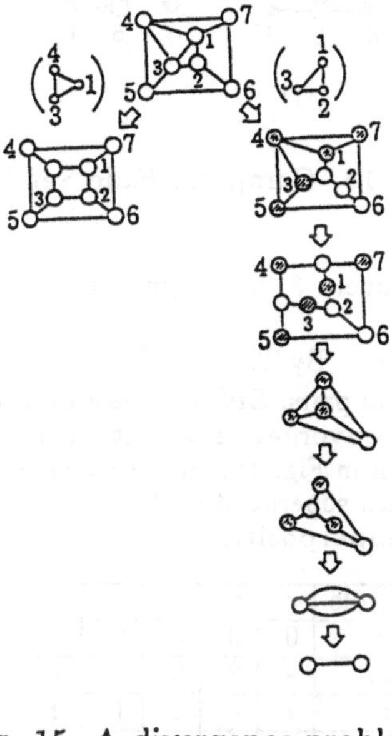

Fig. 15. A divergence problem.

If a *GRS* $\mathcal{R}$ is complete, any graph will have a unique reduction result that does not depend on where and in which order rules are applied. In this sense, we can use a complete $\mathcal{R}$ to perform a strategy-free reduction. We propose a new set of reduction rules $\widetilde{\mathcal{R}}_0$ in (Fig. 16) satisfying with the following properties:

(1) $\widetilde{\mathcal{R}}_0$ is complete.

(2) Any graph reducible to a single edge by $\mathcal{R}_0$ can be reduced to the same single edge by $\widetilde{\mathcal{R}}_0$.

Methods for calculating reliabilities by using the reduction rules $\widetilde{\mathcal{R}}_0$ are shown for r_0 in [26] and for r_1, r_2 and r_3 in [12]. Figure 17 shows an example of the advantages of using the complete rule set (compare with Fig. 15).

The rule r_3 was chosen so that the critical pair obtained by the superposition of $\Delta - Y$ and itself is convergent, and r_0 was chosen to satisfy the convergence of the critical pair obtained by the superposition of $\Delta - Y$ and r_1. In particular, r_4 ensures that $\widetilde{\mathcal{R}}_0$ satisfies property 2 $(see[12])$.

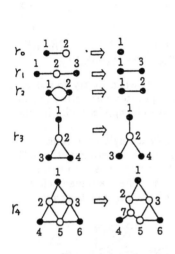

Fig. 16. Complete Rule Set $\widetilde{\mathcal{R}}_0$. 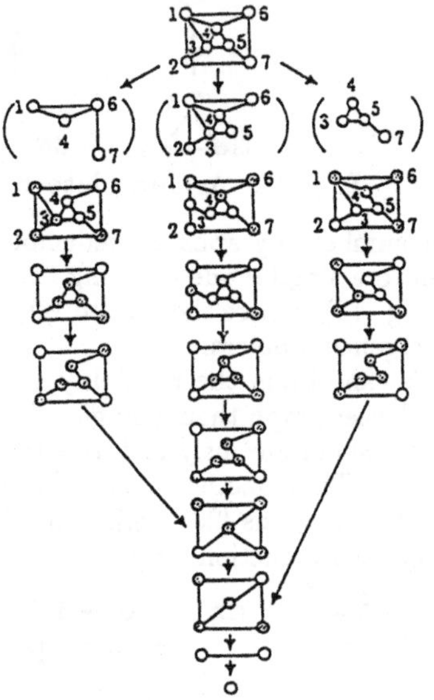 Fig. 17. Reduction using the Complete Rule Set.

Theorem 3 $\widetilde{\mathcal{R}}_0$ *is complete.*

Proof. By Theorem 2, we need only to show that $\widetilde{\mathcal{R}}_0$ is SN and convergence of all its critical pairs. SN: $\widetilde{\mathcal{R}}_0$ has a cycle+edge reduction and therefore, from Theorem 1, is SN.

Convergence of critical pairs: An example $(r_3$ and $r_3)$ of the convergence check is shown in Fig. 18, and other cases can be checked similarly. (Table 2. shows the number of each superposition.)
Superposition

Rule	r_0	r_1	r_2	r_3	r_4
r_0	0	0	0	0	0
r_1	*	2	0	1	1
r_2	*	*	1	1	0
r_3	*	*	*	12	1
r_4	*	*	*	*	1

Table. 2.

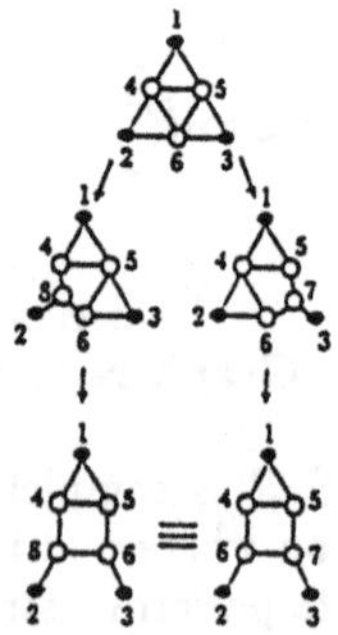

Fig. 18. Convergence of critical pair.

4 Conclusion

This paper proposes Graph Rewriting Systems which are abstract reduction systems for graphs and show how they can be used to analyze the reliability of communication networks.

A boundary graph was introduced to made define rewriting together with redex occurrence and to well-define the convergence of a critical pair. Using a critical pair lemma, we constructed a complete system from a conventional incomplete system.

Acknowledgements

We express our gratitude to Professor B. Courcelle for his encouragement and for his valuable comments on an earlier draft of this paper.

We also thank K. Koyama, H. Katsuno, Y. Toyama, K. Shirayanagi, J. Yamada, K. Mano, Y. Tsukada, T. Abe, and S. Komura for their advice and for many helpful discussions during the preparation of this paper.

References

[1] Arnborg, S., Courcelle, B. , Proskurowski, A. and Seese, D.: *"An Algebraic Theory of Graph Reduction,"* Research Report n° 91-36, Bordeaux-I University, 1991.

[2] Ball, M. O.: *"Complexity of Network Reliability Computations,"* Networks , Vol. 10, pp. 153-165, 1980.

[3] Barendreght, H.P.: *"The Lambda Calculus-Its Syntax and Semantics,"* Revised Edition, North Holland, 1984.

[4] Bauderon, Y. and Courcelle, B.: *"Graph Expression and Graph Rewritings,"* Mathematical Systems Theory, 20, pp. 83-127, 1987.

[5] Brandenburg, F.J.: *"On context-free graph grammars with bounded degrees"* Proceedings of Toyohashi Symposium on Theoretical Computer Science, pp. 23-26 August, 1990.

[6] Courcelle, B. : *"An axiomatic definition of context-free rewriting systems and its application to NLC graph grammars."* Theoretical Computer Science, 55, pp. 141-181, 1987.

[7] Courcelle, B. : *"The monadic second order logic of graphs VI: On several presentations of graphs by relational structures."* Research Report n° 89-99, Bordeaux-I University, 1989.

[8] Ehrig, H., Pfender, M., and Schnider, H. : *"Graph Grammars: An algebraic approach,"* Proceeding 14th IEEE Symposium Switching and Automata Theory, Iowa City, pp. 167-180, 1973.

[9] Ehrig, H. : *"Introduction to the algebraic theory of graphs grammars (A Survry),"* in Proceeding of 1st International Workshop on Graph Grammars, Lecture Notes Computer Science, Vol. 73, Springer Verlag, Berlin, pp. 1-69, 1979.

[10] Ehrig, H., Kreowski, H.J., Maggiolo-Shettini, A., Rosen, B. R., and Winkowski, J. : *"Transformations of structures: an algebraic approach,"* Mathematical Systems Theory, Vol 14, pp. 305-334, 1981.

[11] Harary, F. : *"Graph Theory,"* Addison-Wesley, Reading, MA 1969.

[12] Hayashi, M. and Okada, Y.: *"The Equality of Graphs by Reduction Method and Application,"* IEICE Japan Tech. Rep. CAS and VLD 90-53, pp. 37-44, 1990.

[13] Hindley, J. and Seldin, J: *"Introduction to Combinators and Lambda Calculus,"* London Mathematical Society Student Texts 1, Cambridge University, 1986.

[14] Huet, G.P.: *"Confluent Reductions: Abstract Properties and Applications to Term Rewriting Systems ,"* Journal of the Association for Computing Machinery, Vol. 27, No. 4, pp. 11-21, 1980.

[15] Knuth, D.E. and Bendix, P.B. : *"Simple Word Problems in Universal Algebras,"* Computational problems in abstract algebras, Pergamon Press, Oxford 1970.

[16] Le Chenadec, P. : *"Canonical Forms in Finitely Presented Algebra,"* Research Notes in Theoretical Computer Science, Pitman London, 1986.

[17] Litovsky, I. and Metivier, Y.: *"Computing Trees with Graph Rewriting Systems with Priorities,"* Research Report n° 90-87, Bordeaux-I University, 1990.

[18] Maggiolo-Schettini, A. and Winkowski, J.: *"Process of Transforming Structures"* Journal of Computer and System Sciences 24, pp. 245-282, 1982.

[19] Khan, N.M., Rajamani, K. and Banerjee, S.K.: *"A direct method to Calculate the Frequency and Duration of Failures for Large Networks,"* IEEE Transactions on Reliability Analysis, Vol. R-26, No. 5, December pp.318-321, 1977.

[20] Billaud, M., Lafon, P., Metivier, E. and Sopena, E. : *"Graph Rewriting Systems with Priorities,"* 15th International Workshop WG, Lecture Note in Computer Science 411, pp. 94-106, 1989.

[21] Newman, M.H.A. : *"On theories with a combinatorial definition of 'equivalence' ,"* Ann. Math., Vol. 43, No. 2, pp. 223-243, 1942.

[22] Okada, Y. and M, Hayashi.: *"Graph Rewriting Systems and Application to Network Reliability Analysis."* IEICE Japan Tech. Rep. Comp90-7, pp. 57-66, 1990.

[23] Politof, T.: *"Δ − Y reducible graphs,"* Working Paper 88-12-52, Department of Decision Science and Management Information Systems,"* Concordia University (May 1989).

[24] Politof, T. and Satyanarayana, A. : *"Efficient Algorithms for Reliability Analysis of Planar Networks - A survey,"* IEEE Transactions Reliability, Vol. R-35, No. 3, pp. 252-259, August, 1986.

[25] Politof, T. and Satyanarayana, A. : *"Network Reliability and Inner-Four-Cycle-Free Graph,"* Mathematics of Operation Research, Vol. 11, No. 3, pp. 484-505, August, 1986.

[26] Politof, T. and Satyanarayana, A. : *"A linear time Algorithm to Compute the Reliability of Planar Cube-Free Networks,"* IEEE Transactions on Reliability, Vol. R-39, pp. 557-563, Dec, 1990.

[27] Rosenthal, A. : *"Computing The Reliability of Complex Networks* SIAMJ. APPL. MATH. Vol. 32, No. 2, March, 1977.

[28] Rosenthal, A. and Frisque, D. : *"Transformations for Simplifying Network Reliability Calculations"* Networks, Vol. 7, pp. 97-111, 1977.

[29] Satyanarayana, A. and Chan, M. : *"Network Reliability and the Factoring Theorem,"* Networks, Vol. 13, pp. 107-120, 1983.

[30] Satyanarayana, A. and Wood, R. K. : *"A Linear-Time Algorithm for computing K-terminal reliability of normal graphs in series-parallel networks,"* SIAM J. Computing, Vol. 14, November, 1985.

[31] Sethi, R. : *"Testing for the Church-Rosser theorems, "* J.ACM, Vol. 21, No. 4, pp. 671-679, October, 1974.

Nondeterministic Control Structures for Graph Rewriting Systems

Albert Zündorf[1]   and   Andy Schürr[2]

Lehrstuhl für Informatik III
Aachen University of Technology
Ahornstraße 55, D-5100 Aachen, W-Germany

e-mail: albert@... and andy@...
...@rwthi3.informatik.rwth-aachen.de

Abstract:

The work reported here is part of the IPSEN[3] project whose very goal is the development of an Integrated Project Support ENvironment. Within this project directed, attributed, node- and edge- labeled graphs (diane graphs) are used to model the internal structure of software documents and PROgrammed Graph REwriting SyStems are used to specify the operational behavior of document processing tools like syntax-directed editors, static analyzers, or incremental compilers and interpreters. Recently a very high-level language, named PROGRESS, has been developed to support these activities. This language offers its users a convenient, partly textual, partly graphical concrete syntax and a rich system of consistency checking rules (mainly type compatibility rules) for the underlying calculus of programmed diane-graph rewriting systems.

This paper presents a partly imperative, partly rule-oriented sublanguage of PROGRESS for composing complex graph queries and graph transformations (transactions) out of simple subgraph tests and graph rewriting rules (productions). It also contains a formal definition of this sublanguage by mapping its main control structures onto so-called nondeterministic control flow graphs. We believe that these control structures or at least the underlying flow graph formalism could be incorporated into many graph-/tree-/term- rewriting systems in order to control the nondeterministic selection and application of rewriting rules.

1   Introduction

Modern software systems for application areas like office automation and software engineering are usually highly interactive and deal with *complex, structured objects*. The systematic development of these systems requires precise and readable descriptions of their desired behavior. Therefore, many specification languages and methods have been introduced to produce formal descriptions of various aspects of a software system, such as the design of object structures, the effect of operations on objects, or the synchronization of concurrently executed tasks. Many of these languages use special classes of graphs as their underlying data models. Conceptual graphs [Sowa 84], (semantic) data base models [HK 87], petri nets [GJRT 82], or attributed trees [Reps 84] are well-known examples of this kind.

---

1. Supported by *Deutsche Forschungsgemeinschaft*.
2. Supported by *Stiftung Volkswagenwerk* and the *Bundesministerium für Forschung und Technologie*.
3. cf. [Nagl 80], [EJS 88], [Lew 88]

Within the research project IPSEN a graph grammar based specification method is in use to model the internal structure of software documents and to produce executable specifications of corresponding document processing tools, as e.g. syntax-directed editors, static analyzers, or incremental compilers [ES 85]. The development of such a specification, which is termed '*programmed graph rewriting system*', consists of two closely related subtasks. The first one is to design a graph model for the corresponding complex object structure. The second one is to program object (graph) analyzing and modifying operations by composing sequences of subgraph tests and graph rewriting rules.

Parallel to the continuous evolution of this graph grammar formalism and a graph grammar engineering method, the design of a *graph grammar specification language* is in progress. The outcome of this design process is - as far as we know - the first attempt to combine the advantages of object-oriented database definition languages and attribute grammars (with respect to the description of static data structures) with the advantages of programmed graph rewriting systems (with respect to the definition of nondeterministic graph transformations) within one strongly-typed language (cf. [Zünd 89], [Schürr 91], [Schürr 91a]).

A first version of this language named PROGRESS (for PROgrammed Graph REwriting SyStems) has been fixed a few months ago, and a prototype of a programming environment for this version of the language is now under development (cf. [NaSc 91]).

2 The "ferryman's problem" specified with PROGRESS

For an informal introduction to the way of programming the language PROGRESS supports we will use a small specification called "the ferryman's problem". This specification is a precise description as well as a programmed solution of the following problem:

> There are a cabbage, a goat, and a wolf that must be ferried over a river. But the ferry is so small that it is not able to carry more than one of these objects at the same time. Furthermore, at no time one should leave the goat and the cabbage or the wolf and the goat without any supervision on the same river side.

Figure 1 shows the start situation of "the ferryman's problem" in the form of a directed, attributed, node- and edge labeled graph (diane-graph).

Figure 1: Start situation for "the ferryman's problem"

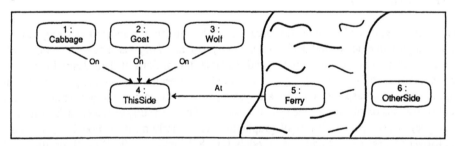

PROGRESS is a *strongly typed language*. Thus, all kinds of graph elements like nodes, edges, and attributes belong to exactly one type describing their properties. Common properties of different node types may be described in an object-oriented/based hierarchy of node classes.

Figure 2 shows a cutout of the PROGRESS data definition part for the graph of Figure 1. This

Figure 2: PROGRESS description for the graph of Figure 1

```
section GlobalGraphscheme;
(*Description of the example's graph model.*)
   node class THING end;
   edge type On: THING -> RIVER_SIDE;
   node class RIVER_SIDE end;
   node class CARGO is a THING end;
   edge type In: CARGO -> BOAT;
   node class BOAT is a THING end;
   edge type At: BOAT -> RIVER_SIDE;
end; (* GlobalGraphscheme *)
```

```
section SpecialObjectTypes;
(* Specialization of the global graph scheme. *)
(* The concrete object/node types are given. *)
   node type ThisSide: RIVER_SIDE end;
   node type OtherSide: RIVER_SIDE end;
   node type Ferry: BOAT end;
   node type Cabbage: CARGO end;
   node type Goat: CARGO end;
   node type Wolf: CARGO end;
end; (* SpecialObjectTypes *)
```

part of the specification determines the following properties for the graph elements of Figure 1:

- ❑ A node of the type Wolf
  - ▮ belongs to the class CARGO and possesses all properties defined in that class and its superclass THING, i.e., it may be a source of an edge of the type On or In.
- ❑ A node of type Ferry belongs to the class BOAT with the following properties:
  - ▮ it may be source of an edge of type On or At and
  - ▮ it may be target of an edge of type In.
- ❑ A node of type ThisSide, belonging to the class RIVER_SIDE, may be target of an edge of type On or At.
- ❑ An edge labeled On
  - ▮ demands a source node of a type that belongs (directly or indirectly) to the class THING. (In our example these are the types Wolf, Goat, Cabbage, and Ferry) and
  - ▮ demands a target node of a type that belongs (directly or indirectly) to the class RIVER_SIDE (ThisSide and OtherSide).
- ❑ An edge labeled In demands a source node class Cargo and a target node class BOAT.
- ❑ An edge labeled At leads from a BOAT to a RIVER_SIDE.

We hope the reader now is able to imagine how the static structure of diane-graphs are described in PROGRESS although our example doesn't include any examples for the definition of attributes and directed attribute dependencies. For a full, detailed description of these features of PROGRESS cf. [Schürr 91], [Zünd 89].

For the definition of basic operations on diane-graphs PROGRESS provides its users with means for the construction of *graph rewriting rules* (Productions and Tests) and *derived relations* on nodes (Pathes and Restrictions).

Figure 3 shows the production LoadFerry. This production applied to the graph of Figure 1

Figure 3: The production LoadFerry

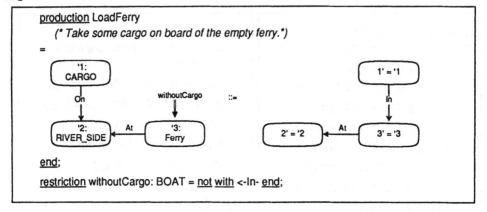

works as follows:

1. First a (partial) subgraph of the current host graph is determined which is isomorphic to the graph on the left hand side of the production. I.e., a subgraph consisting of two nodes with types of the classes CARGO and RIVER-SIDE respectively, one node of the type Ferry, where the CARGO node is connected to the RIVER-SIDE node by an edge with label On and the Ferry is connected to the (same) RIVER-SIDE node by an edge with label At. Additionally, the Ferry node must fulfill the *restriction* withoutCargo, whose definition is also part of Figure 3. It requires the Ferry node not to be the target of an edge of type In. One subgraph of the graph of Figure 1 matching these conditions consists of the nodes 2, 4 and 5.

2. When a matching subgraph of the host graph has been found/selected, this subgraph is removed from the host graph and the graph on the right hand side of the production is inserted. A node inscription like "2' = '2" signals an identical replacement of a node by itself. This means that 2' is the same node as '2 and all its attribute values and all incoming or outgoing edges remain unchanged unless they are explicitly mentioned within the production.

In our example the production LoadFerry replaces all nodes identically and, therefore, neither deletes an old node nor creates a new one. Beside this (identical replacement of three nodes) the production deletes an On edge (between the selected CARGO node and the RIVER_SIDE node) and a new edge with label In is inserted. One possible result graph of an application of LoadFerry to the graph of Figure 1 is shown in Figure 4.

In our running example for this paper we only use productions of this simple form. In general a PROGRESS production may also include[1]:

❑ parameters for node types and attribute values,

❑ application conditions concerning attribute values,

❑ instructions for changing the identical embedding of the new inserted subgraph into the surrounding host graph,

---

1. For a full description of the features of a production cf. [Schürr 91], [Zünd 89].

Figure 4: One possible result of the application of LoadFerry to the graph of Figure 1

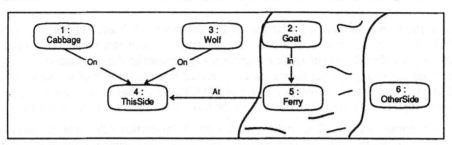

❑ assignments to attributes of new or identically replaced nodes,

❑ assignments to attribute valued return parameters.

At the end of this chapter we will discuss two other sorts of basic operations within PROGRESS, so called (subgraph) tests and pathes. Figure 5 shows the PROGRESS test SupervisionNecessary and two pathes used for its implementation.

Figure 5: The subgraph test SupervisionNecessary

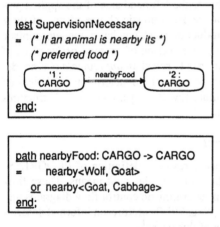

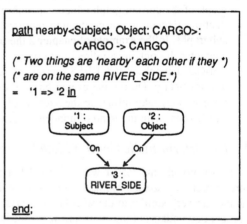

A *test* may be seen as a special form of a production with identical left- and right-hand side and, therefore, never causes any graph modifications (It is executed just by searching for a subgraph in the current host graph that matches the described conditions). The test SupervisionNecessary searches for two pieces of CARGO that are related through the *path* nearbyFood. This path relates two nodes to each other if they are *either* related by nearby<Wolf, Goat> *or* by nearby<Goat, Cabbage>. And two nodes are related to each other by the path nearby<x, y> if they are connected to the *same* RIVER_SIDE by edges with label On and have types equal to the actual parameters x and y.

We hope, that our explanations of pathes, productions and tests were detailed enough to understand the rest of the paper dedicated to . . .

3 The Control Structures of PROGRESS

After a short and informal introduction to the definition of 'basic' operations within PROGRESS we now are able to discuss the language's control structures. These control structures guide the selection of applicable graph rewriting rules/productions and may be used to compose complex graph transformation rules, named transactions, out of simple tests and productions. Designing the control structures of PROGRESS we had to take care to preserve the main properties of tests and productions on the level of transactions. These properties are:

❑ the **atomic** character: The application of a simple production either replaces one subgraph by another one and terminates successfully or fails without causing any graph changes. As a consequence of this fact, the execution of a sequence of graph rewriting rules should also either terminate successfully (if and only if the application of all graph rewriting rules is possible) or fail without any graph changing effects.

❑ the **boolean** character: Simple productions and complex transactions signal their state of termination to the calling environment. This implicit (boolean) return value determines the selection of the next graph rewriting rule.

❑ the **nondeterministic** character: Even in simple cases there may be more than one subgraph of the current graph which is isomorphic to the left hand side of the applied production. Executing the production requires a nondeterministic choice of one of these subgraphs. In the example of chapter 2 the production LoadFerry has a nondeterministic character. It contains no directives concerning the selection of an *appropriate* piece of CARGO. Thus the PROGRESS runtime system has to select an *appropriate* one nondeterministically. During the execution of a transaction such nondeterministic choices (if done wrong) may influence the success of future rewrite steps. Our control structures have to deal with this kind of nondeterminism.

3.1 Nondeterministic Control Flow Diagrams

In order to provide the reader with a formal but easily readable definition of PROGRESS' control structures, *nondeterministic control flow diagrams* are the right means. These flow diagrams are very similar to the well-known class of deterministic control flow diagrams and are proper descendants of the control flow graphs in [Nagl 79]. Therefore, an informal description of their semantics should suffice for the purpose of this paper[1].

Figure 6: Example control diagrams

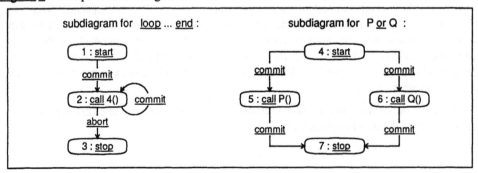

---

1. A formal definition may be found in [Schürr 91].

A node labeled with a production call (P(), Q(), ..., 4(), ...) stands for the execution of the corresponding production, test or another PROGRESS action, or for the execution of a whole subdiagram, whose start node is called. If the call is executed successfully, the execution of the calling diagram proceeds by following one of the outgoing commit edges (nondeterministically chosen). If the call fails, a nondeterministically chosen abort edge has to be followed (if the required type of edge doesn't exist, then another path through the control diagram must be tried). Every subdiagram contains exactly one node labeled start and exactly one stop node. A successful execution path through a subdiagram leads from its start node (through all the called subdiagrams) to its stop node. If there exists no such path the execution of this subdiagram fails, the host graph remains unchanged, and the calling diagram will have to follow an abort edge.

Note: An interpreter which has to find a successful path through such a diagram may have to keep track of its nondeterministic choices. If in one step there exists no outgoing edge of the desired label, the interpreter will have to "backtrack" to the state of its last choice (undoing all changes made to the host graph and the execution environment) and to try another alternative. (Additionally, there may be the problem of nonterminating execution cycles, cf. chapter 3.4.)

Additionally note, that our control diagrams properly reflect the properties of productions because a subdiagram call

❑ has an **atomic** character: only those pathes through a called subdiagram ending at its stop node might have any graph changing effects. All modifications done while following pathes into dead-ends are undone within the "backtracking" steps.

❑ has a **boolean** character: the success of the execution of an action determines if commit or abort edges must be followed.

❑ has a **nondeterministic** character: any Call node may have an arbitrary number of outgoing commit or abort edges. Thus, there may be several successful execution pathes through a subdiagram performing different graph transformations. A successful path through the whole control diagram will be selected nondeterministically. A deterministic working interpreter for PROGRESS will have to use mechanisms like backtracking to simulate this behavior (and for escaping out of dead-ends).

3.2 skip-, fail-, not- and def-statement

This chapter introduces some "pseudo" actions and simple operations, that we will need later on. It also introduces a control diagram for the call of a single PROGRESS action[1].

Figure 7: Control diagram for the call of a single production or test P():

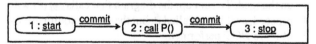

An execution of the above control diagram means: apply the production/test P() to the current host graph. If there are several matching subgraphs choose one which allows the calling diagram to succeed (if possible). If there is no matching subgraph, the host graph will remain unchanged and the execution of the subdiagram fails.

---

1. In the following PROGRESS action means a production, a test, or a compound action built by the proposed control structures.

Figure 8: Control diagram for skip:

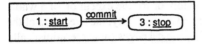

Always successful "pseudo" actions. The host graph remains unchanged. Follow a commit edge in the calling diagram.

Figure 9: Control diagram for fail:

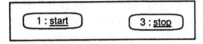

Always failing "pseudo" action. The host graph remains unchanged. Follow an abort edge in the calling diagram.

Figure 10: Control diagram for not P():

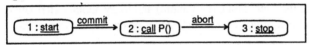

If it is possible to execute P() then the subdiagram for not P() has no legal execution path. Thus the calling diagram has to follow an abort edge. If P() is not executable, then the subdiagram for not P() succeeds and the calling diagram will follow a commit edge.

Note, that not P() never will have a graph changing effect. If P succeeds, then not P fails and the host graph remains unchanged. If P fails, it did not change the graph, thus not P doesn't too. Nevertheless, not LoadFerry is NOT semantically equivalent to skip. If it is possible to apply LoadFerry to the current graph, then not LoadFerry **fails**, while skip is always successful. As the user may be utterly confused by using not for an action that normally modifies the graph, this is forbidden in PROGRESS. The operator not may only be used with *effectless actions* like tests, skip, fail, or composite actions consisting of effectless subactions.

In order to allow the user to apply not to an action with graph changing effects we introduce the def-statement in Figure 11.

Figure 11: Control diagram for def P():

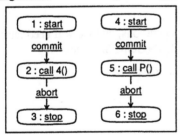

If P is executable, then the right subdiagram (of the figure on the left) fails (the graph remains unchanged) and thus the left subdiagram succeeds. If P fails, then the right subdiagram reaches its stop node by the abort edge and the left subdiagram fails leaving the current graph unchanged. To sum up, the diagram that calls def (P) will proceed, as if it had called P, but the graph modifications of P are NOT performed.

3.3 &-, or-, and and-statement

The first control structure we want to propose is the simple sequence of two or more PROGRESS actions, the &-statement[1].

---

1. Say concatenation statement.

Example: LoadFerry & not SupervisionNecessary & FerryOver & UnloadFerry

General form: P1 & P2 & ... & Pn

Control diagram: Figure 12

Figure 12: Control diagram for a &-statement:

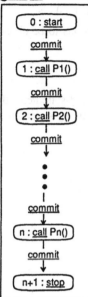

A legal path through the control diagram of a &-statement demands that all of its subactions are executable. The called actions are executed nondeterministically. Theoretically, only those alternatives are considered which allow the subsequent actions to succeed (in practice, backtracking will be necessary). In our example above the action LoadFerry has several matching subgraphs in the graph of Figure 1. It might put a Cabbage, a Goat, or a Wolf into the ferry. Depending on that choice, the test not SupervisionNecessary succeeds or fails. In the current situation, the Goat should be selected in order to be able to complete the path through our control diagram.

For the sequence
LoadFerry & LoadFerry & FerryOver
there exists no successful execution path, because only an unloaded ferry may be loaded. Thus the second call to LoadFerry will fail no matter what piece of CARGO is loaded by the first one. According to the dynamic semantics of control flow diagrams this means that no part of the sequence is executed (resp. everything is undone) and the calling diagram has to continue with an abort edge.

The or-statement combines a number of alternative actions. One of these actions must be chosen nondeterministically and then executed.

Example: LoadFerry or skip

General form: P1 or P2 or ... or Pn

Control diagram: Figure 13

Figure 13: Control diagram for an or-statement:

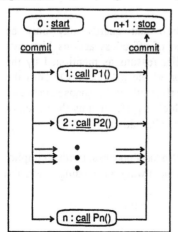

If none of the actions Pi is executable, then the subdiagram for the or-statement has no legal execution path. If several alternatives are possible, then that will be chosen which allows the calling diagram to succeed.

Thus, if there is a number of alternatives like "LoadFerry" or "do nothing" a PROGRESS interpreter has to select one that guarantees a successful termination of the whole execution process.

While the &-statement determines the order for the execution of its subactions, the following and-statement only demands that all its subactions have to be executed (in any order).

Example: not SupervisionNecessary and FerryOver

General form: P1 and P2 and . . . and Pn

Semantic equivalence: (P1 & P2 & . . . & Pn) or (P2 & P1 & P3 & . . . & Pn) or (. . .) . . .

Again PROGRESS' dynamic semantics demand that sequence of basic actions to be chosen which allows the calling diagram to succeed.

3.4   choose- *and* loop-*statement*

In the same way as the &-statement is the 'deterministic' variant of the and-statement, the following choose-statement may be considered as a deterministic variant of the or-statement.

Example: choose ReducePopulation else skip end

General form: choose when Q1 then P1
                          else when Q2 then P2

                          . . .

                          else when Qn then Pn
                  end

Semantic equivalence: choose . . . else P else . . . end
                          is a abbreviation for:
                                  choose . . . else when P then P else . . . end

Control diagram: Figure 14

Figure 14: Control diagram for a choose-statement:

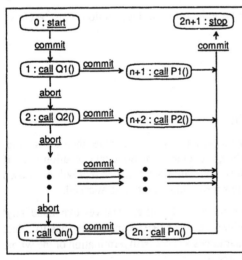

In general, a choose-statement consists of a list of *guarded* alternatives. The guards are examined subsequently until a succeeding one is found. That alternative will be chosen and then be executed. If there is no executable guard or if the chosen call isn't executable then the diagram for the choose-statement has no legal execution path.

We request that guards may only be composed of effectless actions like tests (for similar reasons as mentioned by the discussion of the not operator). Thus the action performed by the choose-statement is that which is performed by the action of the chosen alternative.

Executing our example above means first trying to perform ReducePopulation (with an implicit guard def(ReducePopulation)). Only if this action fails, then the always succeeding alternative skip will be selected.

The last control structure we want to propose is the iteration, the loop-statement.

Example: loop when not ProblemSolved then
               LoadFerry
            or (not SupervisionNecessary & FerryOver)
            or UnLoadFerry
     end (* loop *)

General form: loop   when Q1 then P1
         else when Q2 then P2
          . . .
         else when Qn then Pn
     end

Control diagram: Figure 15

Figure 15: Control diagram for a loop-statement:

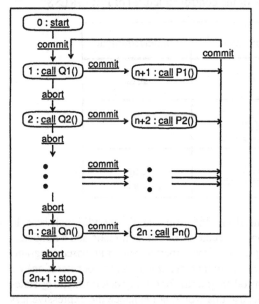

Like a choose-statement a loop-statement consists of a sequence of guarded alternatives. The body of a loop is executed like a choose-statement. The first guard which succeeds determines the action which is executed by the next iteration step. If no guard succeeds, the loop terminates successfully.

Note: If a selected action Pi fails, there is no legal continuation for the execution path through the control diagram. In this case backtracking starts (undoing graph changes) and the PROGRESS interpreter first tries to revise nondeterministic choices during the execution of earlier iteration steps and then at last aborts the execution of the whole loop (undoing all graph changes).

Note: A deterministically working interpreter using backtracking might have a serious termination problem with the example loop-statement above. The interpreter might loop forever just executing FerryOver in every iteration step. Our semantics for the loop-statement defines only how a successful execution looks like. It doesn't determine (like in Prolog) how this execution path has to be found by an deterministically working interpreter. In this case only an interpreter with a fairly (randomly) working algorithm for the selection of possible alternatives will be able to terminate the above loop (but probably not in the minimal number of iteration steps[1]). In general **the user himself has to take care of the termination of the loop**. An always terminating solution for the "ferryman's problem" may be found in Figure 16.

---

1. There is no way to arrange the order of the three 'ferry' actions so that an deterministically choosing interpreter (like a Prolog interpreter which takes the 'first' possible action) terminates the loop. cf. chapter 4.

Figure 16: A solution for the Ferryman's problem:

```
use   StepsLeft : Integer := 1000;
            (*Estimated upper bound for steps needed *)
in loop when not ProblemSolved and StepsLeft > 0 then
            ( LoadFerry or FerryOver or UnloadFerry)
      & (      choose ReducePopulation else skip end
            and StepsLeft := StepsLeft '-' 1 )
      end (* loop *)
```

This loop will find a solution for the Ferryman's problem that needs less (equal) than one thousand steps (if there exists one). For a PROGRESS specification that finds a solution with a minimal number of steps cf. Appendix B.

3.5   Overview

The following table contains a short overview of the control structures of PROGRESS:

	deterministic	nondeterministic
sequence	&	and
branch	choose (optional guards)	or
iteration (tail recursion)	loop (optional guards)	loop (... or ...)

In general, the control structures a language needs are the sequence, the (conditional) branch, and the iteration. PROGRESS has all these control structures in a deterministic and a nondeterministic version[1], thus reaching a smooth integration of the imperative and rule based programming paradigm. For the deterministic version of our branch-statement (and of our iteration) we introduced guarded alternatives. We did not introduce guards for our nondeterministic or-statement like those of [Dij 75]. We did not so, because this would not enhance the expressiveness of our language. A statement like

... or when FerryIsFull then UnloadFerry or ...

would be semantically equivalent to

... or FerryIsFull & UnloadFerry or ...

If in this example the guarded operation UnloadFerry fails the nondeterministic semantics of our or-statement will **not abort** but just try another alternative in order to find a successful execution for the whole statement. In our choose-statement a guarded operation that fails causes the whole statement to fail, because the order of examination of alternatives is prescribed.

In the following chapter 4 we will compare our control structures with those of several other programming languages based on (graph) (rewriting) rules or production systems.

---

1. Using a single or-statement as body of a loop-statement, we yield a nondeterministic choice of alternatives within our iteration statement.

4   Related Work

There are many approaches to define control structures for production systems, (graph) grammars, and (graph) rewriting systems. An exhaustive comparison of PROGRESS with such languages may be found in [Schürr 91], [Zünd 89]. Due to lack of space we have to focus our interest here onto *two typical representatives* of the whole class of (programmed) rule based systems. These are the well-known language Prolog and the Programmed Graph Grammars of [Gött 88].

[Gött 88] describes a graph rewriting system whose basic operations are very similar to productions of PROGRESS (without complex application conditions) and proposes three control structures for his language: wapp, sapp, and capp. sapp stands for Sequential APPlication. The sapp-statement's subactions will be executed one after the other, skipping all inexecutable subactions. Consequently, the execution of the whole sapp construct succeeds as long as at least one of its subactions succeeds. A capp-statement (Case APPlicable) consists of an sequence of alternative subactions. The first subaction that succeeds terminates the capp. The wapp-statement (While APPlicable) executes its body until it fails. To sum up, the control structures of [Gött 88] do respect the *boolean character* of graph rewriting rules but they neither support the definition of *atomic sequences* of graph rewriting rules (already aborting without any graph changing effects in the case of failure of just a single subaction) nor do they pay regard to the *nondeterministic character* of graph rewrite steps (by backtracking out of deadends and thus preventing "wrong" nondeterministic subgraph selections).

Now we will discuss the "control structures" of Prolog. We have chosen this language, because its comparison with PROGRESS shows the benefits of our control structures from another point of view. Figure 17 shows a solution for the Ferryman's problem implemented in Prolog,

Figure 17: Ferryman's problem in Prolog:

```
solveFMP(StepsLeft, StartGraph, TargetGraph)   :- problemSolved(StartGraph, TargetGraph) .
solveFMP(StepsLeft, StartGraph, TargetGraph)
                        :- StepsLeft > 0, StepsLeft2 = StepsLeft - 1,
                           oneStep(StartGraph, StepGraph),
                           tryRemovePopulation(StepGraph, RemoveGraph),
                           solveFMP( StepsLeft2, RemoveGraph, TargetGraph) .

oneStep(OldGraph, NewGraph) :- loadFerry(OldGraph, NewGraph) .
oneStep(OldGraph, NewGraph) :- ferryOver(OldGraph, NewGraph) .
oneStep(OldGraph, NewGraph) :- unloadFerry(OldGraph, NewGraph) .

tryRemovePopulation(InG, OutG)  :- reducePopulation(InG, OutG), ! .
tryRemovePopulation(InG, InG) .
```

which is analogous to our implementation given in Figure 16[1]. The PROGRESS loop has been translated using tail recursion within the definition of the predicate solveFMP[2]. The predicate oneStep is (semantically) equivalent to the following deterministic branch statement:

choose LoadFerry else FerryOver else UnloadFerry end

---

1. Skipping the definition of predicates corresponding to basic PROGRESS tests and productions.
2. The parameters StartGraph and TargetGraph are implicit in PROGRESS.

Note, that the PROGRESS specification in Figure 16 makes use of the nondeterministic or- and and-statements. Thus we were not forced to lay down any arbitrarily chosen precedence order for the selection of executable subactions as the Prolog implementation necessarily does[1]. The predicate tryRemovePopulation corresponds to the choose-statement of Figure 16. Note the cut (!) at the end of the first rule for this predicate. A guarded alternative in PROGRESS:

choose ... when Q1 then P1 ... end

corresponds to the following Prolog rule:

... :- Q1, !, P1 .

If the guard succeeds our semantic rules demand that the corresponding action is executed. If this action fails, then the whole enclosing control structure fails and backtracking to other alternatives is forbidden. To enforce this behavior in Prolog we have to use the cut. (An unguarded alternative is semantically equivalent to one with its action as a guard.)

As one can see, there is a close correspondence between the control structures proposed in this paper and the execution mechanisms for Prolog. We think that our control structures are in some way superior, because we do not only provide the deterministic 'and' and 'or' (like Prolog), but also nondeterministic 'and' and 'or'-statements. And last but not least the unification of (treelike) terms has been replaced by the construction of subgraph isomorphisms and PROGRESS' backtracking mechanism includes undoing of changes to the underlying graph database whereas Prolog does not support undoing of assert and retract of facts during backtracking[2].

5 Conclusions, Future Work

We think that the control structures we have proposed meet the requirements of chapter 3.1. The atomic, boolean, and nondeterministic character of the basic operations of PROGRESS are transferred in a consistent manner to its control structures. And the practical experiences made with PROGRESS and its control structures, e.g. in [Westf 91], and several other specifications are very satisfying.

Within the IPSEN[3] project PROGRESS currently is used to specify the internal behavior and implementation of several software tools. There exists a first prototype of a programming environment for PROGRESS itself. It now includes a syntax-directed editor and an incrementally working analyzing tool performing a large number of checks during editing.

There are two main directions for further development of PROGRESS:

❑ Development of the language PROGRESS: The main problem that remains is to incorporate a module concept into PROGRESS in order to be able to build large, complex specifications by composing simple, abstract, reusable subspecifications. This problem has to be seen together with the introduction of a concept of hierarchical graphs.

---

1. Note, that in Prolog a mechanism like the StepsLeft counter is necessary to guarantee the termination. (Another common mechanism to guarantee termination in Prolog is to keep track of the whole execution history. Then, after each step the new state is compared with all existing states. If it already has been 'visited' the current execution path fails, cf. [SteSha 86]). In PROGRESS a fairly working interpreter could find a solution without such a mechanism because we have used the nondeterministic or-statement, cf. chapter 3.4.
2. We will have to pay the prize when we are building an interpreter for PROGRESS.
3. Interactive Programming Support ENvironment cf. [Nagl 80], [EJS 88], [Lew 88]

❏ Development of software tools for PROGRESS: After the implementation of the PROGRESS editor our main activities now are directed towards the final goal of directly executing PROGRESS specifications and of using PROGRESS specifications as a base for the generation of semantically equivalent source code written in a conventional (imperative) programming language (Modula-2, C(++), . . .). Some basic implementation work is already done. Hopefully, at the end of this year an interpreter and a compiler for the graph scheme definition sublanguage will be available, including the (incremental) evaluation of derived attributes and path definitions. The development of an interpreter for PROGRESS supporting the whole language is under work (but will last longer).

As a result of these activities we hope that the family of PROGRESS users as well as the number of its applications will increase, thus getting more feedback and practical experience for future work.

References

[Dij 75]   Dijkstra, E.W.: Guarded Commands, Nondeterminacy, and Formal Derivation of Programs, CACM Vol. 18, No. 8, acm Press (1975), S. 453-457

[EKR 91]   Ehrig H., Kreowski H.-J., Rozenberg G. (eds.): *Proc. 4th Int. Workshop on Graph-Grammars and Their Application to Computer Science*; (published in) LNCS, Springer Verlag (1991)

[Eng 86]   Engels G.: *Graphen als zentrale Datenstrukturen in einer Software-Entwicklungsumgebung*; Ph.D. Thesis, VDI-Verlag (1986)

[EJS 88]   Engels G., Janning Th., Schäfer W.: *A Highly Integrated Tool Set for Program Development Support*; Proc. ACM SIGSMALL Conf., acm Press(1988), pp.1-10

[ES 85]   Engels G., Schäfer W.: *Graph Grammar Engineering: A Method Used for the Development of an Integrated Programming Support Environment*; in Ehrig et al.(Eds.): Proc. TAPSOFT'85, LNCS 186; Berlin: Springer-Verlag, pp. 179-193

[HK 87]   Hull R., King R.: *Semantic Database Modeling: Survey, Applications, and Research Issues*; in: ACM Computing Surveys, vol. 19, No. 3, acm Press (1987), pp. 201-260

[GJRT 82]   Genrich H.J., Janssens D., Rozenberg G., Thiagarajan P.S.: *Petri nets and their Relation to Graph Grammars*; in [ENR 82], pp. 115-142

[Gött 88]   Göttler H.: *Graphgrammatiken in der Softwaretechnik*; IFB 178, Springer Verlag (1988)

[Lew 88]   Lewerentz C.: *Extended Programming in the Large in a Software Development Environment*; in: Proc. 3rd. Int. ACM SIGPLAN/SIGSOFT Symp. on Practical Software Engineering Environments, SIGSOFT Notes, Vol. 13, No. 5, acm Press (1988), pp. 173-182

[Nagl 79]   Nagl M.: *Graph-Grammatiken*; Vieweg Verlag (1979)

[NaSc 91]   Nagl M., Schürr A.: *A Specification Environment for Graph Grammars*; in [EKR 91]

[Nagl 80]   Nagl M.: *An Incremental Compiler as Part of a System for Software Production*; IFB 25, Springer Verlag (1980), pp. 29-44

[Reps 84]   Reps T.: *Generating Language-Based Environments*; Ph.D. Thesis, MIT Press (1984)

[Schürr 91]   Schürr A.: *Operationales Spezifizieren mit programmierten Graphersetzungssystemen*; Ph.D. Thesis, RWTH Aachen (1991)

[Schürr 91a]   Schürr A.: *PROGRESS: A VHL-Language Based on Graph Grammars*; in: [EKR 91]

[Sowa 84]   Sowa J.F.: *Conceptual Structures: Information Processing in Minds and Machines*; Addison-Wesley (1984)

[SteSha 86]   Sterling, L., Shapiro, E.: The Art of Prolog; The MIT Press, Cambridge, Massachusetts, London, England (1986).

[Westf 91]   Westfechtel B.: *Revisionskontrolle in einer integrierten Softwareentwicklungsumgebung*; Ph.D. Thesis, RWTH Aachen (1991)

[Zünd 89]   Zündorf A.: *Kontrollstrukturen für die Spezifikationssprache PROGRESS*; Master Thesis, RWTH Aachen (1989)

A Language for Generic Graph-Transformations

Marc Andries[1] and Jan Paredaens[2]

Abstract: We define a class of graphs that represent object-oriented databases. Queries, updates and restructurings in such databases are performed by transformations of these graphs. We therefore first define five transformation-operations on this graph-class. Then we define *generic transformations* as a natural generalization of the concept of BP-completeness, originally defined for the relational database model. Our main result states that the language consisting of the five transformation-operations expresses exactly the generic transformations.

1 Introduction

Over the past few years, database research has been characterized by the development of a variety of datamodels. These models can be classified according to their underlying philosophy, the most well-known of which are probably the relational, the deductive and the object-oriented. In an object-oriented database, objects (i.e., representations of real-world entities) are structured in a composition-graph (i.e., a property of some object may be another object) as well as in an inheritance-graph (i.e., some object may be a specialization of another object). Consequently, it has been recognized by many database-researchers over the past decade that graphs are an ideal means for modeling data in an object-oriented database model [1, 2].

If objects are modeled by means of graphs, manipulating objects turns down to transforming graphs. In [3] and [4], a graph-oriented object database model called GOOD is introduced, in which graph theory is used to uniformly define an object-oriented datamodel and data manipulation formalism. Thereby, the advantages of both object-orientation and graphical interfaces are combined.

In both articles, the GOOD-model is shown to be of significant expressive and modeling power. In [3], it is shown how GOOD can simulate arbitrary recursive functions, while in [4] it is illustrated how the most prominent aspects of object-orientation (such as inheritance of both data and methods, encapsulation, extendibility,...) can be incorporated in the model [3]. Since at the same time, its datamodel and manipulation formalism are defined using a limited number of very basic 'building-blocks', GOOD may be regarded as a very general object-oriented database model.

To demonstrate the viability of a newly proposed database manipulation language, its expressive power should be compared to (and shown better than) that of other languages. A possible approach towards such a comparison of database languages is the use of so-called *completeness*-criteria. E.g., in [5, 6] it is shown that a relation S over some domain can be derived from some other relation R over that same domain, by applying relational algebra or calculus, if and only if the following two conditions are satisfied. First, no new values may be added. Second, every domain-permutation that maps R to itself, must

---

[1]The work of this author was supported by the I.W.O.N.L.

[2]Authors' present address: University of Antwerp (UIA), Dept. of Math. & Comp. Science, Universiteitsplein 1, B-2610 Antwerp, Belgium.

[3]Since this paper is concerned with the expressiveness of the GOOD-language, the latter issues, which are primarily concerned with *data-modeling*, will not be considered in the remainder.

also map S to itself, i.e., every *automorphism* of R must be an automorphism of S. An operation satisfying these conditions is often called a *generic* operation.

What does this criterion intuitively signifies ? The presence of a (non-trivial) automorphism for some relation R can be interpreted by saying that for every value in R, there exists another value (namely its image under the automorphism) which can 'take its place' in the relation, because when each value is replaced by its image, we obtain the same relation. The criterion now states that if such a change of place is possible in the input relation of an database operation, it should still be possible in the output relation. Violating this is only possible by manipulating values through more than just their relationships with other values (given in the relations of the database), i.e., by *interpreting* them. Consequently, this criterion is really very natural, since it merely prohibits interpreting values.

Apart from its theoretical importance, the validity of the criterion for a certain query-language can also have a practical usage. If two given relations namely satisfy the criterion, this implies the existence of a transformation between them in the language under consideration.

The above criterion has been called the *BP-completeness* of the relational algebra and calculus [7]. If we want to check this criterion for other languages than those for flat relational databases, we have to find appropriate definitions for the concepts of 'derivation' and 'automorphism'. We do this in Section 2 for the transformation-language of GOOD. The main issue of this paper is Theorem 1, which shows that this transformation-language satisfies the BP-criterion. We will only give an outline of the proof: for full details the reader is referred to [8]. Finally, at the end of Section 3, we show that a weaker variation of the transformation-language is also BP-complete.

2   A Graph-Oriented Object Database Model

2.1   Object base instances

In GOOD, an object base instance is conceptually represented as a directed labeled graph. The nodes of the graph represent the objects of the database. A distinction is made between atomic values (represented in the graph by what we call *printable nodes*) and abstract objects (represented in the graph by what we call *non-printable nodes*). A non-printable node only has the class indicated to which the object it represents belongs (e.g., person, vehicle, ...). A printable node has a double label: one label represents its type (e.g., string, number, picture, ...), the other represents its constant actual value.

The edges of the graph represent relationships between objects and values. A distinction is made between functional and non-functional relationships (resp. edges).

To allow the actual construction of such an instance, we assume the existence of four infinitely enumerable and pairwise disjoint sets, respectively NPOL of *non-printable object labels* (which correspond to class names), POL of *printable object labels* (which correspond to simple types), FEL of *functional edge labels* and NFEL of *non-functional edge labels*. Furthermore, we assume there is a domain-function π which associates a set of constants (e.g., strings, numbers, ...) to each printable object label. Formally:

Definition 2.1 (Object base instance) *An object base instance $\mathcal{I}$ is a labeled graph (N,E) such that*

- N is a finite set of labeled nodes; if n is a node in N such that its label, denoted by $\lambda_{\mathcal{I}}(n)$, is in NPOL (resp. in POL), then n is called a non-printable node (resp. a printable node);

- each printable node n in N has an additional label, denoted by $\text{print}(n)$, which is called its printlabel; this must be an element of $\pi(\lambda_{\mathcal{I}}(n))$;

- E is a set of labeled edges; if e is a labeled edge in E, then $e = (m, \alpha, n)$ with $m, n \in N$, and its label $\lambda_{\mathcal{I}}(e) = \alpha \in FEL \cup NFEL$. If $\lambda_{\mathcal{I}}(e)$ is in FEL (resp. in NFEL), then e is called a functional edge (resp. a non-functional edge);

- if (m, α, n_1) and $(m, \alpha, n_2) \in E$, then $\lambda_{\mathcal{I}}(n_1) = \lambda_{\mathcal{I}}(n_2)$; moreover, if $\alpha \in FEL$, then $n_1 = n_2$;

- if $\lambda_{\mathcal{I}}(n_1) = \lambda_{\mathcal{I}}(n_2)$ is in POL and $\text{print}(n_1) = \text{print}(n_2)$, then $n_1 = n_2$. ∎

The set N is often indicated with $Nodes(\mathcal{I})$, while E is indicated with $Edges(\mathcal{I})$.

To be able to formulate and prove the main issue of this paper, we introduce the following additional concepts.

Definition 2.2 (Isomorphism) Let $\mathcal{I} = (N, E)$ and $\mathcal{I}' = (N', E')$ be object base instances. An isomorphism i from $\mathcal{I}$ to $\mathcal{I}'$ is a one to one function of N onto N', such that:

1. $\forall n \in N : \lambda_{\mathcal{I}'}(i(n)) = \lambda_{\mathcal{I}}(n)$

2. $\forall n \in N : \lambda_{\mathcal{I}}(n) \in POL \Rightarrow n = i(n)$

3. $\forall n, n' \in N, \forall \alpha \in FEL \cup NFEL : (n, \alpha, n') \in E \Rightarrow (i(n), \alpha, i(n')) \in E'$.

An automorphism of $\mathcal{I}$ is an isomorphism of $\mathcal{I}$ onto itself. $Aut(\mathcal{I})$ is the set of all automorphisms of $\mathcal{I}$. ∎

We now define a special group-homomorphism between the automorphism-groups of two instances, that naturally extends an automorphism of one instance to an automorphism of the other instance.

Definition 2.3 (Genericity-morphism) Let $\mathcal{I}$ and $\mathcal{I}'$ be object base instances. We call a group-homomorphism $h : Aut(\mathcal{I}) \to Aut(\mathcal{I}')$ such that

$$\forall n \in N \cap N', \forall a \in Aut(\mathcal{I}) : a(n) = h(a(n))$$

a genericity-morphism of type $(\mathcal{I}, \mathcal{I}')$. ∎

Such a group-homomorphism does not necessarily exist for two arbitrary instances. A transformation between two instances $\mathcal{I}$ and $\mathcal{I}'$ for which there exists a genericity-morphism of type $(\mathcal{I}, \mathcal{I}')$, is what we called a *generic* transformation in the Introduction.

In the next section we show that the transformation-language of GOOD (which we will define next) expresses exactly the generic transformations. We wish to stress however that the above definition is not a mere translation of the concept of genericity as stated for flat relations in the Introduction. We draw the readers attention to the extra condition that the mapping that extends every automorphism of $\mathcal{I}$ into an automorphism of $\mathcal{I}'$ must be a group-*homomorphism*. We claim that the fact that an additional constraint has to be imposed on database-transformations in order to be called generic, is a consequence of the difference between value-based models (such as the relational) and pure object-based models (such as GOOD).

2.2 Instance Transformation

The GOOD-language for transforming object-base instances consists of five basic operations. An arbitrary sequence of such operations is called a GOOD-*transformation*. Every operation is based on the notion of *pattern*, which describes the parts of the instance for which the operation is to be executed.

Informally, when an operation is applied to some instance, the instance is scanned, and each time the pattern can be *embedded* somewhere, the operation under consideration is executed for the objects which correspond to this embedding. Formally:

Definition 2.4 (Pattern) *A* pattern *is an object base instance, with the exception that it may contain printable nodes without printlabels.* ■

Definition 2.5 (Embedding) *Let* $\mathcal{I} = (N, E)$ *be an object base instance and let* $\mathcal{J} = (M, F)$ *be a pattern. An* embedding *of* $\mathcal{J}$ *in* $\mathcal{I}$ *is a total mapping* $i : M \rightarrow N$ *such that*

1. $\forall n \in M : \lambda_{\mathcal{I}}(i(n)) = \lambda_{\mathcal{J}}(n)$

2. $\forall n \in M : \lambda_{\mathcal{J}}(n) \in POL \Rightarrow \text{print}(n) = \text{print}(i(n))$ *(if* n *has a printlabel)*

3. $\forall n, n' \in M, \forall \alpha \in FEL \cup NFEL : (n, \alpha, n') \in F \Rightarrow (i(n), \alpha, i(n')) \in E.$ ■

Lemma 1 *The composition of an automorphism of an instance and an embedding of a pattern in that instance, is itself an embedding of that pattern in the instance.* ■

We now discuss the five basic operations of the GOOD manipulation language. Concretely, this language allows the addition and deletion of objects and relationships. This gives us five operations: one for the addition of nodes with only functional outgoing edges, one for the addition of nodes with only non-functional outgoing edges, one for the addition of edges, one for the deletion of nodes and one for the deletion of edges. [4]

First we formally define and illustrate two addition- and two deletion-operations.

Definition 2.6 (Node Addition) *Let* $\mathcal{I} = (N, E)$ *be an object base instance and let* $\mathcal{J} = (M, F)$ *be a pattern. Let* $m_1, \ldots, m_n \in M$, $K \in NPOL$ *and* $\alpha_1, \ldots, \alpha_n \in FEL$. *The* node addition $NA[\mathcal{J}, \mathcal{I}, K, \{(\alpha_1, m_1), \ldots, (\alpha_n, m_n)\}]$ *results in a new instance* $\mathcal{I}'$, *defined as the minimal object base instance (up to isomorphism) for which:*

1. $\mathcal{I}$ *is a subinstance of* $\mathcal{I}'$;

2. for each embedding i *of* $\mathcal{J}$ *in* $\mathcal{I}$, *there exists a K-labeled node* n *of* $\mathcal{I}'$ *such that* $(n, \alpha_1, i(m_1)), \ldots, (n, \alpha_n, i(m_n))$ *are edges of* $\mathcal{I}'$;

3. each edge of $\mathcal{I}'$ *leaving a node of* $\mathcal{I}$ *is also an edge of* $\mathcal{I}$. ■

Definition 2.7 (Edge Addition) *Let* $\mathcal{I} = (N, E)$ *be an object base instance and let* $\mathcal{J} = (M, F)$ *be a pattern. Let* m, m' *be nodes in* M *and let* α *be an edge label. The* edge addition $EA[\mathcal{J}, \mathcal{I}, (m, \alpha, m')]$ *results in a new instance* $\mathcal{I}'$, *defined as he minimal object base instance of which* $\mathcal{I}$ *is a subinstance, and such that for each embedding* i *of* $\mathcal{J}$ *in* $\mathcal{I}$, $(i(m), \alpha, i(m'))$ *is an edge of* $\mathcal{I}'$. ■

---

[4] For all these operations, the model offers a graphical representation. Uniformly, patterns are indicated in plain line, what is added is indicated in boldface, and what is deleted is indicated in outline.

Definition 2.8 (Node deletion) *Let $\mathcal{I} = (N, E)$ be an object base instance and let $\mathcal{J} = (M, F)$ be a pattern. Let m be a non-printable node in M. The node deletion $ND[\mathcal{J}, \mathcal{I}, m]$ results in a new instance $\mathcal{I}'$, defined as the maximal object base instance for which $\mathcal{I}'$ is a subinstance of $\mathcal{I}$, and such that for each embedding i of $\mathcal{J}$ in $\mathcal{I}$, i(m) is not a node of $\mathcal{I}'$.* ∎

Definition 2.9 (Edge deletion) *Let $\mathcal{I} = (N, E)$ be an object base instance and let $\mathcal{J} = (M, F)$ be a pattern. with $(m, \alpha, m') \in F$. The edge deletion $ED[\mathcal{J}, \mathcal{I}, (m, \alpha, m')]$ results in a new instance $\mathcal{I}'$, defined as the maximal object base instance which is a subinstance of $\mathcal{I}$, and such that for each embedding i of $\mathcal{J}$ in $\mathcal{I}$, $(i(m), \alpha, i(m'))$ is not an edge of $\mathcal{I}'$.* ∎

Consider an object-base instance, representing some colored parts, each of which consists of a set of other parts. Suppose we want to remove from this instance the parts that have no subparts. Figure 1 shows how this can be accomplished in GOOD. In the first three operations we group all those parts in a node labeled **Atomic Parts**. First we create this node, by means of an operation with an empty pattern. Then we group *all* parts in this node, after which we remove from this set the parts that *do* consist of one or more subparts. This is accomplished by means of an edge deletion. Finally we delete the parts that are still in the set, as well as the set itself.

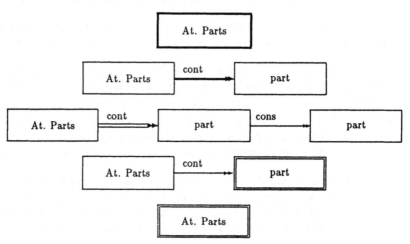

Figure 1: Removing parts without subparts

We now define and illustrate the third addition-operation.

Definition 2.10 (Abstraction) *Let $\mathcal{I} = (N, E)$ be an object base instance and let $\mathcal{J} = (M, F)$ be a pattern. Let n be a non-printable node in M. Let $K \in NPOL$ and $\alpha, \beta \in NFEL$. The abstraction $AB[\mathcal{J}, \mathcal{I}, n, K, \alpha, \beta]$ results in a new instance $\mathcal{I}'$, defined as the minimal object base instance (up to isomorphism) for which*

1. $\mathcal{I}$ is a subinstance of $\mathcal{I}'$;

2. for each embedding i of $\mathcal{J}$ in $\mathcal{I}$, there exists a K-labeled node p of $\mathcal{I}'$ such that $(p, \beta, i(n))$ is an edge of $\mathcal{I}'$;

3. if (p, β, q_1) *and* (p, β, q_2) *are both edges of* $\mathcal{I}'$, *then for each node* r *of* $\mathcal{I}'$, $(q_1, \alpha, r) \in E' \Leftrightarrow (q_2, \alpha, r) \in E'$;

4. each edge of $\mathcal{I}'$ *leaving a node of* $\mathcal{I}$ *is also an edge of* $\mathcal{I}$. ∎

Suppose we want to group the parts in our example object base instance according to the subparts they consist of. Figure 2 shows how this can be accomplished in GOOD. By this one operation, a new node labeled `Common Subparts` is created for each set of parts with common subparts. The label `Common Subparts` corresponds to K in the formal definition. The edge according to which parts are grouped (i.e. `consists_of`, corresponding to α in the formal definition) is indicated with a dashed line.

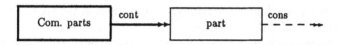

Figure 2: Grouping parts by their subparts

Theorem 1 of this paper will characterize the expressive power of the language consisting of arbitrary sequences of these five primitive operations *on instance-level.* This means that we will state a necessary and sufficient condition for derivability of one given instance from another by means of a sequence of primitive GOOD-operations.

However, on an instance-independent level, the language as introduced above is not powerful enough. Therefore, in [4] a method-construct is added to the language, allowing the naming of arbitrary fixed sequences of primitive GOOD-operations and calls to other methods. With this addition, the language is capable of simulating arbitrary Turing-machines. For a *given* input-instance, however, we can always check whether a method-call in some GOOD-transformation will still have any effect. Consequently, such method-calls can always be replaced by the corresponding method-body (if the call will still have an effect) or removed from the program (if the call will have no effect at all). Therefore, GOOD-methods are outside the scope of this paper.

To conclude this section, we introduce a notation to indicate that an instance is the result of applying a GOOD-transformation (i.e., a sequence of basic GOOD-operations) to another instance.

Definition 2.11 (GOOD-implication) *Let* $\mathcal{I} = (N, E)$ *and* $\mathcal{I}' = (N', E')$ *be object base instances.* $\mathcal{I} \stackrel{GOOD}{\Longrightarrow} \mathcal{I}'$ *indicates that there is an instance* $\hat{\mathcal{I}} = (\hat{N}, \hat{E})$ *such that:*

1. there is a GOOD-transformation that maps $\mathcal{I}$ *to* $\hat{\mathcal{I}}$;

2. there is an isomorphism from $\hat{\mathcal{I}}$ *to* $\mathcal{I}'$ *for which the elements of* $N \cap \hat{N}$ *are fixpoints.* ∎

We often call the instance $\hat{\mathcal{I}}$ the *intermediate* instance of the GOOD-implication. This definition may look a bit unnecessarily complicated. However, it would be rather meaningless just to consider instances $\mathcal{I}'$ being the result of applying some GOOD-transformation to $\mathcal{I}$, since the result of the two node addition-operations is defined *up to isomorphism.* Consequently, it is more meaningful to consider instances, isomorphic to the result of applying a transformation to another instance.

3   The Algebraic Characterization

We now characterize the expressiveness of the GOOD-transformation language by stating a property which is equivalent to definition 2.11.

Theorem 1 *Let $\mathcal{I}$ and $\mathcal{I}'$ be instances. Then the following properties are equivalent.*

1. $\mathcal{I} \overset{GOOD}{\Longrightarrow} \mathcal{I}'$.

2. *There exists a genericity-morphism h of type $(\mathcal{I}, \mathcal{I}')$.*

Sketch of Proof   To prove that GOOD-implication implies the existence of a genericity-morphism h, we use induction on the number of operations in the GOOD-transformation from the definition of GOOD-implication.

The other implication is first proved for the special case in which $\mathcal{I}$ is a subinstance of $\mathcal{I}'$ (cfr. Lemma 3). In Lemma 4 we then prove that the implication is valid for two arbitrary instances. ■

Before we can sketch the proof of the second implication for the special case of an instance and its superinstance, we have to adapt some concepts, known from group-theory and graph-theory [9], to the context of genericity-morphisms.

Definition 3.1 (Coset)   *Let G be a subgroup of $Aut(\mathcal{I})$ and let $a \in Aut(\mathcal{I})$. We define a coset of G as $a \circ G = \{a \circ b \mid b \in G\}$. $CosetAut(\mathcal{I})$ is the set of all cosets of all subgroups of $Aut(\mathcal{I})$.* ■

Definition 3.2 (Orbit)   *Let $\mathcal{I}$ be a subinstance of $\mathcal{I}'$ such that there exists a genericity-morphism h of type $(\mathcal{I}, \mathcal{I}')$. We call the orbit of a node $n \in \mathcal{I}' - \mathcal{I}$ w.r.t. h the set*

$$orb_h(n) = \{n' \in Nodes(\mathcal{I}' - \mathcal{I}) \mid \exists a \in Aut(\mathcal{I}) : h(a)(n) = n'\}$$

In each orbit we choose an arbitrary but fixed node, called the representative *of the orbit. $Orbits_h(\mathcal{I}' - \mathcal{I})$ is the set of all the orbits of nodes of $\mathcal{I}' - \mathcal{I}$ w.r.t. h.* ■

It can easily be seen that $Orbits_h(\mathcal{I}' - \mathcal{I})$ is a partition of $Nodes(\mathcal{I}' - \mathcal{I})$.

Definition 3.3 (Stabilizer)   *Let $\mathcal{I}$ be a subinstance of $\mathcal{I}'$ such that there exists a genericity-morphism h of type $(\mathcal{I}, \mathcal{I}')$. Let $m \in Nodes(\mathcal{I}' - \mathcal{I})$. The stabilizer of m w.r.t. h is the set*

$$st_h(m) = \{a \in Aut(\mathcal{I}) \mid h(a)(m) = m\}$$ ■

Obviously, any $st_h(m)$ is a subgroup of $Aut(\mathcal{I})$.

Definition 3.4 ($\mathcal{I}_{\text{diff}}$)   *Let $\mathcal{I} = (N, E)$ be an object base instance. We define $\mathcal{I}_{\text{diff}}$ as the instance (N, E') with*

$$E' = E \cup \{(n, equal, n) \mid n \in N\} \cup \{(n, diff, m) \mid n, m \in N, n \neq m\}$$

We assume that 'diff' and 'equal' are no edge labels of $\mathcal{I}$. ■

We now describe an extension of the instance $\mathcal{I}'$, based on the genericity-morphism h (see Figure 3).

Definition 3.5 *Let $\mathcal{I}$ be a subinstance of $\mathcal{I}'$, such that there exists a genericity-morphism h of type $(\mathcal{I}, \mathcal{I}')$. We define the extension $\langle \mathcal{I}' \rangle$ of $\mathcal{I}'$ w.r.t. $\mathcal{I}$ and h as follows.*

Consider $Orbits_h(\mathcal{I}' - \mathcal{I})$ as a set of non-printable nodes, one for each element, labeled by a unique name for that orbit. Consider $Aut(\mathcal{I})$ as a set of non-printable nodes, one for each element, labeled by AUT. Consider $CosetAut(\mathcal{I})$ as a set of non-printable nodes, one for each element, labeled by a unique name for their associated subgroup. We assume that all these labels are new. We then define:

$$Nodes(\langle \mathcal{I}' \rangle) = Nodes(\mathcal{I}') \cup Orbits_h(\mathcal{I}' - \mathcal{I}) \cup Aut(\mathcal{I}) \cup CosetAut(\mathcal{I})$$
$$Edges(\langle \mathcal{I}' \rangle) = Edges(\mathcal{I}') \cup Edges(\mathcal{I}_{diff}) \cup$$
$$\{(a, n, a(n)) \mid a \in Aut(\mathcal{I}), n \in Nodes(\mathcal{I})\} \cup$$
$$\{(C, \xi, a) \mid C \in CosetAut(\mathcal{I}), a \in C\} \cup$$
$$\{(m, \mu, A) \mid A \in Orbits_h(\mathcal{I}' - \mathcal{I}), m \in A\} \cup$$
$$\{(h(b)(m_0), \nu, b \circ st_h(m_0)) \mid m_0 \text{ an orbit-representative,}$$
$$b \in Aut(\mathcal{I})\}$$

We assume that n ($n \in Nodes(\mathcal{I})$), μ and ν are functional edge labels [5] and that ξ is a non-functional edge label, all not occurring in $\mathcal{I}'$. We also assume that $Orbits_h(\mathcal{I}' - \mathcal{I})$, $Aut(\mathcal{I})$ and $CosetAut(\mathcal{I})$ are disjoint sets of nodes not occurring in $\mathcal{I}'$. ∎

The following lemma is of critical importance to the proof of Theorem 1. It shows that nodes of $\mathcal{I}' - \mathcal{I}$ are determined uniquely by the coset and the orbit they are associated with in $\langle \mathcal{I}' \rangle$.

Lemma 2 *Let $m_1, m_2 \in Nodes(\mathcal{I}' - \mathcal{I})$, $(m_1, \mu, A), (m_2, \mu, A), (m_1, \nu, C), (m_2, \nu, C) \in Edges(\langle \mathcal{I}' \rangle)$. Then $m_1 = m_2$.* ∎

We can now state and prove the second implication of Theorem 1 for the special case of an instance and its superinstance.

Lemma 3 *Let $\mathcal{I}$ be a subinstance of $\mathcal{I}'$ such that there exists a genericity-morphism h of type $(\mathcal{I}, \mathcal{I}')$. Then $\mathcal{I} \overset{GOOD}{\Longrightarrow} \mathcal{I}'$.*

Sketch of Proof We gradually construct a superinstance $\langle \hat{\mathcal{I}} \rangle$ of $\mathcal{I}$. Step by step, the various components as described in definition 3.5 are added to $\mathcal{I}$ by means of a GOOD-transformation, each time resulting in an intermediate instance $\hat{\mathcal{K}}_j$, where j indicates the number of the step. The isomorphism between $\langle \hat{\mathcal{I}} \rangle$ and $\langle \mathcal{I}' \rangle$ and between their auto-morphism-groups is also described 'gradually'. At each step of the construction, a superinstance $\hat{\mathcal{K}}_j$ of $\mathcal{I}$, a subinstance $\mathcal{K}'_j$ of $\langle \mathcal{I}' \rangle$ and a mapping i are defined such that $i : \hat{\mathcal{K}}_j \to \mathcal{K}'_j$ is an isomorphism, fixing the nodes of $\mathcal{I}$.

Due to lack of space, and since proving that a mapping is a GOOD-isomorphism or a group-isomorphism simply requires checking the different items in the respective definitions, we omit these proofs from the remainder of this sketch, and just give the GOOD-transformation that adds the various components of $\langle \hat{\mathcal{I}} \rangle$.

---

[5] It can be shown that the edges labeled by n ($n \in Nodes(\mathcal{I})$), μ and ν are indeed functional.

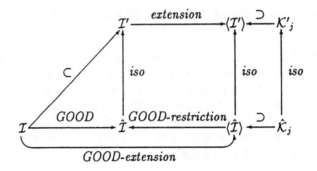

Figure 3: An overview of instances used in the proof

Step 1 : Initially, $\hat{\mathcal{K}}_1 = \mathcal{I}$.

Step 2 : We now add *diff-* and *equal*-edges. First we add an *equal*-edge between each non-printable node and itself. Then we add a *diff*-edge between each two non-printable nodes. Finally, if between two nodes there is both an *equal-* and a *diff*-edge, we delete the *diff*-edge. This way we obtain $\hat{\mathcal{K}}_2$.

Step 3 : We now add AUT-labeled nodes and their outgoing edges, labeled by nodes of $\mathcal{I}$. This operation is accomplished with a single node addition. We use $\hat{\mathcal{K}}_2$ itself as pattern, and add a node labeled by AUT with outgoing edges to each node n of $\hat{\mathcal{K}}_2$, labeled by n. To see that this operation does what it is supposed to do, reconsider Lemma 1. The resulting instance is called $\hat{\mathcal{K}}_3$.

Step 4 : We next add nodes corresponding to the elements of $CosetAut(\mathcal{I})$, and their outgoing ξ-edges. Therefore, for each subgroup D of $Aut(\mathcal{I})$, the following five operations are applied consecutively to $\hat{\mathcal{K}}_3$.

- a node addition with $\hat{\mathcal{K}}_3$ as pattern, of a node labeled by D' (which we assume to be a new label), with outgoing functional edges, all with different new labels, to all the AUT-nodes that correspond to an automorphism in D;

- an edge addition of non-functional edges labeled by ξ', for each functional edge added in the previous operation;

- abstraction of the ξ'-edges, resulting in the creation of D-nodes, with outgoing ξ''-edges;

- an edge addition of non-functional edges labeled by ξ, each time there is a ξ''-edge, followed by a ξ'-edge;

- a node deletion of all D'-nodes.

The first operation of this step results in the creation of a node $\hat{D}$ for each subgroup D (with outgoing edges to its members), but also in the creation of a node $a \circ D$ for each $a \in Aut(\mathcal{I})$, which is (group-)isomorphic to $Aut(\hat{\mathcal{K}}_3)$. By the use of functional edges with all different labels, however, a lot of redundant nodes were also created. Recalling section 2, the abstraction-operation allows one to group objects according to common non-functional properties. Therefore, the following four operations group D'-nodes that represent the same set, thereby creating a unique D-node for each coset. The resulting instance is called $\hat{\mathcal{K}}_4$.

Step 5 :   We now add nodes corresponding to some orbit. This operation is accomplished by means of one node addition for each element O of $Orbits_h(\mathcal{I}' - \mathcal{I})$, each time with $\hat{\mathcal{K}}_4$ as pattern, of a node $\hat{O}$ labeled by a unique name for that orbit. Since the result of this node addition is defined as the *minimal* superinstance of $\hat{\mathcal{K}}_4$ that also contains an orbit-node, such an operation creates exactly one node. Consequently, exactly one node is added for each orbit. The resulting instance is called $\hat{\mathcal{K}}_5$.

Step 6 :   We now add nodes corresponding to the nodes of $\mathcal{I}' - \mathcal{I}$, together with their outgoing μ- and ν-edges. Therefore, one node addition is applied to $\hat{\mathcal{K}}_5$ for each element of $Orbits_h(\mathcal{I}' - \mathcal{I})$. Given the representative m_0 of an orbit, the operation has $\hat{\mathcal{K}}_5$ as pattern, and adds a node with the same label as m_0, and two outgoing functional edges. One μ-labeled edge arrives at the node that corresponds to the orbit O of m_0, while the other edge with label ν arrives at the node that corresponds to the coset $st_h(m_0)$. The resulting instance is called $\hat{\mathcal{K}}_6$. The fact that this single node addition really adds one node for each node of $\mathcal{I}' - \mathcal{I}$ is proved using Lemma 2.

Step 7 :   Finally, we add the edges of $\mathcal{I}'$ that are not edges of $\mathcal{I}$. For each such edge, say (n, α, m), an edge addition is applied with $\hat{\mathcal{K}}_6$ as pattern, of an edge $(\hat{n}, \alpha, \hat{m})$. Obviously, for each such edge and for each $b \in Aut(\hat{\mathcal{K}}_6)$, an edge $(b(\hat{n}), \alpha, b(\hat{m}))$ is also created. The result of this seventh step of the construction is the superinstance of $\mathcal{I}$ which we previously called $\langle \hat{\mathcal{I}} \rangle$.

We now have to restrict $\langle \hat{\mathcal{I}} \rangle$ to an instance, isomorphic to the given instance $\mathcal{I}'$. This can be done very easily by deleting all nodes and edges added in steps 2 through 5. This way we obtain $\hat{\mathcal{I}}$. By restricting the instance $\langle \mathcal{I}' \rangle$ to the instance $\mathcal{I}'$, and the isomorphism i to the nodes of $\hat{\mathcal{I}}$, it can be seen that $\hat{\mathcal{I}}$ and $\mathcal{I}'$ are indeed isomorphic. ■

Lemma 4 *Let $\mathcal{I}$ and $\mathcal{I}'$ be instances such that there exists a genericity-morphism h of type $(\mathcal{I}, \mathcal{I}')$. Then $\mathcal{I} \overset{GOOD}{\Longrightarrow} \mathcal{I}'$.*

Sketch of Proof   To be able to apply Lemma 3, we first construct a superinstance of $\mathcal{I}$ and $\mathcal{I}'$. Since in general, it is not possible to construct a superinstance of two arbitrary instances (e.g., functionality might be violated), we first have to 'rename' the instance $\mathcal{I}$ to an instance $\bar{\mathcal{I}}$ with all different new edge labels. Next we show the existence of a genericity-morphism between the automorphism-groups of $\mathcal{I}$ and the superinstance. Application of Lemma 3 now shows the existence of a GOOD-transformation from $\mathcal{I}$ to the superinstance. Finally, we restrict this superinstance by means of deletions to an instance isomorphic to $\mathcal{I}'$. ■

One might wonder why we have to place such a strong restriction on printlabels (i.e., the constants of the object-base) in the definition of isomorphisms, i.e., why the second item of definition 2.2 is necessary. To get some insight into this problem, let us see what happens to Theorem 1 if we relax this restriction. Therefore we change definitions 2.2, 2.3 and 2.11 as follows. The newly defined concepts are called *weak*, whereas their old counterparts will be called *strong*.

Definition 3.6 (Weak Genericity-morphism) *Let* $\mathcal{I} = (N, E)$ *and* $\mathcal{I}' = (N', E')$ *be object base instances. A* weak isomorphism i *from* $\mathcal{I}$ *to* $\mathcal{I}'$ *is a one to one function of N onto N', such that:*

1. $\forall n \in N : \lambda_{\mathcal{I}'}(i(n)) = \lambda_{\mathcal{I}}(n)$

2. $\forall n, n' \in N, \forall \alpha \in FEL \cup NFEL : (n, \alpha, n') \in E \Rightarrow (i(n), \alpha, i(n')) \in E'.$

A weak automorphism *of* $\mathcal{I}$ *is a weak isomorphism of* $\mathcal{I}$ *onto itself.* $Aut_w(\mathcal{I})$ *is the set of all weak automorphisms of* $\mathcal{I}$. *A group-homomorphism* $h : Aut_w(\mathcal{I}) \to Aut_w(\mathcal{I}')$ *such that*
$$\forall n \in N \cap N', \forall a \in Aut_w(\mathcal{I}) : a(n) = h(a(n))$$
is called a weak genericity-morphism *of type* $(\mathcal{I}, \mathcal{I}')$. ∎

Obviously, with this new definition, Lemma 1 would no longer hold if we allowed arbitrary patterns in GOOD-operations. Therefore we adapt definition 2.4 in a way that patterns are no longer allowed to contain printlabels. Such patterns are called weak patterns. When the five basic operations of the GOOD manipulation language are defined using these weak patterns, we call them weak operations, which can be sequenced to obtain weak GOOD-transformations. This leads us to the following new definition of GOOD-implication.

Definition 3.7 (Weak GOOD-implication) *Let* $\mathcal{I} = (N, E)$, $\mathcal{I}' = (N', E')$ *be object base instances.* $\mathcal{I} \overset{GOOD}{\Longrightarrow}_w \mathcal{I}'$ *indicates that there is an instance* $\hat{\mathcal{I}} = (\hat{N}, \hat{E})$ *such that:*

1. *there is a weak GOOD-transformation that maps* $\mathcal{I}$ *to* $\hat{\mathcal{I}}$;

2. *there is a strong isomorphism from* $\hat{\mathcal{I}}$ *to* $\mathcal{I}'$ *for which the elements of* $N \cap \hat{N}$ *are fixpoints.* ∎

If Theorem 1 is modified according to the new definitions, it still holds:

Theorem 2 *Let* $\mathcal{I}$ *and* $\mathcal{I}'$ *be instances. Then the following properties are equivalent.*

1. $\mathcal{I} \overset{GOOD}{\Longrightarrow}_w \mathcal{I}'$.

2. *There exists a weak genericity-morphism* h *of type* $(\mathcal{I}, \mathcal{I}')$. ∎

4 Research Directions

Following the suggestions of two of the anonymous referees of this paper, we have recently started an investigation into the possible connections with the area of Graph-Grammars. It immediately became apparent that the semantics of GOOD-transformations is very closely related to programmed applications of graph-grammar productions. The question is now whether results from both research-areas are mutually applicable.

As mentioned before, a limitation of the theorems of this paper is that they consider only the relationship between two given instance-graphs at a time. Therefore, another issue which we are currently investigating is the study of certain mappings between two *sets* of instance-graphs, thereby trying to characterize the set of GOOD-transformations itself.

References

[1] Mariano Consens and Alberto Mendelzon. GraphLog: a visual formalism for real life recursion. In *Proceedings of the Ninth ACM SIGACT-SIGMOD-SIGART Symposium on Principles of Database Systems*, pages 404–416, April 1990.

[2] C. Beeri. A formal approach to object-oriented databases. *Data & Knowledge Engineering*, 5(4):353–382, 1990.

[3] Marc Gyssens, Jan Paredaens, and Dirk Van Gucht. A graph-oriented object database model. In *Proceedings of the Ninth ACM SIGACT-SIGMOD-SIGART Symposium on Principles of Database Systems*, pages 417–424, April 1990.

[4] Marc Gyssens, Jan Paredaens, and Dirk Van Gucht. A graph-oriented object model for end-user interfaces. In *Proceedings of the 1990 ACM SIGMOD International Conference on Management of Data*, pages 24–33, May 1990.

[5] Jan Paredaens. On the expressive power of the relational algebra. *Information Processing Letters*, 7(2):107–111, february 1978.

[6] Francois Bancilhon. On the completeness of query languages for relational data bases. In *Proceedings of the 7th Symposium on Mathematical Foundations of Computer Science*, volume 64 of *Lecture Notes in Computer Science*, pages 112–123, Berlin, 1978. Springer-Verlag.

[7] Ashok K. Chandra and David Harel. Computable queries for relational databases. *Journal of Computer and System Sciences*, 21:156–178, 1980.

[8] Marc Andries and Jan Paredaens. An algebraic characterization for the GOOD-transformation language. Technical Report 91-11, University of Antwerp (U.I.A.), 1991.

[9] Béla Bollobás. *Graph Theory (An Introductory Course)*. Number 63 in Graduate Texts in Mathematics. Springer Verlag, New York Heidelberg Berlin, 1979.

Attributed Elementary Programmed Graph Grammars

Rudolf Freund   **Brigitte Haberstroh**

Institut für Computersprachen   Institut für Informationssysteme

Technische Universität Wien, Karlsplatz 13, Vienna, Austria

Abstract. A new mechanism for generating graph languages is introduced which is based on the controlled rewriting of graphs using only six elementary types of graph productions, namely the addition, the deletion and the renaming of a node or an edge. Although these elementary graph productions are acting strictly locally and no embedding transformations are needed, in the unrestricted and monotone case, elementary programmed graph grammars have the same generative power as expression graph grammars. As a graph with attributes assigned to its nodes and edges is an even more useful tool for the description of certain data structures than a directed graph, attributed elementary programmed graph grammars turn out to be adequate graph rewriting systems applicable in many areas of computer science.

0. Introduction

In the last decade many generating mechanisms for graph languages have been investigated, most of them emphasizing the role of the embedding transformations, i.e. the methods how to embed a rewritten subgraph into the remaining rest graph. For a survey of some of these graph generating systems see [LNCS 153] and [LNCS 291]. Our approach minimizes the role of embedding and uses only six elementary types of strictly locally operating graph productions, namely the adding of a new node or a new edge with a given label, the deleting of a node or an edge with the required label and the changing of the label of a node or of an edge between two nodes with specified labels.

Various concepts of attributed directed graphs can be found in literature as the underlying data model in applications like pattern recognition (e.g. see [Kaplan et al.]). Attributed graph grammars as the operational specification method for manipulating these attributed graphs and several applications in the domain of software development are described in [Engels et al.] and [Göttler].

In the first section of this paper, the necessary definitions for introducing elementary programmed graph grammars and some explanatory examples are cited. Moreover the equivalence of our approach and the expression approach (see [Nagl]) in the unrestricted and the monotone case is exhibited, a result that is based on the model of sequential programmed graph grammars proposed in [Bunke].

Furthermore the method of analysing graphs by using elementary programmed graph grammars for practical purposes is discussed. In the second section attributed elementary programmed graph grammars are introduced, and by several selected examples we show the applicability of our approach in different areas of computer science. A discussion of future research topics concludes the paper.

1. Definitions and preliminary results

Additional definitions and results of formal language theory can be looked up in [Hopcroft Ullman] or [Salomaa]. For a detailed survey of regulated rewriting in formal language theory the interested reader is referred to [Dassow Păun].

The structures to be considered in the following are finite directed graphs with labelled edges and nodes.

Definition 1.1. A *graph* over (V, W) is a tuple $g = (\mathcal{N}, \mathit{n}, \mathcal{E})$, where V and W are alphabets for labelling the nodes and edges, $\mathcal{N}$ is a finite set of *nodes*, $\mathcal{E} \in \mathcal{P}(\mathcal{N} \times \mathcal{N} \times W)$ ($\mathcal{P}(M)$ denotes the powerset of the set M) and $\mathit{n}: \mathcal{N} \to V$ is the labelling function for the nodes in $\mathcal{N}$. Any tuple (k, k', w) with $k, k' \in \mathcal{N}$ and $w \in W$ can be interpreted as a (*directed*) *edge* from the node k to the node k' and labelled by w. According to these definitions, loops are allowed in the graphs considered throughout the paper, whereas multiple edges are not.

Definition 1.2. Let $\gamma (V, W)$ denote the set of all graphs over (V, W) and g_ε the *empty graph*, i. e. the graph with an empty set of nodes.

Definition 1.3. Let $g, g' \in \gamma (V, W)$; then g and g' are called *isomorphic* – abbreviated $g \cong g'$ – iff there exists a bijective mapping $f: \mathcal{N}(g) \to \mathcal{N}(g')$ with $(k, k', w) \in \mathcal{E}(g) \iff (f(k), f(k'), w) \in \mathcal{E}(g')$ for any $k, k' \in \mathcal{N}(g)$ and $w \in W$; g and g' are called *equivalent* – abbreviated $g \equiv g'$ – iff $g \cong g'$ and, moreover, $\mathit{n}(g) = \mathit{n}(g') \cdot f$, where "$\cdot$" denotes the composition of functions or relations, e. g. $(h \cdot f)(x) = h(f(x))$.

As equivalent graphs exactly have the same structure and differ only in the denotations of the nodes, we usually shall consider equivalence classes of graphs:

Definition 1.4. For any $g \in \gamma(V,W)$ let $[g]$ denote the equivalence class of all graphs of $\gamma(V,W)$ equivalent to g. A class $[g]$ of graphs is represented as a labelled graph without denotations of the nodes.
Furthermore we define $\Gamma(V,W) := \gamma(V,W)/\equiv$, i. e. $\Gamma(V,W) = \{ [g] \mid g \in \gamma(V,W) \}$.

Example 1.1. A word $w = w_1 \dots w_m$ over an alphabet V can be represented as a graph $g \in \gamma(V, \{ e \})$ with $\mathcal{N}(g) = \{1, \dots, m \}$, $\mathcal{E}(g) = \{ (n, n + 1, e) \mid 1 \le n \le m - 1 \}$ and $\mathit{n}(g)(n) = w_n$ for $1 \le n \le m$. Representing a node by a circle with its denotation outside and its label inside, g can be described by the graph in figure 1, whereas the corresponding class $[g]$ is shown in figure 2:

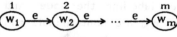

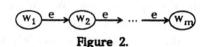

Figure 1.   **Figure 2.**

Let us now, in terms of the definitions above, define **elementary graph productions** over (V, W) and the results of their applications, where g denotes the original graph and g' the graph after having applied the elementary graph production.

Definition 1.5.

(1) $\underline{E(X)}$: add a new node with label X.
$|\mathcal{N}(g')| = |\mathcal{N}(g)| + 1$, $\mathcal{N}(g) \subseteq \mathcal{N}(g')$, $\mathcal{N}(g) - \mathcal{N}(g') = \{u\}$, $\mathcal{E}(g') = \mathcal{E}(g)$,
$n(g') | \mathcal{N}(g) = n(g)$, $n(g')(u) = X$ (where f | C denotes the restriction of the function $f: A \to B$ to the subset $C \subseteq A$).

(2) $\underline{C(X,Y)}$: change the label of a node from X to Y.
$\mathcal{N}(g') = \mathcal{N}(g)$, $\mathcal{E}(g') = \mathcal{E}(g)$, and for some $u \in \mathcal{N}(g)$
$n(g') | (\mathcal{N}(g) - \{u\}) = n(g) | (\mathcal{N}(g) - \{u\})$, $n(g)(u) = X$ and $n(g')(u) = Y$.

(3) $\underline{D(X)}$: delete a node with label X.
$|\mathcal{N}(g')| = |\mathcal{N}(g)| - 1$, $\mathcal{N}(g) - \mathcal{N}(g') = \{u\}$, $n(g) | (\mathcal{N}(g) - \{u\}) = n(g')$,
$n(g)(u) = X$, and $\mathcal{E}(g') = \mathcal{E}(g)$, i. e. we demand
$\forall v \in \mathcal{N}(g)$, $\forall w \in W$: $(v,u,w) \notin \mathcal{E}(g)$ and $(u,v,w) \notin \mathcal{E}(g)$ assuring that no edges are leaving from or ending in the node u with label X of the graph g.

(4) $\underline{E(X,Y;w)}$: add an edge with label w between a node with label X and another node with label Y.
$\mathcal{N}(g') = \mathcal{N}(g)$, $n(g') = n(g)$, and for some $u, v \in \mathcal{N}(g)$
$n(g)(u) = X$, $n(g)(v) = Y$, $(u,v,w) \notin \mathcal{E}(g)$ (which prohibits multiple edges) and $\mathcal{E}(g') = \mathcal{E}(g) \cup \{(u,v,w)\}$.

(5) $\underline{C(X,Y;w,z)}$: change the label of an edge between a node with label X and another node with label Y from w to z.
$\mathcal{N}(g') = \mathcal{N}(g)$, $n(g') = n(g)$, and for some $u, v \in \mathcal{N}(g)$
$n(g)(u) = X$, $n(g)(v) = Y$, $(u,v,z) \notin \mathcal{E}(g)$, $(u,v,w) \in \mathcal{E}(g)$, and
$\mathcal{E}(g') = (\mathcal{E}(g) - \{(u,v,w)\}) \cup \{(u,v,z)\}$.

(6) $\underline{D(X,Y;w)}$: delete an edge with label w between a node with label X and another node with label Y.
$\mathcal{N}(g') = \mathcal{N}(g)$, $n(g') = n(g)$, and for some $u, v \in \mathcal{N}(g)$
$n(g)(u) = X$, $n(g)(v) = Y$, $(u,v,w) \in \mathcal{E}(g)$ and $\mathcal{E}(g') = \mathcal{E}(g) - \{(u,v,w)\}$.

An **elementary graph production** described under (i) is called to be **of type i** for $1 \leq i \leq 6$. The set of all possible elementary graph productions over (V, W) is denoted by $\mathcal{P}(V,W)$.

Definition 1.6. If the graph $g' \in \gamma(V,W)$ can be obtained by applying an elementary graph production p – which is of some type i as described above – to the graph $g \in \gamma(V,W)$, then g' is said to be **directly derivable** from g by p – abbreviated $g \xrightarrow{p} g'$; $h' \in \Gamma(V,W)$ is directly derivable from $h \in \Gamma(V,W)$ by p – abbreviated $h \xrightarrow{p} h'$ – iff there is a $g \in h$ and a $g' \in h'$ with $g \xrightarrow{p} g'$.

Example 1.2. ⓐ is directly derivable from $[g_\varepsilon]$ by E(a).

Definition 1.7. An **elementary programmed graph grammar** is a 7-tuple $G = (V_N, W_N, V_T, W_T, P, P_0, P_f)$ such that

(1) $V = V_N \cup V_T$ and $W = W_N \cup W_T$ are alphabets for labelling the nodes and edges, respectively, where the elements of V_N and W_N are nonterminals and the elements of V_T and W_T are terminals; $V_N \cap V_T = W_N \cap W_T = \emptyset$.

(2) P is a finite set of triples (**rules**) (r: p, σ(r), φ(r)) where r: p is an elementary graph production in $\mathfrak{L}(V,W)$ labelled by r and σ(r) and φ(r) are two sets of labels of such core rules in P, i. e. σ and φ are two mappings from Lab(P) to $\mathfrak{P}(Lab(P))$, Lab(P) denoting the set of labels of the productions in P: Lab(P) = { r | (r: p, σ(r), φ(r)) ∈ P }

(3) $P_0 \subseteq$ Lab(P) is a set of initial labels (of initial rules);

(4) $P_F \subseteq$ Lab(P) is a set of final labels (of final rules).

A **rule** (r: p, σ(r), φ(r)) in P is of **type** m – with $1 \leq m \leq 6$ – iff the elementary graph production p is of type m. Moreover, we define $\mathfrak{p}(r) := p$.

G is called to be of **type** $n_1 n_2 n_3 n_4 n_5 n_6$ – with $n_m \in \{0,1\}$ for $1 \leq m \leq 6$ – if no rule (r: p, σ(r), φ(r)) ∈ P is of type m for some m with $n_m = 0$.

Definition 1.8. Let G = ($V_N, W_N, V_T, W_T, P, P_0, P_F$) be an elementary programmed graph grammar, r,r' ∈ Lab(P), (r: p, σ(r), φ(r)) ∈ P and g,g' ∈ γ(V,W). The pair (g',r') is **directly derivable** from (g,r) **in** G – abbreviated (g,r) $\vdash_G$ (g',r') – iff either

(1) $g \underset{p}{\Rightarrow} g'$ and r' ∈ σ(r) (which means that the production p can be applied to the graph g, yielding the new graph g') and we also write (g,r) $\overset{Y}{\vdash_G}$ (g',r') – or

(2) g' = g, the production p cannot be applied to the graph g, and r' ∈ φ(r); in this case we also write (g,r) $\overset{N}{\vdash_G}$ (g',r').

For h,h' ∈ Γ(V,W) we define (h,r) $\vdash_G$ (h',r') (respectively (h,r) $\overset{Y}{\vdash_G}$ (h',r') and (h,r) $\overset{N}{\vdash_G}$ (h',r')) iff there exists a g ∈ h and a g' ∈ h' such that (g,r) $\vdash_G$ (g',r') (respectively (g,r) $\overset{Y}{\vdash_G}$ (g',r') and (g,r) $\overset{N}{\vdash_G}$ (g',r')).

As usual, $\overset{*}{\vdash_G}$ denotes the reflexive and transitive closure of the relation $\vdash_G$.

Definition 1.9. The **graph language generated by G** is defined by
$\mathcal{L}(G)$ = { h | h ∈ Γ(V_T,W_T) and ([g_ε], r_0) $\overset{*}{\vdash_G}$ (h, r_F) for some $r_0 \in P_0$ and some $r_F \in P_F$ }.

Remark 1.1. We sometimes shall make use of the fact that also
$\mathcal{L}(G)$ = { [g] | g ∈ γ(V_T,W_T) and (g_ε, r_0) $\overset{*}{\vdash_G}$ (g, r_F) for some $r_0 \in P_0$ and some $r_F \in P_F$ }.

Example 1.3. Let G_1 = ({A}, $\emptyset$, {a}, {e}, P_1, {r_0}, {r_5}) be the elementary programmed graph grammar with P_1 = { (r_0: E(A), {r_4}, $\emptyset$), (r_1: E(a), {r_4}, $\emptyset$), (r_2: E(a,a;e), {r_2}, {r_5}), (r_3: C(a,A), {r_3}, {r_4}), (r_4: C(A,a), {r_1}, {r_2,r_3}), (r_5: C(A,A), $\emptyset$, $\emptyset$)}. Then $\mathcal{L}(G_1)$ exactly contains the equivalence classes [g] of complete graphs g ∈ γ({a},{e}) with $|\mathcal{N}(g)| = 2^n$ for some n ∈ **N** (where **N** denotes the set of positive integers 1,2,... and $\mathbf{N}_0 := \mathbf{N} \cup \{0\}$, and a graph g ∈ γ(V,W) is called **complete** iff $\mathcal{E} = \mathcal{N} \times \mathcal{N} \times W$).

An elementary programmed graph grammar can be represented as a graph itself that looks like the graph of a (nondeterministic) finite automaton, the edges of which are labelled by Y(es) or N(o) and the nodes by the elementary graph productions of the rules; the attributes "START" and "STOP" denote initial respectively final rules. Thus the elementary programmed graph grammar G_1 is described by the following *control graph* (in the sequel we shall omit the circles round the rules):

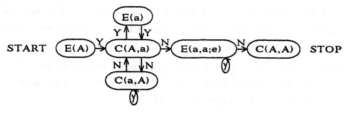

Definition 1.10. $\mathfrak{L}(P, n_1n_2n_3n_4n_5n_6)$ is the **family of graph languages** generated by elementary programmed graph grammars *of type* $n_1n_2n_3n_4n_5n_6$ with $n_m \in \{0,1\}$ for $1 \le m \le 6$. Moreover, graph languages respectively elementary programmed graph grammars of type 111111, 110111, and 110110 are called **unrestricted, monotone** and **strictly monotone,** respectively.

For instance, $\mathcal{L}(G_1) \in \mathfrak{L}(P,110100)$, because the elementary programmed graph grammar G_1 of Example 1.3 is of type 110100.

In the following we exhibit that the graph languages in $\mathfrak{L}(P,111111)$ and $\mathfrak{L}(P,110111)$ coincide with the recursively enumerable respectively monotone graph languages of equivalence classes of graphs generated by unrestricted respectively monotone sequential programmed graph grammars ([Bunke]) or expression graph grammars ([Nagl]) - abbreviated by $\mathfrak{L}$(rec enum) resp. $\mathfrak{L}$(mon).

As a derivation in an elementary programmed graph grammar starts from the empty graph in contrast to other approaches we need the following definitions:

Definition 1.11. $\mathfrak{L}(P,110111 - \varepsilon)$ denotes the family of graph languages in $\mathfrak{L}(P,110111)$ not containing $[g_\varepsilon]$.

Definition 1.12. Two *elementary programmed graph grammars* G and G' are called **equivalent** iff $\mathcal{L}(G) = \mathcal{L}(G')$.

Lemma 1.1. For any elementary programmed graph grammar G of type 110111 there exists an equivalent elementary programmed graph grammar G' of type 110111 with $\mathcal{L}(G') = \mathcal{L}(G) - \{[g_\varepsilon]\}$.

We now are in the position to show how our approach fits into the framework of other systems of graph grammars:

Theorem 1.1. $\mathfrak{L}$(mon) $= \mathfrak{L}(P,110111 - \varepsilon)$ and $\mathfrak{L}$(rec enum) $= \mathfrak{L}(P,111111)$.

For a formal proof of this theorem the reader must be referred to [Freund].

Remark 1.2. The definition of a graph language by an elementary programmed graph grammar G can also be used for deciding the question $[g] \in \mathcal{L}(G)$ for any given graph g, because $[g] \in \mathcal{L}(G)$ iff g can be reduced to g_ε by applying the complementary productions p^c of the productions p in the control graph representing G, thus simulating a derivation of g in G in the reversed direction, where we define the *complementary productions* by
$(E(X))^c := D(X)$, $(C(X,Y))^c := C(Y,X)$, $(D(X))^c := E(X)$,
$(E(X,Y;w))^c := D(X,Y;w)$, $(C(X,Y;w,z))^c := C(X,Y;z,w)$, $(D(X,Y;w))^c := E(X,Y;w)$.
Because of the lack of embedding rules an efficient backtracking algorithm based on these considerations can be implemented; the computer program described in [Haberstroh] shows the applicability of our approach for practical purposes.

In order to get clearly arranged examples, for convenience also an additional kind of graph productions is used within the (attributed) elementary graph grammars described in the following section:

(7) <u>*C(A,B,C,D;w,z):*</u> *change the label of an edge between a node with label A and another node with label B from w to z and also relabel these nodes by C respectively D.*

$\mathcal{N}(g') = \mathcal{N}(g)$, and for some $u, v \in \mathcal{N}(g)$

$n(g)(u) = A$, $n(g)(v) = B$, $n(g')(u) = C$, $n(g')(v) = D$,

$n(g) \mid (\mathcal{N}(g) - \{u,v\}) = n(g') \mid (\mathcal{N}(g') - \{u,v\})$,

$(u,v,z) \notin \mathcal{E}(g)$, $(u,v,w) \in \mathcal{E}(g)$, $\mathcal{E}(g') = (\mathcal{E}(g) - \{(u,v,w)\}) \cup \{(u,v,z)\}$.

Lemma 1.2. The following control graph shows how a production $C(A,B,C,D;w,z)$ of type 7 can be simulated using only productions of type 2 and 5:

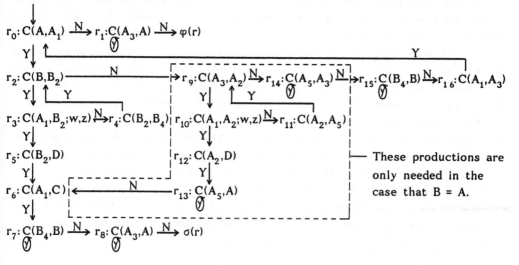

These productions are only needed in the case that B = A.

If $\varphi(r) = \emptyset$ the following simple simulation is sufficient for our purposes:

$$\longrightarrow r_0 : C(A,A_1) \xrightarrow{Y} C(B,B_2) \xrightarrow{Y} C(A_1,B_2;w,z) \xrightarrow{Y} C(A_1,C) \xrightarrow{Y} C(B_2,D) \xrightarrow{Y} \sigma(r).$$

2. Attributed elementary programmed graph grammars

In order to be able to keep more than the structural informations described by a directed graph we assign additional attributes to the nodes and the edges:

Definition 2.1. An **attributed graph** over (V,W,A) is a tuple $g_a = (\mathcal{N}, n, \mathcal{E}, a)$, where $g = (\mathcal{N}, n, \mathcal{E})$ is the corresponding graph over (V, W) and $a: \mathcal{N} \cup \mathcal{E} \to A$ is the attribution function assigning attributes of the set A to each of the nodes and edges of g thus yielding g_a. The set of all attributed graphs over (V,W,A) is denoted by $\gamma_a(\mathbf{V,W,A})$. In general we shall not impose any restrictions on the set A such as finiteness or restrictions on the function a except that for practical purposes we shall assume that a is a computable function yielding values in A that can be expressed by a computer within a finite amount of time.

Definition 2.2. Let $g_a, g_a' \in \gamma_a(V,W,A)$; then g_a and g_a' are called *isomorphic* iff $g \cong g'$; g_a and g_a' are called *equivalent* ($g_a \equiv g_a'$) iff $g \cong g'$, $n(g) = n(g') \cdot f$, and $a(g_a) = a(g_a') \cdot f$, where $f: \mathcal{N}(g_a) \to \mathcal{N}(g_a')$ is the bijective function establishing the equivalence of g and g'. Moreover, $\Gamma_a(\mathbf{V,W,A}) := \gamma_a(V,W,A)/\equiv$, and $[g_a] \in \Gamma_a(V,W,A)$ denotes the equivalence class of all graphs of $\gamma_a(V,W,A)$ equivalent to g_a.

Example 2.1. $g_\varepsilon \in \gamma_a(V,W,A)$ for arbitrary sets V,W,A.

Example 2.2. The attributed graph $g_a \in \gamma_a(\{a\},\{w\},\{1,2\})$ with $\mathcal{N}(g_a) = \{1,2\}$, $n(g_a)(1) = n(g_a)(2) = a$, $a(g_a)(1) = 1$, $a(g_a)(2) = 2$, $\mathcal{E}(g_a) = \{(1,2,w)\}$, and $a(g_a)(1,2,w) = 1$ is represented in a more descriptive way as follows: 1 (a,1) $\xrightarrow{(w,1)}$ (a,2) 2

Definition 2.3. An **attributed elementary programmed graph grammar** is an 8-tuple $G = (V_N,W_N,V_T,W_T,A,P,P_0,P_f)$, where V_N,W_N,V_T,W_T,P_0,P_f have the same meaning as for an elementary programmed graph grammar and, moreover, A is the set of **attributes** for the nodes respectively edges; P is a finite set of rules $(r; p,\sigma(r),\varphi(r))$ with $p \in \mathcal{P}(V,W)$, $\varphi(r) \subseteq Lab(P)$, and $\sigma(r) \subseteq Lab(P) \times \mathfrak{F}$, $\mathfrak{F}$ being a finite set of partial recursive functions with arguments in A. Like in definition 1.8 the pair (g_a',r') is said to be *directly derivable* from (g_a,r) in G – abbreviated $(g_a,r) \xrightarrow{} (g_a',r')$ – iff either

(1) $g_a' = g_a$, the production $p(r)$ cannot be applied to the graph g, and $r' \in \varphi(r)$ – we also write $(g_a,r) \xrightarrow[G]{N} (g_a',r')$ – or

(2) $g \xrightarrow{p(r)} g'$, $(r',f) \in \sigma(r)$, and the attributes of the nodes resp. the edges affected by the application of the production $p(r)$ to g_a (resp. its underlying graph g) are changed according to the function f thus yielding g_a' from g_a.

Because of the special features of the different kinds of elementary graph productions $p(r)$ in the rules of P we have to distinguish between the different functions that can be assigned to each type of productions and the result of a derivation $(g_a,r) \xrightarrow[G]{(Y,f)} (g_a',r')$ in case that $g \xrightarrow{p(r)} g'$:

a) For the deleting productions D(X) and D(X,Y;w) no functions need to be assigned because of the nature of these productions, i.e. we may assume $\sigma(r) \subseteq Lab(P)$ for $p(r)$ being of type 3 or 6.

b) For an entering production E(X) only constant functions are of any importance, i.e. we may assume $\sigma(r) \subseteq Lab(P) \times A$ for $p(r)$ being of type 1, and for a rule r with $p(r)$ of type 4, i.e. of the form E(X,Y;w), we shall also allow only constant attributes to be assigned to a new edge between two nodes, i.e. $\sigma(r) \subseteq Lab(P) \times A$.

c) For $p(r) = C(X,Y)$ we demand $\sigma(r) \subseteq Lab(P) \times \mathfrak{F}(A,A)$ – where $\mathfrak{F}(M,K)$ denotes the subset of $\mathfrak{F}$ containing all functions $f \in \mathfrak{F}$ such that f: $M \to K$ –, i.e. if $u \in \mathcal{N}(g)$ with $n(g_a)(u) = X$, $n(g_a')(u) = Y$ is the node the label of which is changed by $p(r)$, then $a(g_a') | (\mathcal{N}(g_a) - \{u\}) = a(g_a) | (\mathcal{N}(g_a) - \{u\})$ and $a(g_a')(u) = f(a(g_a)(u))$.

d) For $p(r) = C(X,Y;w,z)$ we allow $\sigma(r) \subseteq Lab(P) \times (\mathfrak{F}(A,A))$, i.e. the attribute of the edge the label of which is changed by $p(r)$ as well as the attributes of the two nodes u,v adjacent to this edge can be altered according to the function $f = (f_1,f_2,f_3)$, where $f_i: A^3 \to A$, $1 \le i \le 3$, such that

$a(g_a')(u) = f_1(a(g_a)(u), a(g_a)(v), a(g_a)(u,v,w))$,
$a(g_a')(v) = f_2(a(g_a)(u), a(g_a)(v), a(g_a)(u,v,w))$,
$a(g_a')(u,v,z) = f_3(a(g_a)(u), a(g_a)(v), a(g_a)(u,v,w))$.

G is called to be of *type* $n_1n_2n_3n_4n_5n_6$, if for no rule $r \in P$ $p(r)$ is of type m for some m with $n_m = 0$.

Definition 2.4. The **language of attributed graphs generated by G** is $\mathcal{L}(G) = \{ h \mid h \in \Gamma(V_T,W_T,A)$ and $([g_\varepsilon],r_0) \xrightarrow[G]{*} (h,r_f)$ for some $r_0 \in P_0$ and some $r_f \in P_f\}$.

Moreover, languages of attributed graphs respectively attributed elementary programmed graph grammars of type 111111, 110111, and 110110 are called **unrestricted, monotone** and **strictly monotone.**

Additionally, for $\mathfrak{p}(r) = C(A,B,CD;w,z)$, i.e. a production of type 7, $\sigma(r)$ and $\varphi(r)$ can be chosen as for productions of type 5, and we can prove the following result corresponding to lemma 1.2:

Lemma 2.1. The following control graph shows how a derivation step in G by means of a rule r with $\mathfrak{p}(r) = C(A,B,C,D;w,z)$ and $\sigma(r) = \{(r_i,h_i) \mid 1 \leq i \leq n\}$ can be simulated in G' only using rules r' with $\mathfrak{p}(r')$ being of type 2 or 5:

Simulation of the production $C(A,B,A_6,B_7;w,z)$ like in lemma 1.2 with identical atttribution functions leaving all attributes unchanged.

$$\downarrow N \qquad \downarrow (Y,id) \qquad \xrightarrow{(Y,h_1)} r_1': C(A_6,C) \xrightarrow{(Y,id)} r_1'': C(B_7,D) \xrightarrow{(Y,id)} r_1$$
$$\varphi(r) \quad r_0: C(A_6,B_7;z,z) \qquad \vdots$$
$$\xrightarrow{(Y,h_n)} r_n': C(A_6,C) \xrightarrow{(Y,id)} r_n'': C(B_7,D) \xrightarrow{(Y,id)} r_n$$

where id is the identical attribution function leaving all attributes unchanged.

If $\varphi(r) = \emptyset$ the following simple simulation is sufficient:

$$\xrightarrow{(Y,h_1)} r': C(A_6,C) \xrightarrow{(Y,id)} r_1'': C(B_7,D) \xrightarrow{(Y,id)} r_1$$
$$r_0: C(A,A_6) \xrightarrow{(Y,id)} r_0': C(B,B_7) \xrightarrow{(Y,id)} r_0'': C(A_6,B_7;w,z) \quad \vdots$$
$$\xrightarrow{(Y,h_n)} r_n': C(A_6,C) \xrightarrow{(Y,id)} r_n'': C(B_7,D) \xrightarrow{(Y,id)} r_n$$

Definition 2.5. For $m,n,k \in \mathbf{N}_0$ with $n + k \leq m + 1$ let $I_{n,k}^m: \mathbf{N}_0^m \to \mathbf{N}_0^k$ denote the *projection* $I_{n,k}^m(x_1, \ldots, x_m) = (x_n, \ldots, x_{n+k-1})$.

The following example shows how attributed elementary programmed graph grammars can be used for evaluating expressions (see [Kaplan et al.] and [Göttler]):

Example 2.3. (*Dataflow graphs*)
Let op be a commutative operator with two arguments, e.g. '+'. Then the attributed elementary programmed graph grammar given by the control graph

$$\longrightarrow r_1: C(op,op') \xrightarrow{(Y,id)} r_2: C(op',arg;is\_arg1,is\_arg1)$$
$$\downarrow (Y,F)$$
$$r_3: C(op',arg,arg,arg;is\_arg2,is\_arg2) \xrightarrow{(Y,G)}$$

$$F = (I_{2,1}^2, I_{2,1}^2, I_{0,0}^2), \quad G = (\underline{op} \cdot I_{1,2}^2, I_{2,1}^2, I_{0,0}^2),$$

performs the evaluation of the corresponding expression, provided suitable attributes having been assigned to the arguments. Starting with the values 2 and 9, the attributed graph in figure 3 finally is transformed to the attributed graph in figure 6.

(op,0)	(op',0)	(op',2)	(arg,11)
is_arg1 / \ is_arg2	is_arg1 / \ is_arg2	is_arg1 / \ is_arg2	is_arg1 / \ is_arg2
(arg,2) (arg,9)	(arg,2) (arg,9)	(arg,2) (arg,9)	(arg,2) (arg,9)
Figure 3.	**Figure 4.**	**Figure 5.**	**Figure 6.**

Remark 2.1. Our approach again yields the possibility of applying the elementary graph productions in the reverse direction according to the control graph in the case of *strictly monotone* attributed elementary programmed graph grammars provided that all attribution functions appearing in the control graph can be reverted in an appropriate way, which makes our system very well suited for some applications in software development as are described in [Kaplan et al.] or [Goettler].

Example 2.4. (*Pascal's triangle of binomial coefficients*)

The following attributed elementary programmed graph grammar generates graphs like that shown in figure 7 being parts of a more complex infinite planar structure, namely Pascal's triangle of binomial coefficients.

In figure 8 we have omitted the labels of the nodes and edges in order to show the attributes more explicitly.

$r_0: E(p) \xrightarrow{(Y,(1))} r_1: E(A) \xrightarrow{(Y,(1))} r_2: E(B)$ $r_6: C(A,B,p,p;r,r) \longrightarrow$ STOP

$\qquad\qquad\xrightarrow{(Y,(1,1,2,3))}$ $\uparrow(Y,(2))$ $(Y,(I^6_{1,1},I^6_{2,1},I^6_{6,1}))$

$r_3: E(p,A;u) \xrightarrow{(Y,(0))} r_4: E(p,B,u) \xrightarrow{(Y,(1))} r_5: E(A,B;r) \xrightarrow{(Y,(2))} r_7: C(A,L) \longleftarrow$

$\qquad\qquad\qquad\qquad\qquad\qquad\qquad\qquad\qquad\qquad (Y,h_a) \qquad\qquad\qquad (Y,I^4_{1,4})$

$r_8: E(A) \xrightarrow{(Y,(1))} r_9: E(D) \xrightarrow{(Y,(1))} r_{10}: E(L,A;u) \xrightarrow{(Y,(0))} r_{11}: E(A,D;r) \xrightarrow{(Y,(2))} r_{12}: E(L,D;u)$

$\qquad\qquad\xrightarrow{(Y,(1))}$

$\qquad\qquad\qquad\qquad\qquad\qquad\qquad\qquad N$

$r_{13}: C(L,B,p,L;r,r) \xrightarrow{\hspace{4cm}} r_{14}: C(L,p) \qquad r_{18}: C(D,B)$

$\downarrow (Y,(I^9_{1,1},I^9_{5,4},I^9_{9,1})) \qquad\qquad\qquad\qquad (Y,I^4_{1,1}) \qquad\qquad \uparrow N$

$r_{19}: E(L,D,u) \qquad\qquad\qquad r_{15}: C(C,D,D,p;r,r) \longleftarrow r_{17}: C(C,D,D,B;r,r)$

$\downarrow (Y,(1)) \qquad (Y,(I^9_{1,4},I^9_{5,1},I^9_{9,1})) \bigcirc \downarrow N \qquad\qquad \bigcirc (Y,(I^9_{1,4},I^9_{5,4},I^9_{9,1}))$

$r_{20}: C(L,D,L,E;u,u) \qquad\qquad r_{16}: C(A,D,p,p;r,r)$

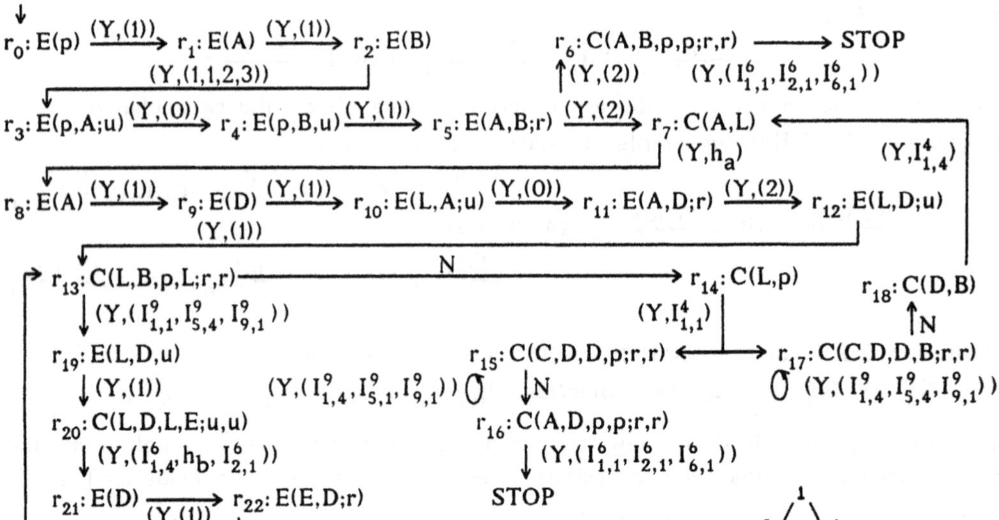

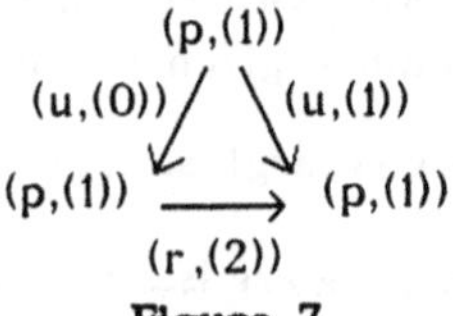

$\downarrow (Y,(I^6_{1,4},h_b,I^6_{2,1})) \qquad\qquad\downarrow (Y,(I^6_{1,1},I^6_{2,1},I^6_{6,1}))$

$r_{21}: E(D) \xrightarrow[\;(Y,(1))\;]{} r_{22}: E(E,D;r) \qquad$ STOP

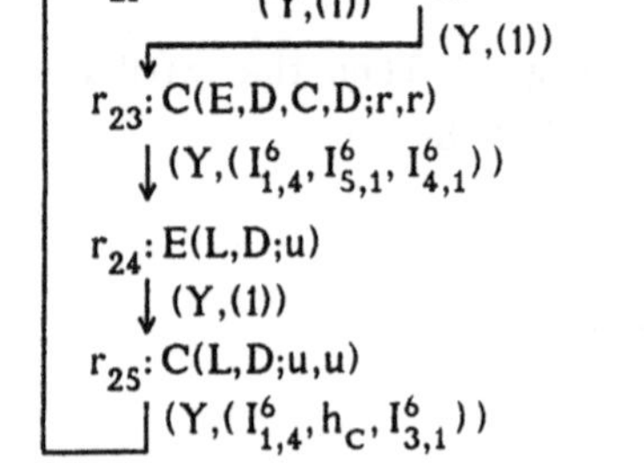

$\qquad\qquad\qquad\qquad \downarrow (Y,(1))$

$r_{23}: C(E,D,C,D;r,r) \qquad\qquad (p,(1))$

$\downarrow (Y,(I^6_{1,4},I^6_{5,1},I^4_{4,1})) \qquad (u,(0))\diagup \quad \diagdown (u,(1))$

$r_{24}: E(L,D;u) \qquad\qquad (p,(1)) \longrightarrow (p,(1))$

$\downarrow (Y,(1)) \qquad\qquad\qquad (r,(2))$

$r_{25}: C(L,D;u,u) \qquad\qquad$ **Figure 7.**

$\downarrow (Y,(I^6_{1,4},h_c,I^6_{3,1}))$

$h_a(x_1) = (x_1,0,1,2)$

$h_b(x_1,\dots,x_6) = (x_5 + x_1,x_2,x_3,x_4)$

$h_c(x_1,\dots,x_6) = (x_1,x_2 + 1,x_3 + 1,x_4 + 1)$

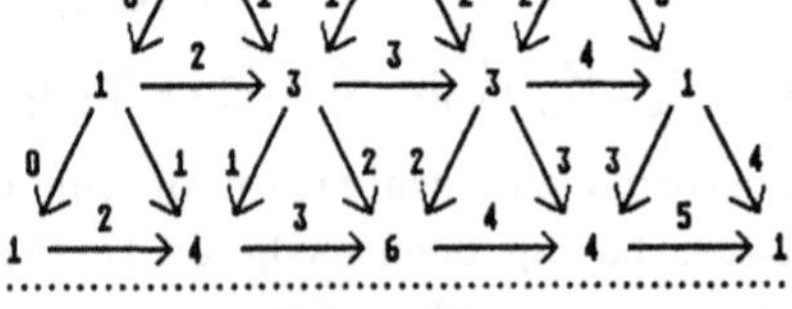

Figure 8.

The construction of the attributed elementary programmed graph grammar for Pascal's triangle may look rather complicated, but one should be aware of the fact that for the first triangle shown in figure 7 being handled with a computer program, the nodes and the edges as well as the corresponding attributes have to be entered step by step – and that is just what the rules r_0 to r_6 are standing for.

3. Conclusion

We have exhibited that elementary programmed graph grammars are powerful enough to generate any recursively enumerable set of equivalence classes of directed graphs, although in our approach only a few elementary graph productions are used and the role of embedding, which is of central importance for other approaches, is reduced to a minimum.

Moreover, attributed elementary programmed graph grammars have been shown to be useful in many areas of computer science especially as a ground system of graph rewriting other models may be built on. Only a more general concept for the applicability of our mechanism has been presented in this paper, but, for example, we have already used attributed elementary programmed graph grammars for the description of various models of neural networks within a unique formal framework (see [Haberstroh Freund]).

Acknowledgements

We gratefully appreciate the fruitful discussions with Gregor Engels and Annegret Habel as well as the useful hints of some of the referees.

References

[Bunke]: Bunke, Horst: On the Generative Power of Sequential and Parallel Programmed Graph Grammars. Computing 29, 89-112, 1982.

[Dassow Păun]: Dassow, Jürgen, Păun, Gheorghe: Regulated Rewriting in Formal Language Theory. Springer-Verlag, Berlin, New York, 1989.

[Engels et al.]: Engels, Gregor, Lewerentz, Claus, Schäfer, Wilhelm: Graph Grammar Engineering: A Software Specification Method. In: [LNCS 291],186-291.

[Freund]: Freund, Rudolf: Elementary Programmed Graph Grammars, Technical report 185/2/F/GG1/91, Technical University of Vienna, 1991.

[Göttler]: Göttler, Herbert: Attributed Graph Grammars for Graphics. In: [LNCS 153], 130-142.

[Haberstroh]: Haberstroh, Brigitte: Analyseprogramm für elementare programmierte Graphgrammatiken. Report 180/3/H/GG2/90, Technical University of Vienna, 1990.

[Haberstroh Freund]: Haberstroh, Brigitte, Freund, Rudolf: Describing Neural Networks by Using Attributed Elementary Programmed Graph Grammars. To be presented at the Workshop Connectionism of the Seventh Austrian Conference on Artificial Intelligence 1991.

[Hopcroft Ullman]: Hopcroft, John E., Ullman, Jeffrey D.: Introduction to Automata Theory, Languages and Computation. Addison-Wesley Publishing Company, Reading Massachussetts, 1979.

[Kaplan et al.]: Kaplan, Simon M., Goering, Steven K., Campbell, Ray H.: Supporting the Software Development Process with Attributed NLC Graph Grammars. In: [LNCS 291], 309-325.

[LNCS 153]: Lecture Notes in Computer Science, Vol. 153: Graph Grammars and Their Application to Computer Science. Springer-Verlag, Berlin, New York, 1983.

[LNCS 291]: Lecture Notes in Computer Science, Vol. 291: Graph Grammars and Their Application to Computer Science. Springer-Verlag, Berlin, New York, 1987.

[Nagl]: Nagl, Manfred: Formal Languages of Labelled Graphs. Computing 16, 113-137, 1976.

[Salomaa]: Salomaa, A.: Formal Languages. Academic Press, New York, London, 1973.

The Complexity of Approximating the Class Steiner Tree Problem

Edmund Ihler

Institut für Informatik, Universität Freiburg

Rheinstraße 10 - 12, W - 7800 Freiburg, Germany

Abstract

Given a connected, undirected distance graph with required classes of nodes
and optional Steiner nodes, find a shortest tree containing at least one node of each
required class. This problem called CLASS STEINER TREE is NP-hard and therefore
we are dependent on approximation.

In this paper, we investigate various restrictions of the problem comparing their
complexities with respect to approximability. A main result is that for an input of
trees without Steiner nodes and unit edges only, CLASS STEINER TREE is as hard
to approximate as MINIMUM SET COVER, for which no constant approximation is
known, too. Further we prove that if this restricted version has an approximation
scheme, all members of the optimization problem class MAX SNP do.

1 Introduction

A large number of important optimization problems have been shown to be NP-hard,
and this number is increasing constantly. Therefore, good approximation algorithms are
of great importance. Moreover, there is the question for a general theory, classifying
optimization problems by their complexity with respect to approximation. Different approaches have been made toward such a general theory (see e.g. [PY88,PR90,Ihl90]).

In this paper, we want to investigate a minimization problem with respect to the
following properties: Is it polynomially solvable or - if not - is an approximation scheme
known or at least some constant approximation, or none of the three. Here, *constant
approximation* means a polynomial time approximation algorithm whose relative error is
bounded by a constant. An *approximation scheme* is a family of constant approximations,
one for each arbitrary relative error bound $\epsilon > 0$. We study the frontiers between the
mentioned types of complexity by investigating a certain minimization problem that is
important in VLSI and network design.

We consider the problem of finding a shortest network, connecting classes of nodes of
a graph. To be more precise: Given an undirected connected distance graph where its set
of nodes is partitioned into a class of optional *Steiner nodes* and *required classes* of nodes,
compute a tree of minimum total length that contains at least one node of each required
class. We will call this problem CLASS STEINER TREE.

Various applications to this problem are possible. For example in the routing phase of VLSI design after the placement of components on a chip, pins of different components have to be connected by an electrical network of minimum total length [RW90]. Using the remaining freedom of rotating and mirroring the components leads to several positions for each required pin. So we can consider in a natural way the possible positions of such a pin as a required class of nodes.

As another possible application, consider the problem of connecting local networks by a minimum global network, independent of the local networks [IRW91]. Assume that we have e.g. existing local networks of computers and several possibilities for new connections with specific costs between computers and/or optional branching points. We can model this problem using the local networks as required classes of nodes with one node for each of their computers. The optional branching points are Steiner nodes.

Clearly, CLASS STEINER TREE is a natural generalization of the famous STEINER TREE. As STEINER TREE is NP-hard [Kar74], CLASS STEINER TREE is too. Therefore we depend on approximation. For STEINER TREE, polynomial time algorithms are known that compute an approximate Steiner tree less than twice as long as the minimum Steiner tree (e.g. [KMB81,Meh88]). But it is still an open question, whether there is an approximation scheme.

Some progress has been made toward this question in general. In [PY88], a complexity class MAX SNP for optimization problems has been introduced. Many well known problems like MAX 3SAT or MAXIMUM INDEPENDENT SET and NODE COVER for bounded node degree have been proved to be complete in this class under a certain kind of transformation that preserves approximation schemes. All these complete problems share the property that for none of it an approximation scheme has been found yet. But such a scheme for one of it would imply an approximation scheme for each problem in MAX SNP. Hence, for a problem proved to be complete or hard in this sense, there is little hope for an approximation scheme.

STEINER TREE has been shown MAX SNP-hard [BP89]. Similar to NP-hardness, this implies MAX SNP-hardness for CLASS STEINER TREE. MAX SNP-hard means that the problem is not in MAX SNP but that the reduction required for completeness is possible. There are many results showing that CLASS STEINER TREE is a harder problem than STEINER TREE. For certain restrictions, e.g., where no Steiner nodes are allowed or where an input graph has to be a tree, the complexities differ considerably: STEINER TREE gets polynomially solvable whereas CLASS STEINER TREE remains NP-hard [RW90]. Another indication for the higher complexity of CLASS STEINER TREE is that approximations with bounds for their relative error have been found [Ihl91], but none of the bounds is constant, i.e., independent of the input graph.

In Section 2, we introduce transformations between minimization problems that preserve these approximation properties like the existence of an approximation scheme or of a constant approximation. The main tools are *C-reductions*, transforming polynomial approximation algorithms with a constant bound for their relative error, and *S-reductions* that transform a whole approximation scheme.

In Section 3, we present a strong restricted version of CLASS STEINER TREE that is polynomially solvable, and how we can modify this problem to an NP-hard but constant approximable version. In Section 4, we work in the other direction: How strong can we restrict CLASS STEINER TREE preserving the complexity of the general version. We prove in Section 4 that CLASS STEINER TREE for VLSI oriented graphs where the edge lengths are integers polynomially bounded by the number of nodes (*similarity assumption*, see e.g. [Gab85]) is not easier to approximate if we consider only graphs without Steiner nodes and with all edges of length 1. We show that it is not more complex to find an approximation scheme for graphs where the edge lengths are integers, the similarity assumption holds, and the node degree is bounded by some integer, than for graphs with node degree bounded by 3, without Steiner nodes and with unit edges only.

In Section 5, we give strong evidence that an approximation scheme or at least a constant approximation even for strongly restricted versions is not achievable. There we prove that CLASS STEINER TREE for trees without Steiner nodes and unit edges only is as hard to approximate as MINIMUM SET COVER. In MINIMUM SET COVER, a cover for a set is given by a collection of subsets and we have to find a minimum number of subsets, covering the same original set. The power of this result depends on the fact that it is an old open question whether there exists a constant approximation for MINIMUM SET COVER and that MINIMUM SET COVER has been shown to be MAX SNP-hard [PY88].

2   Basic definitions

We describe a *minimization problem* Π by a set $\mathbf{I}$ of *instances*, a collection of finite nonempty sets containing a set $\mathbf{S}(I)$ of *solutions* for each instance $I \in \mathbf{I}$, and a *solution value function* $/\cdot, \cdot/$, mapping the pairs (I, S) with $I \in \mathbf{I}$ and $S \in \mathbf{S}(I)$ into the rational numbers. The problem asks for each instance $I \in \mathbf{I}$ for a minimal solution $S^* \in \mathbf{S}(I)$, i.e., a solution minimizing the solution value function (see e.g. [GJ79]).

A polynomial time algorithm A is an *approximation algorithm (PTA)* for a minimization problem Π, if A finds for each instance $I \in \mathbf{I}$ a solution $A(I) \in \mathbf{S}(I)$. A PTA is called a $(1 + \epsilon)$-*CPTA* or just a *CPTA* for Π, if for each $I \in \mathbf{I}$: $/I, A(I)/ \leq (1 + \epsilon)/I, S^*(I)/$, and $\epsilon > 0$ is a constant.

A *polynomial time approximation scheme (PTAS)* for a minimization problem Π is a family $(A_\epsilon)_{\epsilon>0}$ of CPTAs, containing a $(1 + \epsilon)$-CPTA for Π for each $\epsilon > 0$.

We will now define transformations between minimization problems, that preserve CPTAs or whole PTASs. For two minimization problems Π, Π', we call a pair of polynomial transformations (f, g) with an *instance transformation* $f : \mathbf{I} \to (\mathbf{I}')^m$, and a *solution transformation* $g : \mathbf{I} \times (\mathbf{I}')^m \times (\mathbf{S}'(\mathbf{I}'))^m \to \mathbf{S}(\mathbf{I})$ a *(multi-) C-reduction*, denoted by $\Pi \leq_C \Pi'$, if whenever A' is a CPTA for Π', then $g \circ (\cdot, f, A' \circ f)$ is a CPTA for Π. Here $A' \circ f$ is defined by $(A' \circ f_1, \ldots, A' \circ f_m)$ where $f = (f_1, \ldots, f_m)$.

If $(f_\epsilon, g_\epsilon, e)_{\epsilon>0}$, where $(f_\epsilon, g_\epsilon)_{\epsilon>0}$ is a family of C-reductions, one for each $\epsilon > 0$, and e, called the *error transformation*, is a polynomial transformation $e : (0, \infty) \to (0, \infty)^m$, then this triple is called a *(multi-) S-reduction*, denoted by $\Pi \leq_S \Pi'$, if whenever $(A'_{\epsilon'})_{\epsilon'>0}$ is a PTAS for Π', then $g_\epsilon \circ (\cdot, f_\epsilon, A'_{e(\epsilon)} \circ f_\epsilon)_{\epsilon>0}$ is a PTAS for Π. Here $A'_{e(\epsilon)} \circ f_\epsilon$ is defined for each $\epsilon > 0$ by $(A'_{e_1(\epsilon)} \circ f_{\epsilon 1}, \ldots, A'_{e_m(\epsilon)} \circ f_{\epsilon m})$ where $f_\epsilon = (f_{\epsilon 1}, \ldots, f_{\epsilon m})$ and $e = (e_1, \ldots, e_m)$. If e is the identity mapping (in all components), then we call $(f_\epsilon, g_\epsilon)_{\epsilon>0}$ a *(multi-) S^{id}-reduction*. Usually $(f_\epsilon, g_\epsilon, e)_{\epsilon>0}$ is given by a triple (f, g, e), independent of ϵ, and $m = 1$. If we have $\Pi \leq_{S^{id}} \Pi'$ and $\Pi' \leq_{S^{id}} \Pi$, then we write $\Pi \equiv_{S^{id}} \Pi'$. It is an important consequence of our definitions that C-reductions as well as S-reductions compose. To see this, note that the transformations f, e can have at most a polynomial number of polynomial components, because f, e itself have to be polynomially computable. This implies that the corresponding $\leq$-relations are transitive.

For distance graphs (V, E, l), we use standard notation, where V is the set of *nodes*, E is the set of *edges*, and l maps the edges to nonnegative *edge lengths*. We describe *simple paths* between two nodes $v, w \in V$ by a sequence of nodes $(v, v_1, \ldots, v_k, w)$, where each node of the sequence appears once. The *(total) length* of a graph, a subgraph, or a path is given by the sum of the lengths of all its edges.

We define the problem CLASS STEINER TREE as follows. Let $G = (V, E, l)$ be a connected, undirected distance graph, where $V = S \cup \bigcup_{i=1}^{c} R_i$ is partitioned into disjoint sets: the *required classes* R_i, $1 \leq i \leq c$, and the *Steiner nodes* S. A *class Steiner tree* is a tree that contains at least one node for each required class. We ask for a class Steiner tree of minimum total length. If we restrict CLASS STEINER TREE to classes where $|R_i| = 1$ for $1 \leq i \leq c$, this defines the classical STEINER TREE problem.

After all this definitions, let us first give two simple versions of CLASS STEINER TREE in the next section: A polynomial solvable one and a NP-hard one, together with a CPTA.

3 A polynomial and a constant approximable version

For many restrictions of the input graph, CLASS STEINER TREE remains NP-hard, whereas other classical NP-complete problems become polynomially solvable. This holds e.g. for input graphs that are trees without Steiner nodes, having unit edges only, a node degree and a cardinality for the required classes both bounded by 3 [RW90,IRW91]. Also versions of CLASS STEINER TREE embedded in the plane and in the unit grid have been proved NP-hard [IRW91]. Further restrictions are possible.

But where is the frontier between NP-hard and polynomial versions? Consider a restriction where the node degree is bounded by 2. Because the graph has to be connected, this implies a linear order to the nodes. So we can compute a minimum class Steiner tree as follows. For each node, we add recursively the next node and the edge connecting these nodes until all classes are represented. Then we just have to take a shortest of these subgraphs, and we have found a (degenerated) minimum class Steiner tree.

There is another less trivial polynomial version given in [Ihl91], where the number of classes is bounded. Slightly modified, this version becomes NP-hard, but we can give a CPTA. Here, not the total number of required classes is bounded by k, but the number of required classes that contain more than one node. This new version CLASS STEINER TREE$(|\{R_i : |R_i| > 1\}| \leq k)$ is a generalization of STEINER TREE and therefore is NP-hard. But there is a simple way to apply any CPTA known for STEINER TREE to this problem.

Theorem 1 CLASS STEINER TREE$(|\{R_i : |R_i| > 1\}| \leq k) \equiv_{Sid}$ STEINER TREE.

Proof: Let the graph G have n nodes and let the required classes R_i have n_i nodes, then $\prod_{i=1}^{c} n_i \leq n^k$ different tuples $(r_1, \ldots, r_c) \in R_1 \times \cdots \times R_c$ are possible. A transformation f can now create by its component f_i for each such tuple i a graph G_i' for STEINER TREE, by using the original graph, and interpreting the nodes of the tuple as required nodes and the remaining nodes as Steiner nodes. If a CPTA computes a Steiner tree for each of the graphs G'_i, a transformation g can pick a shortest one of the computed trees and reinterpret it for G. Clearly, this describes a CPTA for G, because the number of graphs G'_i is bounded by $O(n^k)$, and there is a tuple i where the length of the minimal Steiner tree for G_i equals this of the minimal class Steiner tree for G. The relative error bound is the same. Hence (f, g) is a multi S^{id}-reduction from CLASS STEINER TREE to STEINER TREE, and by a trivial S^{id}-reduction in the other direction we have proved the theorem. $\square$

It can easily be seen that the same version without Steiner nodes gets polynomially solvable using $O(n^k)$ minimal spanning tree computations.

In this section, we considered very restricted versions. Slightly more general cases of CLASS STEINER TREE become considerably more complex. In the next section, we give a restricted version that is as hard to approximate as the general one.

4   Versions of equal complexity

How strong can we restrict CLASS STEINER TREE preserving the complexity with respect to approximation of the general version? Sometimes, it is easier to find a CPTA or a PTAS for a restricted version, but can we apply this to more general cases? The following theorems aim at this questions. First, we show that a PTAS for CLASS STEINER TREE restricted to graphs without Steiner nodes gives a PTAS for the general CLASS STEINER TREE and vice versa.

Theorem 2 CLASS STEINER TREE $\equiv_{S^{id}}$ CLASS STEINER TREE$(S = \emptyset)$.

Proof: We have to construct S^{id}-reductions in both directions.

"$\leq$": For our instance transformation f, we start with the graph $G = (S \cup \bigcup_{i=1}^c R_i, E, l)$ where the number of Steiner nodes is s. Introduce for each $v \in S$ a new class and call these classes $R_{c+1}, \ldots, R_{c+s}$. Interpret each Steiner node as a node of its corresponding new required class. We call all nodes in this slightly modified graph *original nodes*. Now replace each edge $(v, w) \in E$ by a simple path $p(v, w) = (v, r_1, \ldots, r_s, w)$. The nodes $r_1, \ldots, r_s$ should be new nodes with $(r_1, \ldots, r_s) \in R_{c+1} \times \cdots \times R_{c+s}$ for each edge in E, the length of each of the new edges be $l((v, w))$. Hence, each of the classes $R_{c+i}, 1 \leq i \leq s$, consists of $|E| + 1$ nodes, and the number of edges in the new graph G' is $|E|(s + 1)$ (see Figure 1).

The solution transformation g works as follows. Given a class Steiner tree T' for G', take the original nodes that are in T' and reinterpret it. Whenever two original nodes are connected in T' by a simple path of $s + 1$ edges, then choose the corresponding edge in E. Clearly, these edges and the reinterpreted nodes build a class Steiner tree T for G, and $(s + 1)l(T) \leq l(T')$.

Assume that we have given a $(1 + \epsilon')$-CPTA A' that computes a class Steiner tree $T'_{A'}$ for each input graph G' of CLASS STEINER TREE$(S = \emptyset)$, then this implies a $(1 + \epsilon')$-CPTA $A = g \circ (\cdot, f, A' \circ f)$ computing T_A for G, because of the following: A minimum class Steiner tree T_{min} for G leads to a class Steiner tree T' for G' by taking for each edge (v, w) of T_{min} the corresponding simple path $p(v, w)$ in G'. Hence, for a minimum class Steiner tree T'_{min} of G' this yields $(s + 1)l(T_{min}) = l(T') \geq l(T'_{min})$, and altogether

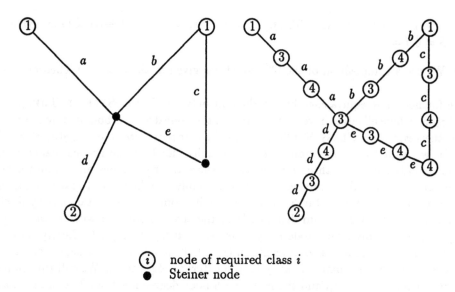

$\textcircled{i}$ node of required class i
● Steiner node

Figure 1: An example graph G and its image G' transformed by f of the proof for Theorem 2.

$\frac{l(T_A)}{l(T_{min})} \leq \frac{(s+1)}{(s+1)} \frac{l(T'_{A'})}{l(T'_{min})} \leq 1 + \epsilon'$. Note that both, f and g can be computed in polynomial time.

"$\geq$": Obvious by a trivial S^{id}-reduction. □

The effect of this theorem is, that all results that we can derive for the restricted version where no Steiner nodes are allowed, also hold for the general version.

A similar theorem is possible for CLASS STEINER TREE(l : integers, sim. ass.), where the lengths of all edges are integers and bounded by a polynomial function in the number of nodes, and CLASS STEINER TREE($l = 1$), a restriction to graphs with unit edges only. This polynomially bounding of the edge lengths is called *similarity assumption* (see e.g. [Gab85]). Both assumptions are very natural and hold, e.g., for graphs derived from the VLSI layout geometry, where the edge lengths are multiples of a certain layout unit.

Theorem 3 CLASS STEINER TREE(l : integers, sim. ass.) $\equiv_{S^{id}}$ CLASS STEINER TREE($l = 1$).

Proof: Again, we have to construct S^{id}-reductions in both directions.

"$\leq$": We give a new transformation $f : G \mapsto G'$ for a graph $G = (S \cup \bigcup_{i=1}^{c} R_i, E, l)$. Replace each $(v, w) \in E$ by a simple path $p(v, w) = (v, s_1, \ldots, s_{l(v,w)-1}, w)$ where all nodes $s_1, \ldots, s_{l(v,w)-1}$ are new Steiner nodes and all edges have length 1.

Our solution transformation g starts with a given class Steiner tree T' for G'. For each edge $(v, w) \in E$ test, whether the corresponding path $p(v, w)$ is completely contained in T'. All such edges determine the tree T.

Analogously to the proof of Theorem 2, we can prove that (f, g) is an S^{id}-reduction. It is important that f and g are polynomial transformations, as the similarity assumption holds.

"$\geq$": This direction is obvious. □

Corollary 1 CLASS STEINER TREE*(l : integers, sim. ass.)* $\equiv_{S^{id}}$ CLASS STEINER TREE*(S = $\emptyset$, l = 1).*

Proof: We use a composition of the S^{id}-reductions given in the proves of Theorem 3 and 2. $\square$

We further show equivalence by S-reduction between CLASS STEINER TREE*(l = 1, deg(v) $\leq$ k)*, where all edges have unit length and the node degree is bounded by a constant *k*, and the restriction CLASS STEINER TREE*(l : integers, sim. ass., deg(v) $\leq$ 3)*. To this end, we want to use a certain family of graphs. Each of the graphs, called *i-star*, consists roughly spoken of three binary trees, where the root nodes are connected to a common *center node*. We recursively define this family as follows. A 1-star consists of a single center node, and an *i*-star S_i for $i \geq 2$ is built from an $(i-1)$-star S_{i-1} by finding a node v of S_{i-1} with minimum distance from the center node and with a node degree $deg(v) < 3$, and adding a new node w by a new edge (v, w) of length 1. Clearly, an *i*-star S_i is a tree, has i nodes and a total length of $(i-1)$. If we allow a node degree of 3 for all the nodes, then we can connect exactly $(2+i)$ external edges to S_i. We call this number the *connectivity degree* of S_i, and the nodes with node degree less than 3 *connector nodes* (see Figure 2).

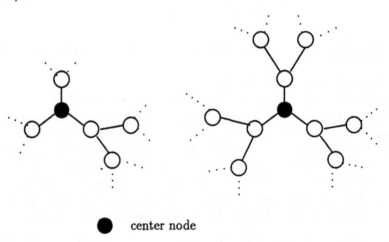

● center node

Figure 2: A 6-star and a 10-star.

Theorem 4 CLASS STEINER TREE*(l = 1, deg(v) $\leq$ k)* $\equiv_S$ CLASS STEINER TREE*(l : integers, sim. ass., deg(v) $\leq$ 3)*, for $k > 3$.

Proof: "$\leq$" We transfer an instance graph G for an arbitrary $\epsilon > 0$ by the following f_ϵ to G'. Each node has a degree not exceeding k. Hence, we replace each node v of G by a $(k-2)$-star. Hereby we can connect each edge incident to v to a connector node of the $(k-2)$-star, because its connectivity degree is k. If v is a required node, all nodes, especially the center node, of the $(k-2)$-star are added to the required class of its corresponding node v; otherwise all nodes are additional Steiner nodes. At last we change the lengths of the old unit edges to $2m(k-3)$, where m is an arbitrary positive integer with $m \geq \frac{1}{\epsilon}$. Hence, we have integer lengths for all edges and the similarity assumption holds for any constant ϵ, too.

A solution transformation g independent of ϵ takes all center nodes that are contained in a class Steiner tree T' of G'. A class Steiner tree T of G then consists of all nodes of G corresponding to these center nodes, and of the corresponding edges of T'. We can verify $2m(k-3)l(T) \leq l(T')$.

For the error transformation we use $e(\epsilon) = \frac{\epsilon}{2+\epsilon}$. All given transformations are polynomial in time by construction for any constant ϵ.

Assume a minimum class Steiner tree T_{min} for G, then we can build a class Steiner tree T' for G' by using for a leaf of T_{min} the corresponding connector node as a leaf for T', for the remaining nodes the whole corresponding $(k-2)$-star, and the corresponding edges of T_{min}. We have $2m(k-3)l(T_{min}) + (l(T_{min}) + 1 - 1)(k-3)) \geq l(T')$, because T_{min} consists of unit edges and therefore has $l(T_{min}) + 1$ nodes, each tree has at least one leaf, and the total length of a $(k-2)$-star is $(k-3)$. For a minimum class Steiner tree T'_{min} for G' this implies $(2m+1)(k-3)l(T_{min}) \geq l(T'_{min})$. The chosen e guarantees that a given PTAS $(A'_{\epsilon'})_{\epsilon'>0}$ computing a class Steiner tree $T'_{A'_{\epsilon'}}$ changes by $(f_\epsilon, g, e)_{\epsilon>0}$ to a PTAS $(A_\epsilon)_{\epsilon>0}$ for G computing T_{A_ϵ}:

$$\frac{l(T_{A_\epsilon})}{l(T_{min})} \leq \frac{(2m+1)(k-3)}{2m(k-3)} \frac{l(T'_{A'_{e(\epsilon)}})}{l(T'_{min})} \leq (1 + \frac{\epsilon}{2})(1 + \frac{\epsilon}{2+\epsilon}) = 1 + \epsilon.$$

"$\geq$" Use the S-reduction given in the proof of Theorem 3. $\square$

Corollary 2 CLASS STEINER TREE(l : integers, sim. ass., $deg(v) \leq k$) $\equiv_S$ CLASS STEINER TREE($S = \emptyset$, $l = 1$, $deg(v) \leq 3$), for $k > 3$.

Proof: Compose the S-reductions in the proofs of Theorems 3,4,3,2 in this order. $\square$

Remember that an equivalence $\equiv_S$ guarantees that a PTAS for one of the problems implies a PTAS for the other problem. If a single $(1 + \epsilon)$-CPTA is known for some ϵ for one of the problems, we have a CPTA for the other problem.

In the next section, we relate the complexity of approximating the CLASS STEINER TREE to the one of an old problem. This explains the lack of progress in finding a constant approximation for our problem.

5 Class Steiner Tree is hard to approximate

We saw that CLASS STEINER TREE remains NP-hard for strong restrictions, especially when the input graph is a tree, free of Steiner nodes, and all edges have length 1. In this section, we strengthen the conjecture that there is no CPTA for CLASS STEINER TREE even for the mentioned restrictions. We prove that such an algorithm would imply a CPTA for MINIMUM SET COVER, and exactly for this problem the question for a CPTA is still open.

MINIMUM SET COVER is the following minimization problem: Given a finite nonempty collection $C = \{C_1, \ldots, C_n\}$ of nonempty sets C_i, find a subcollection $S = \{S_1, \ldots, S_s\} \subseteq C$ of minimum cardinality $|S|$ being a cover, i.e. $\bigcup_{i=1}^s S_i = \bigcup_{i=1}^n C_i$.

To demonstrate the kind of reduction we will use, first consider CLASS STEINER TREE($S = \emptyset, tree, l = 0, 1$) where only trees without Steiner nodes and edge lengths 0 and 1 are allowed for input.

Theorem 5 MINIMUM SET COVER $\leq_{S^{id}}$ CLASS STEINER TREE($S = \emptyset, tree, l = 0, 1$).

Proof: We have to give polynomial transformations for the S^{id}-reduction. For the instance transformation f, create for each set C_i in the collection C a *set-node* r_i, for each member μ of each set C_i a *member-node* $r_{i\mu}$, and one *root-node* r_0. All these nodes form the set V. $\{r_0\}$, $\{r_i : 1 \leq i \leq n\}$ are two required classes. For each member $\mu \in \bigcup_{i=1}^n C_i$, we create a class $R_\mu = \{r_{i\nu} : \nu = \mu; 1 \leq i \leq n; r_{i\mu} \in V\}$. Connect each set-node r_i by an edge of length 1 to the root-node and each member-node $r_{i\mu}$ by an edge of length 0 to the corresponding set-node r_i. These edges form E, $G = (V, E, l)$ is a tree without Steiner nodes, and the constructed edges define l (see Figure 3).

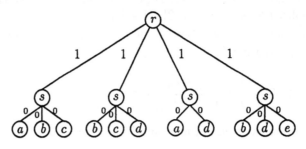

$\textcircled{i}$ node of required class i
collection $C = \{\{a, b, c\}, \{b, c, d\}, \{a, d\}, \{b, d, e\}\}$

Figure 3: An example collection C and its image graph G transformed by f of the proof for Theorem 5.

The solution transformation g is obvious. Given a class Steiner tree T for G, pick all set-nodes r_i of T, then the corresponding sets S_i build a subcollection S which is really a cover for C and $|S| = l(T)$.

Both, f and g, are polynomial time transformations and from a given $(1 + \epsilon')$-CPTA A' for G computing a class Steiner tree $T_{A'}$, we derive a $(1 + \epsilon')$-CPTA A computing a cover S_A using f and g: A minimum cover S_{min} for C implies a class Steiner tree T for G, if we use the set-nodes and member-nodes related to the sets of S_{min}, together with the root-node and all edges between these nodes. Clearly, for a minimum class Steiner tree T_{min}, we have $|S_{min}| = l(T) \geq l(T_{min})$, and therefore $\frac{|S_A|}{|S_{min}|} \leq \frac{l(T_{A'})}{l(T_{min})} \leq 1 + \epsilon'$. $\square$

A simple but interesting corollary results from the fact that MINIMUM SET COVER, restricted to collections where for each set its cardinality is bounded by a constant, as well as the number of appearances of each member, was proved MAX SNP-complete [PY88].

Corollary 3 CLASS STEINER TREE($S = \emptyset, tree, l = 0, 1$) *is MAX SNP-hard.*

Proof: MAX SNP-hard means that the problem is not in MAX SNP but that the reduction required for completeness is possible. The given S^{id}-reduction in the proof of Theorem 5 works for this bounded case of MINIMUM SET COVER, too, and has the form of the required reduction (see [PY88]). $\square$

As edges of the length 0 are not adequate for many of the applications of CLASS STEINER TREE, further restriction to trees built of unit edges are of interest. We show similar approximation results for this NP-hard version.

Theorem 6 MINIMUM SET COVER $\leq_S$ CLASS STEINER TREE$(S = \emptyset, tree, l = 1)$.

Proof: To describe the desired instance transformation f_ϵ for an arbitrary $\epsilon > 0$, recall the transformation f given in the proof of Theorem 5. As before, we create a tree with nodes r_0, r_i, $r_{i\mu}$, and edges of length 0 and 1, and with the classes $\{r_0\}$, $\{r_i : 1 \leq i \leq n\}$, and R_μ for $\mu \in \bigcup_{i=1}^n C_i$ by a given collection C. Then we choose an arbitrary integer $k \geq \frac{m}{\sqrt{1+\epsilon}-1}$, where m is the cardinality of $\bigcup_{i=1}^n C_i$. Replace each edge (r_0, r_i) of length 1 between the root-node r_0 and a set-node r_i by a simple path $p(r_0, r_i) = (r_0, v_{i1}, \dots, v_{ik-1}, r_i)$ where all the v_{ij} are new nodes for each replaced edge and belong to the class of r_i. After this, replace all edges in the current graph by edges of length 1. In this way, we get a graph G that is a tree with the desired properties.

The solution transformation g_ϵ is given by g in the proof of Theorem 5, independent of ϵ. Pick the sets S_i related to the set-nodes r_i contained in a class Steiner tree T for G to generate a cover S. It is easy to verify that

$$k|S| + m \leq l(T). \tag{1}$$

We give the error transformation e by $e(\epsilon) = \sqrt{1+\epsilon} - 1$. Note, that each of $f_\epsilon, g_\epsilon, e$ can be computed in polynomial time.

It remains to show that our reduction $(f_\epsilon, g, e)_{\epsilon>0}$ transforms each given PTAS $(A'_{\epsilon'})_{\epsilon'>0}$ for CLASS STEINER TREE$(S = \emptyset, tree, l = 1)$ to a PTAS $(A_\epsilon)_{\epsilon>0}$ for MINIMUM SET COVER. Start with a minimum cover S_{min} and take for each set S_{min_i} of this cover the corresponding set-node r_i of G and each simple path $p(r_0, r_i)$. Each member of $\bigcup_{i=1}^n C_i$ lies in at least one set S_{min_i}, hence, we can easily contact a member node for each such member by an edge of length 1 to one of the chosen set-nodes. This implies for the constructed class Steiner tree T and a minimum class Steiner tree T_{min} of G

$$k|S_{min}| + m = l(T) \geq l(T_{min}). \tag{2}$$

In addition we know that $m \geq 1$, therefore $|S_{min}| \geq 1$. Remember that our k guarantees $k(\sqrt{1+\epsilon} - 1) \geq m$, hence the following holds:

$$\begin{aligned}
k|S_{min}| &= \frac{1}{\sqrt{1+\epsilon}}(k|S_{min}| + k(\sqrt{1+\epsilon} - 1)|S_{min}|) \\
&\geq \frac{1}{\sqrt{1+\epsilon}}(k|S_{min}| + m)
\end{aligned} \tag{3}$$

Let S_{A_ϵ} be the cover derived by a given PTAS $(A'_{\epsilon'})_{\epsilon'>0}$ for G using $(f_\epsilon, g, e)_{\epsilon>0}$, then the inequalities (1),(2),(3), and the fact $k|S_{min}| > 0$ imply

$$\begin{aligned}
\frac{|S_{A_\epsilon}|}{|S_{min}|} &< \frac{|S_{A_\epsilon}|}{|S_{min}|} + \frac{m}{k|S_{min}|} = \frac{k|S_{A_\epsilon}| + m}{k|S_{min}|} \\
&\leq \frac{l(T_{A'_{e(\epsilon)}})}{\frac{1}{\sqrt{1+\epsilon}}(k|S_{min}| + m)} \leq \frac{l(T_{A'_{e(\epsilon)}})}{l(T_{min})}\sqrt{1+\epsilon} \\
&\leq (1 + e(\epsilon))\sqrt{1+\epsilon} = (1 + \sqrt{1+\epsilon} - 1)\sqrt{1+\epsilon} \\
&= 1 + \epsilon \qquad \square
\end{aligned}$$

Note that the instance transformations given in the proofs of Theorems 5 and 6 serve also as Turing reductions from MINIMUM SET COVER to the corresponding CLASS STEINER TREE version for an alternative proof of its NP-hardness. We can derive a similar corollary to Corollary 3.

Corollary 4 *If a PTAS for* CLASS STEINER TREE$(S = \emptyset, tree, l = 1)$ *is found, then a PTAS is found for each problem of MAX SNP.* $\square$

6 Conclusion

We investigated the complexity of finding a constant approximation or an approximation scheme for CLASS STEINER TREE, a natural generalization of the STEINER TREE problem, where we have required classes of nodes instead of simply required nodes. We gave a polynomial time algorithm for a strong restricted version of CLASS STEINER TREE and a polynomial time approximation algorithm with constant error bound for an NP-hard version. But most nontrivial restrictions are NP-hard. We proved the equivalence of the complexity of approximation for some restrictions and the general version, and we proved that most of it are at least as hard to approximate as the well known problem MINIMUM SET COVER, for which no constant approximation is known yet, too. Moreover, an approximation scheme would imply such a scheme for all problems of the complexity class MAX SNP.

We have been still unsuccessful in proving this kind of hardness for NP-hard restrictions where the graph is embedded in the plane or in the unit grid. Maybe this gives some hope for constant approximability in these cases.

Acknowledgement

Thanks are due to P. Widmayer and G. Reich for helpful discussions and critical comments.

References

[BP89]   Marshall Bern and Paul Plassmann. The Steiner problem with edge lengths 1 and 2. *Information Processing Letters*, 32:171–176, 1989.

[Gab85] H.N. Gabow. Scaling algorithms for network problems. *J. Comp. Sys. Sc.*, 31:148–168, 1985.

[GJ79]   Michael R. Garey and David S. Johnson. *Computers and Intractability: A Guide to the Theory of NP-Completeness*. W. H. Freeman and Company, 1979.

[Ihl90] Edmund Ihler. Approximation and existential second-order logic. Technical report, Institut für Informatik, Universität Freiburg, 1990.

[Ihl91] Edmund Ihler. Bounds on the quality of approximate solutions to the Group Steiner Problem. In *Graph-Theoretic Concepts in Computer Science, WG90*, volume 484 of *Lecture Notes in Computer Science*, pages 109–118. Springer, 1991.

[IRW91] Edmund Ihler, Gabriele Reich, and Peter Widmayer. On shortest networks for classes of points in the plane. Technical report, Institut für Informatik, Universität Freiburg, 1991.

[Kar74] Richard W. Karp. Reducibility among combinatorial problems. In R. E. Miller and J. W. Thatcher, editors, *Complexity of Computer Computations*, pages 85–103. Plenum Press, 1974.

[KMB81] L. Kou, G. Markowsky, and L. Berman. A fast algorithm for Steiner trees. *Acta Informatica*, 15:141–145, 1981.

[Meh88] Kurt Mehlhorn. A faster approximation algorithm for the Steiner problem in graphs. *Information Processing Letters*, 27:125–128, 1988.

[PR90] Alessandro Panconesi and Desh Ranjan. Quantifiers and approximation. In *Proc. 22th Annual ACM Symp. on Theory of Computing*, pages 446–456, 1990.

[PY88] Christos H. Papadimitriou and Mihalis Yannakakis. Optimization, approximation, and complexity classes. In *Proc. 20th Annual ACM Symp. on Theory of Computing*, pages 229–234, 1988.

[RW90] Gabriele Reich and Peter Widmayer. Beyond Steiner's problem: A VLSI oriented generalization. In *Graph-Theoretic Concepts in Computer Science, WG89*, volume 411 of *Lecture Notes in Computer Science*. Springer, 1990.

On Complexity of Some Chain and Antichain Partition Problems

Zbigniew Lonc

Institute of Mathematics

Warsaw University of Technology

00-661 Warsaw, Poland

Abstract. In the paper we deal with computational complexity of a problem C_k (respectively A_k) of a partition of an ordered set into minimum number of at most k-element chains (resp. antichains). We show that C_k, $k \geq 3$, is NP-complete even for N-free ordered sets of length at most k, C_k and A_k are polynomial for series-paralel orders and A_k is polynomial for interval orders. We also consider related problems for graphs.

1. Introduction and Motivation

Consider the following transportation scheduling problem (c.f. Rival [8]): What is the least number of planes needed in a fleet to carry out all of a set of journeys with specified origin, destination, departure time, and arrival time? This problem can be easily expressed as a chain partition problem for an ordered set. Let P be the sets of journeys. Assign to each journey $x \in P$ two numbers $d(x)$ and $a(x)$ representing the departure time and the arrival time for the journey x, respectively. Moreover, for each pair $x, y \in P$, $x \neq y$, let $t(x, y)$ be the transition time which is needed to prepare for journey y after the completion of journey x. For a pair of journeys $x, y \in P$, we say that $x \leq y$ if $a(x) + t(x, y) \leq d(y)$. Under a reasonable assumption that $t(x, y) \leq t(x, z) + t(z, y)$, for every $x, y, z \in P$, P is an ordered set. The least number of planes required to undertake all the journeys is simply the minimum number of pairwise disjoint chains whose union is the whole ordered set P. Establishing this number as well as a construction of the appropriate chain partition is known to be solvable by an algorithm running in $O(n^3)$ time, where $n = |P|$ (c.f. Möhring [3]).

Let us now assume additionally that each of the planes can not, for some technological reasons, say, undertake more than k journeys, where k is a fixed positive integer.

Our problem is now to find the minimum number of pairwise disjoint chains of size at most k whose union is the whole ordered set P.

Let us consider another famous scheduling problem. It is well-known (c.f. Rival [8]) that the k-machine unit execution time problem can be modeled by a partition of an ordered set P into antichains $A_1, A_2, ..., A_m$ such that $|A_i| \leq k$ and

(*) each A_i consists of some minimal elements of the ordered set $P - \bigcup_{j<i} A_j$

for $i = 1, ..., m$.

Establishing the computational complexity of the problem of finding such a partition with the least possible m is a well-known open problem for $k \geq 3$. Möhring [5] proposed a related problem. He asked about computational complexity of the problem with the condition (*) dropped, i.e. of finding the minimum number of pairwise disjoint antichains of size at most k whose union is the whole ordered set P. If the sizes of antichains are not bounded the problem can be solved in $O(n^2)$ time, where $n = |P|$ (see Möhring [3]).

In this paper we deal with algorithmic aspects of the problems of partition of ordered sets into minimum number of chains and antichains with a bounded size.

Let us formulate these problems precisely.

Problem C_k (respectively A_k):
Instance : an ordered set P and a positive integer c.
Question : Is there a partition of P into at most c chains (resp. antichains) of sizes at most k?

In Section 2 we prove that the problem C_k , $k \geq 3$, is NP-complete even for N-free ordered sets of length at most k (the *length* of an ordered set P is equal to the number of elements in a longest chain in P minus 1). In Section 3 we show that A_k is polynomial for interval ordered sets and we sketch the proof that both C_k and A_k are polynomial for series-parallel ordered sets.

The problems C_k and A_k can be generalized in the following natural way into graph partition problems.

Problem CL_k (resp. I_k):
Instance : a graph G and a positive integer c.
Question : Is there a partition of the vertex set of G into at most c cliques (resp. independent sets) of sizes at most k?

Our problems C_k and A_k are just the problems CL_k and I_k for comparability graphs. While both CL_k and I_k are NP-complete for arbitrary graphs it would be interesting to establish the complexity status of them for some classes of perfect graphs other than comparability graphs. (The problems of partition of a perfect graph into minimum number of cliques (resp. independent sets) with unbounded sizes is known to be polynomial (see Grötschel, Lovász and Schrijver [2])). Corneil [1] reports that D.G. Kirkpatrick has shown NP-completeness of a related problem of partition of a chordal graph into k-element cliques, where $k \geq 3$. In Section 4 we prove that for split graphs both CL_k and I_k are polynomial. (A *split* graph is one where the vertex set may be partitioned into a clique and a void set.) We conclude this paper with some open problems.

For definitions of all notions not defined in this paper we refer the reader to Möhring [4].

2. Complexity of the General Chain Partition Problem

Notice that C_1 is obviously polynomial and C_2 is equivalent to the maximum matching problem for comparability graphs so it is polynomial too.

For ordered sets of length at most $k-1$ our problem C_k coincides with the problem of partition of an ordered set into minimum number of chains with no size restriction, which is known to be polynomial (c.f. Möhring [3]).

Theorem 1. *The problem C_k , for $k \geq 3$, is NP-complete even for N-free ordered sets of length at most k.*

Proof. We shall prove NP-completeness of the following problem C'_k which immediately implies NP-completeness of C_k.

Problem C'_k:
Instance: an ordered set P of size being a multiple of k.
Question: Is there a partition of P into chains of size k?

Clearly C'_k belongs to the class NP. First we prove NP-completeness of C'_3. We reduce the problem of exact cover by 3-element sets into C'_3. Let $(X, \mathcal{A})$ be a pair consisting of a finite set X and some family $\mathcal{A}$ of its 3-element subsets. For every 3-element set $A = \{x, y, z\} \in \mathcal{A}$ denote by P_A the ordered set depicted in Figure 1. Assume that $(P_A - A) \cap X = \emptyset$ and $(P_A - A) \cap (P_B - B) = \emptyset$ for all $A, B \in \mathcal{A}, A \neq B$.

Let $P = \bigcup_{A \in \mathcal{A}} P_A$. Clearly P can be constructed in polynomial time. Obviously P is N-free and its length is 3.

We shall show that an exact cover of X by 3-element sets exists if and only if P has a partition into 3-element chains.

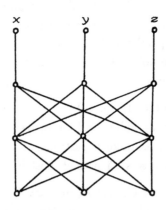

Figure 1.

Let $A_1, A_2, ..., A_p$ be an exact cover of X by 3-element sets. We partition each of the ordered sets P_{A_i}, $i = 1, ..., p$, into 3-element chains (which is obviously feasible). The ordered set $Q = P - \bigcup_{i=1}^{p} P_{A_i}$ is a collection of a certain number of ordered sets being a series composition of three 3-element antichains. Clearly, Q can be partitioned into 3-element chains.

Conversely, let C be a collection of disjoint 3–element chains whose union is the whole ordered set P. For every $x \in X$, let C_x be the chain in C such that $x \in C_x$. Clearly each C_x is contained in P_A for some $A = \{x, y, z\} \in \mathcal{A}$. Suppose that $C_y \subseteq P_B$ and $C_z \subseteq P_C$ for some $B, C \in \mathcal{A}$. If $B \neq A$ or $C \neq A$ then what remains from P after deleting the chains C_x, C_y and C_z is a 7- or 5-element ordered set whose elements are incomparable to the other elements of $P - \{C_x, C_y, C_z\}$. This contradicts to existence of a partition of P into 3-element chains. Thus $A = B = C$. It is evident now that the collection $\{A = \{x, y, z\} \in \mathcal{A} : C_x, C_y, C_z \subseteq P_A\}$ is an exact cover of X by 3-element sets.

Now, we shall reduce C'_3 into C'_k, for $k \geq 4$. Let P be an ordered set of length at most 3 and $|P| \equiv 0 \pmod 3$. Denote by P_k the ordered set being the series composition $Q_1 * Q_2 * ... * Q_{k-3} * P$, where the ordered sets Q_i are $|P|/3$-element antichains for $i = 1, 2, ..., k - 3$. A partition of P into 3-element chains exists if and only if a partition of P_k into k-element chains exists. Indeed, a partition of P into $t = |P|/3$ 3-element chains can be easily extended into a partition of P_k into t k-element chains. To prove the converse assume that some chains $C_1, C_2, ..., C_t$ form a partition of P_k into k-element chains. Notice that $|C_i \cap P| \geq 3$ for $i = 1, ..., t$, since otherwise C_i would have less than k elements. If $|C_i \cap P| \geq 4$ for some i then $|C_i - P| < k - 3$ so

$$t(k-3) = |\bigcup_{j=1}^{t} C_j - P| = \sum_{j=1}^{t} |C_j - P| < t(k-3),$$

a contradiction. Thus $|C_i \cap P| = 3$ for $i = 1, ..., t$ so the collection $C_1 \cap P, ..., C_t \cap P$ is a partition of P into 3-element chains. Obviously P_k is N-free and it has length at most k. ∎

The problem C'_k can be generalized in the following direction. Let N be a subset of the set of all positive integers.

Problem C'_N:

Instance: an ordered set P.

Question: Is there a partition of P into chains of sizes belonging to N?

We are able to prove the following theorem establishing the computational complexity of C'_N. Its proof will appear elsewhere.

Theorem 2. *The problem C'_N is NP-complete if $1 \notin N$ and $2 \notin N$. Otherwise it is polynomial.* ∎

3. Series-parallel and Interval Orders

In this section we prove that the problem A_k is linear for interval orders and we sketch the proof that both C_k and A_k are polynomial for series parallel orders.

Lemma 3. Let $\mathcal{P} = \{A_1, A_2, ..., A_m\}$ be any partition of an interval order I into antichains. The family $\mathcal{P}$ can be linearly ordered into a sequence $A'_1, A'_2, ..., A'_m$ such that for every i, A'_i consists of some minimal elements of $I - \bigcup_{j<i} A'_j$.

Proof. It was noticed by Reuter [7] that maximal antichains of an interval order P can be ordered into a sequence $B_1, B_2, ..., B_s$ such that for every $i, j = 1, ..., s$ if $i < j$ then for each $a \in B_i$ there is $b \in B_j$ such that $a \leq b$ in P.

Denote by p_i the largest p such that $A_i \subseteq B_p$, for $i = 1, ..., m$. We define a linear order $\preceq$ on $\mathcal{P}$. Let $A_i \preceq A_j$ if $p_i < p_j$ or $p_i = p_j$ and $i < j$. Denote by $A'_1, A'_2, ..., A'_m$ the members of $\mathcal{P}$ ordered according to this order. Suppose that our theorem does not hold and denote by l the least i such that A'_i contains a non-minimal element x in $I - \bigcup_{j<i} A'_j$.

Since x is not minimal in $J = I - \bigcup_{j<l} A'_j$, there is $y \in J$ such that $y < x$ in I. Let q be the largest i such that $y \in B_i$. Suppose $q \geq p_l$. Then since $x \in B_{p_l}$, there is $z \in B_q$ such that $x \leq z$. We get a contradiction because $y < x \leq z$ and $y \in B_q$. Thus $q < p_l$ and y belongs to some A'_j, where $j < l$. A contradiction because $y \in J$. ∎

The Lemma 3 shows that the problem A_k is equivalent for interval orders to the k-machine unit execution time problem. Since the latter problem is solvable by a linear time algorithm (see Papadimitriou and Yanakakis [6]) so is A_k.

Theorem 4. There is a linear time algorithm solving A_k for interval orders. ∎

Let us pass on to the series-parallel orders. Since a complement of a series-parallel order is a series-parallel order, the problems C_k and A_k coincide in this case.

Theorem 5. The problems C_k and A_k are polynomial for series-parallel orders.

We are able to show that much more general problems than C_k and A_k can be solved by a polynomial algorithm for series-parallel orders. Call a sequence $(c_1, c_2, ..., c_m)$ k-feasible for an ordered set P if there exists a partition of P into m chains $C_1, C_2, ..., C_m$ such that $|C_i| = c_i \leq k$ for $i = 1, ..., m$. Denote by $\mathcal{F}_P$ the set of all k-feasible sequences for P. We are able to prove the following theorem.

Theorem 6. Let k be fixed. There exists a polynomial algorithm constructing the set $\mathcal{F}_P$ of all k-feasible sequences for a series-parallel ordered set P. ∎

Proof of Theorem 5. This theorem is an immediate consequence of Theorem 6. To see this let us estimate the largest possible number of elements in $\mathcal{F}_P$. The elements of $\mathcal{F}_P$ can be identified with partitions of the integer $n = |P|$ into summands not exceeding k. It is known that the number of such partitions is equal to the number of partitions of n into at most k summands. Thus $|\mathcal{F}_P| \leq n^{k-1}$ so we can find in $\mathcal{F}_P$ the sequence with the smallest number of elements in polynomial time which together with Theorem 5 completes the proof. ∎

We will not prove Theorem 6 here. Instead, we will describe the main idea of the proof. The complete proof contains some boring and rather routine technical details.

Denote by S_k the set of all nondecreasing sequences of positive integers less than or equal to k. Let $\mathbf{c} = (c_1, ..., c_t)$ and $\mathbf{d} = (d_1, ..., d_s)$ be some members of S_k. By the parallel composition of $\mathbf{c}$ and $\mathbf{d}$, denoted by $\mathbf{c} + \mathbf{d}$, we mean the sequence obtained from the sequence $(c_1, ..., c_t, d_1, ..., d_s)$ by rearranging its terms in the nondecreasing way. By the series composition of $\mathbf{c}$ and $\mathbf{d}$, denoted by $\mathbf{c} * \mathbf{d}$, we mean the set of sequences of the form $(c'_1 + d'_1, ..., c'_r + d'_r) \in S_k$, where $(c'_1, ..., c'_r)$ (resp. $(d'_1, ..., d'_r)$) is a sequence obtained from the r-term sequence $(c_1, ..., c_t, 0, ..., 0)$ (resp. $(d_1, ..., d_s, 0, ..., 0)$) by some rearrangement of the terms. It can be shown that the construction of $\mathbf{c} * \mathbf{d}$, given $\mathbf{c}$ and $\mathbf{d}$, requires polynomially many steps.

Let Q and R be ordered sets and let $Q + R$ denote the parallel composition of Q and R. Then it is evident that $\mathcal{F}_{Q+R} = \bigcup_{\mathbf{c} \in \mathcal{F}_Q} \bigcup_{\mathbf{d} \in \mathcal{F}_R} \{\mathbf{c} + \mathbf{d}\}$. For the series composition $Q * R$ of Q and R, it can be shown in a rather straightforward way that $\mathcal{F}_{Q*R} = \bigcup_{\mathbf{c} \in \mathcal{F}_Q} \bigcup_{\mathbf{d} \in \mathcal{F}_R} \mathbf{c} * \mathbf{d}$

In view of the inequalities $|\mathcal{F}_Q| \leq n^{|Q|-1}$ and $|\mathcal{F}_R| \leq n^{|R|-1}$, it is clear that the constructions of $\mathcal{F}_{Q+R}$ and $\mathcal{F}_{Q*R}$ require polynomially many steps.

Every series-parallel order P can be constructed from the one-element orders recursively (using the binary decomposition tree for P) by less than $|P|$ parallel or series compositions so the set $\mathcal{F}_P$ can be found in polynomially many steps.

4. Split Graphs

Theorem 7. *The problem CL_k is polynomial for split graphs.*

Proof. Let S be a split graph and let $S = K \cup V$, where K is a clique and V is a void set. Denote by G the bipartite graph obtained from S by deleting all the edges joining two vertices in K, substituting each vertex $v \in V$ by $k-1$ vertices $v_1, v_2, ..., v_{k-1}$ and joining them to the neighbors of v in S. Clearly the construction of G requires polynomially many steps.

We shall prove that the minimum number m of cliques of size at most k into which S can be partitioned is equal to $|V| + \lceil (|K| - \nu)/k \rceil$, where ν is the maximum size of a matching in G ($\lceil x \rceil$ stands for the least integer not less than x). This will complete our proof since the maximum matching problem is polynomial.

Consider a partition of S into the minimum number m of cliques of size at most k. Let, for every $v \in V$, $C[v]$ be the clique in this partition containing v. Denote the elements of $C[v] - \{v\}$ by $u_1^v, u_2^v, ..., u_{t_v}^v$. Obviously, $t_v \leq k - 1$. Moreover $m = |V| + \lceil (|K| - d)/k \rceil$, where $d = \sum_{v \in V} t_v$. It is clear that the set of edges $\{v_i u_i^v : v \in$

$V, i = 1, ..., t_v\}$ induces a matching of size d in G. Thus $m = |V| + \lceil (|K| - d)/k \rceil \geq |V| + \lceil (|K| - \nu)/k \rceil$.

Conversely, consider a maximum-sized matching M in G and denote $U[v] = \{u \in K : v_i u$ is an edge in M for some $i = 1, ..., k-1\}$, for every $v \in V$. Clearly $|U[v]| \leq k-1$. Obviously the sets $\{v\} \cup U[v]$ are pairwise disjoint cliques of size at most k in S. The graph $S - \bigcup_{v \in V} (\{v\} \cup U[v])$ is a $(|K| - \nu)$-element clique so it can be partitioned into $\lceil (|K| - \nu)/k \rceil$ cliques of size at most k. Thus the whole graph S has a partition into $|V| + \lceil (|K| - \nu)/k \rceil$ cliques of size at most k. Consequently, $m \leq |V| + \lceil (|K| - \nu)/k \rceil$. ∎

Since the complement of a split graph is a split graph too we get the following corollary.

Corollary 8. *The problem I_k is polynomial for split graphs.* ∎

5. Conclusions and Problems

It seems that establishing the complexity of A_k for arbitrary orders is the central open problem in the area considered in this paper. Solving this problem may, however, be hard in view of its close relationship to the k-machine unit execution time problem. Finding the complexity of A_k for 2-dimensional orders and $k \geq 3$ could be a step towards its solution. For these orders C_k and A_k coincide because the complement of the comparability graph of a 2-dimensional order is still the comparability graph of a 2-dimensional order. The complexity of C'_k (and consequently of A'_k) is not known for 2-dimensional orders either. Let us formulate this problem for $k = 3$ in the following equivalent form.

Problem S_3:
Instance: a sequence of $3n$ distinct positive integers.
Question: Is there a partition of the sequence into n 3-term increasing subsequences?

References

1. Corneil D. G., The complexity of generalized clique packing, *Discrete Applied Mathematics* 12(1985), 233-239.

2. Grötschel M., Lovász L. and Schrijver A., The ellipsoid method and its consequences in combinatorial optimization, *Combinatorica* 1(1981), 169-197.

3. Möhring R.H., Algorithmic aspects of comparability graphs and interval graphs, in Graphs and Order (ed. I. Rival), Reidel Publishing Co., Dordrecht 1985, 41-102.

4. Möhring R.H., Computationally tractable classes of ordered sets, in Algorithms and Order (ed. I. Rival), Reidel Publishing Co., Dordrecht 1989, 105-194.

5. Möhring R.H., Problem 9.10, in Graphs and Order (ed. I. Rival), Reidel Publishing Co., Dordrecht 1985, 583.

6. Papadimitriou C.H. and Yanakakis M., Scheduling interval-ordered tasks, *SIAM J. Comp.* 8(1979), 405-409.

7. Reuter K., The jump number and the lattice of maximal antichains, Report 1197, Technische Hochschule Darmstadt, 1-16.

8. Rival I., Some order theoretical ideas about scheduling, IX symposium on operations research (Osnabrück, 1984), Methods of Operations Research 49 (1985).

TIGHT BOUNDS FOR THE RECTANGULAR ART GALLERY PROBLEM

J. Czyzowicz[1], E. Rivera-Campo[2], N. Santoro[3],
J. Urrutia[4] and J. Zaks[5]

1. U. Quebec, Hull, Quebec, Canada.
2. U. Autonoma Metropolitana-I, Mexico.
3. Carleton U., Ottawa, Canada.
4. U. Ottawa, Canada.
5. U. Haifa, Israel.

Abstract

Consider a rectangular art gallery, subdivided into n rectangular rooms; any two adjacent rooms have a door connecting them. We show that $\lceil n/2 \rceil$ guards are always sufficient to protect all rooms in a rectangular art gallery; furthermore, their positioning can be determined in O(n) time. We show that the optimal positioning of the guards can be determined in linear time. We extend the result by proving that in an arbitrary orthogonal art gallery (not necessarily convex, possibly having holes) with n rectangular rooms and k walls, $\lceil (n+k)/2 \rceil$ guards are always sufficient and occasionally necessary to guard all the rooms in our gallery. A linear time algorithm to find the positioning of the guards is obtained.

1. Rectangular galleries

Consider a rectangular art gallery, subdivided into n rectangular rooms; any two adjacent rooms have a door connecting them. (See Figure 1.) We want to protect all of the rooms in the gallery stationing the smallest number of guards possible. Two problems arise: 1. determining the number of guards, and 2. determining their positioning in the gallery.

An equivalent formulation of the first problem is the following. Suppose a rectangle T is decomposed into n rectangles, having mutually disjoint interiors. How few points in T are needed such that every subrectangle will contain at least one of the points? Trivially, n points will suffice.

The example of n rectangles in a row shows that as many as $\lceil n/2 \rceil$ points might be needed.

The main aim in this section is to show that $\lceil n/2 \rceil$ guards are always sufficient to protect all rooms in a rectangular art gallery; furthermore, their positioning can be determined in O(n) time.

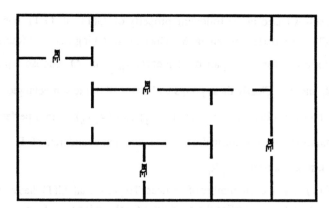

An example of an Art Gallery with eight
rooms for which four guards suffice,
showing the positions of their chairs.

Figure 1

Some definitions will be needed before we procee. A polygonal region is called **orthogonal**
[5], if all its edges are parallel to either the x-axis or to the y-axis. Given a decomposition of a
rectangle T into n rectangles, we can obtain the dual graph $G(T)$ of the decomposition of T by
representing each subrectangle of T by a vertex in $G(T)$, two vertices being adjacent if their
corresponding rectangles share a line segment in their common boundary. See Figure 2.

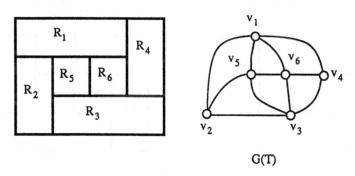

$G(T)$

Figure 2

Notice that the union of the rectangles corresponding to the vertices of any connected subgraph of
$G(T)$ form an orthogonal polygon. We are now ready to prove the following Theorem.

Theorem 1. If a rectangle T is decomposed into n rectangles, having mutually disjoint interiors,
then there exists a set S consisting of at most $\lceil n/2 \rceil$ points, such that every subrectangle meets S;
moreover, all the points of S can be chosen so that each one of them belongs to at most two of the
subrectangles.

Proof. Our result is based in the following property of G(T): if G(T) has an even number of nodes, then G(T) has a perfect matching M. Once the matching has been found, the n/2 points can be chosen as follows: for every pair of elements $\{v_i,v_j\}$ in M choose a point in the common boundary of R_i and R_j. Clearly these points will cover all of the subrectangles of T. For instance for the graph shown in Figure 2, $\{v_1,v_5\}$, $\{v_2,v_3\}$ and $\{v_4,v_6\}$ form a perfect matching M for G(T). Three points can now be located one in the common boundaries of R_1,R_5 another in that of R_2,R_3 and the last one between R_4,R_6.

Assume that G(T) has an even number of vertices. To prove that G(T) has a perfect matching, we proceed to prove that G(T) satisfies the conditions for the existence of a perfect matching stated in the well-known result of Tutte; namely, for any subset S of the vertices of G(T), the number of odd components of G(T)–S does not exceed ISI. Let k be the number of connected components of G(T)–S. Each connected component C_i of G(T)–S is represented by an orthogonal polygon P_i of T. Each such polygon has at least four corner points. The total number of corner points generated by the k components in G(T)–S is at least 4k. In other words, when we delete from T the rectangles representing vertices in S, we obtain a family of k disjoint orthogonal polygons contained in T with a total number of at least 4k corner points.

Our next observation is essential to our proof: when a rectangle represented by a point in S is now replaced, at most four corner points will disappear. (See Figure 3).

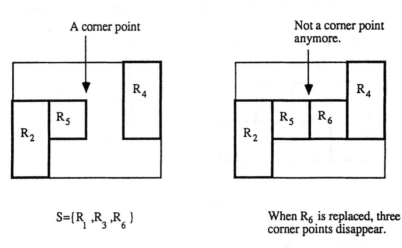

Figure 3

Once all rectangles in S are replaced, all the corner points generated by the components of G(T)–S will disappear, except for the four corner points of T; it follows that $k\le ISI+1$. The reader may verify that if k=ISI+1, then at least one of the components of G(T)–S is even, otherwise the number of vertices of G(T) would be odd, which contradicts the assumption that the number of vertices of G(T) is even.

For the case when n is odd, add an extra rectangle along one side of T and apply the previous arguments.

[]

The result of Theorem 1 shows that $\lceil n/2 \rceil$ guards are sufficient. It also gives a constructive way of determining their positioning in the gallery: find a perfect matching and position the guards on the corresponding doors; the complexity of this algorithm is clearly not linear. A linear (and, thus, optimal) algorithm can be derived exploiting the property established in the following

Theorem 2. If a rectangle T is decomposed into n rectangles, having mutual disjoint interiors, then the dual graph of ·T has a Hamiltonian path.

Proof. Let a rectangle T be arbitrarily decomposed into n subrectangles. Let T* be the rectangle, obtained from the rectangle T by surrounding it with four additional subrectangles, as shown in Figure 4:

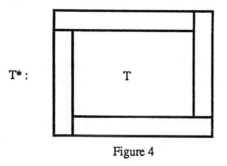

Figure 4

Main claim: The dual graph of the decomposition of T*, in which the outer face is included, is 4-connected.

Proof of the claim: It suffices to show that the deletion of any three or fewer vertices from the dual graph of T* yields a connected subgraph. Equivalently, it suffices to show that if three or fewer subrectangles are to be avoided, then there is always a path from an interior point of any of the remaining subrectangles via the remaining subrectangles to the (remaining part of the) boundary of T*. If the three subrectangles are all in the interior of T*, they cannot block all the *four* directions to the boundary of T*, available from any one of the remaining subrectangles; here a *direction* means up, down left or right. Similar arguments are used in case some of the three subrectangles are in the boundary of T*. Therefore the dual graph of T* is 4-connected.

To complete our proof, let $T^\#$ denote the decomposition of a rectangle into three copies of the decomposition T, joined by subrectangles, and surrounded by four rectangles, as shown in **Figure 5**.

$T^\#$:
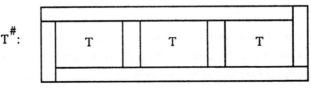

Figure 5

According to the previous part, the dual graph of $T^\#$, where the outer face is included, is planar and 4-connected, hence it is Hamiltonian, by a theorem of Tutte [6]. It follows easily that every Hamiltonian circuit of $T^\#$ must visit one of the copies of T in $T^\#$ exactly once; thus the dual graph of T has a Hamiltonian path.
[]

Theorem 2 implies Theorem 1, since every other edge on the Hamiltonian path yields a matching in the dual graph, having $\lfloor n/2 \rfloor$ edges. Our proof of Theorem 2 actually proves a slightly stronger statement, as follows.

Corollary 1: If a rectangle T is decomposed into rectangles, then the dual graph of T has a Hamiltonian path which starts at one side of T and ends up at the opposite side.

Algorithmically, finding the Hamiltonian path in the dual of T can be done by finding a Hamiltonian circuit in the dual of $T^\#$; this can be done in linear time, using [1].

2. Non-Rectangular Orthogonal Galleries

If our art gallery is not rectangular, our result is obviously false. In the worst case as many as $n-1$ points might be needed to meet all the subregions, as shown in Figure 6.

Figure 6

For arbitrary orthogonal regions, we have the following.

Theorem 3. If **F** is an arbitrary orthogonal region in the plane, having k vertices, and if **F** is decomposed into n subrectangles having mutually disjoint interiors, then, for sufficiently large n, there exists a set S of at most $\lceil (n+k)/2 \rceil$ points, so that every subrectangle meets S. Moreover, the points of S can be so chosen that each one of them meets at most two of the subrectangles.

The proof is similar to the proof of Theorem 1, except that here F is wrapped with one layer consisting of k quadrangles; the rest of the proof is the same as in the proof of Theorem 2, hence it is omitted. Algorithmically, also in this case, S can be found in $O(n+k)$ time using the result by [1].
   Our next result is an extension of Theorem 2 to the case when the region can have holes

Theorem 4. For every orthogonal region **F**, possibly having holes, there exists a constant k, k=k(**F**), depending linearly on the number of vertices of **F**, such that if **F** is decomposed into n subrectangles, having mutually disjoint interiors, then there exists a set S consisting of $\lceil (n+k)/2 \rceil$ points, such that every subrectangle meets S; moreover, the points of S can be chosen so that each one of them meets at most two of the subrectangles.

Proof. Let **F** be an arbitrary orthogonal region, possibly having holes. Let B be the smallest rectangle which contains **F**; let the region inside B and outside F be decomposed by a *fixed* number k of rectangles. Every decomposition of F into n rectangles can be extended to a decomposition of the rectangle B into n+k rectangles; the rest follows from Theorem 1.

   In a similar way, we have the following extension of Theorem 2.

Theorem 5. For every orthogonal region D in the plane, possibly having holes, there exists a constant k=k(D), depending linearly on the number of vertices of D, such that the dual graph of every decomposition of D into rectangles has a path number which is at most k.

3. General polygonal regions

   In the general case, where a polygonal region is decomposed into n convex subregions, it is known [3] that about $\lceil (2/3)n \rceil$ points will be required; if the entire configuration is 3-connected, we have the following result.

Theorem 6. If F is an arbitrary polygonal region in the plane, having k vertices, and if F is decomposed into n convex subregions, then, for sufficiently large n, there exists a set S consisting of at most $\lceil (2/3)(n+k) \rceil$ points, such that every subregion meets S. Moreover, S can be so chosen that every point of S meets at most two of the subregions.

Proof. Let **F** be an arbitrary polygonal region in the plane, having k vertices, and let **F** be decomposed into n convex subregions. Wrap **F** with a layer of quadrangles, as shown in Figure 7, and call the new decomposition **F***.

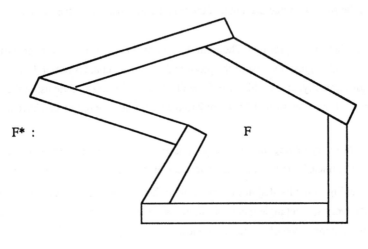

Figure 7

Claim: The dual graph of **F*** is 3-connected.

Proof. Let **A, B** and **C** be any three subregions of **F***, and let x be any point in the interior of **C**. Not all the rays from x to the part of bd(F*) which is in **F*** - (A ∪ B) can be blocked by A ∪ B, since the projection of **A** (and **B**) on a small circle inside **C**, centred at x, is an arc which is less than half the circle, since x ∉ A . Thus, two such arcs add up to less than the full circle. Therefore the dual graph of **F*** - (A ∪ B) is connected. It follows that the dual graph of **F*** is 3-connected.

The dual graph of **F*** is, in addition, planar; therefore, by a theorem of Nishizeki [4], it has a matching which has ⌈(n+k)/3⌉ edges. Thus, ⌈(n+k)/3⌉ points of the set **S** can be chosen, in correspondence with the edges of this matching, lying (in the dual setting) on twice as many subregions. For the remaining one-third of the regions, take one point per region (on a proper boundary point of that subregion).

[]

To see that about (2/3)n points are needed for a convex domain, which is decomposed into n subdomains, consider the dual graph of the m-th iterated Kleetope over the bi-pyramid, for large values of m (for details, see [2]).

Using the idea of the proof of Theorem 5, we get the following.

Theorem 7. For every polygonal region F in the plane, possibly having holes, there exists a constant k=k(F), such that for every decomposition of F into n convex subregions, there exists a set S consisting of at most $\lceil(2/3)(n+k)\rceil$ points, such that every subregion meets S; moreover, S can be so chosen that every point of S meets at most two of the subregions.

Acknowledgment

This work has been supported in part by the Natural Sciences and Engineering Research Council of Canada.

References

[1] N. Chiba and T. Nishizeki, The Hamiltonian cycle problem is linear-time solvable for 4-connected planar graphs, *J. Algorithms*, V.**10** (2) (1989), 187-212.

[2] B. Grunbaum, *Convex polytopes*, J. Weily & Sons, 1967.

[3] B. Grunbaum and J. O'Rourke, 1983 (see [5].).

[4] T. Nishizeki, Lower bounds on cardinality of the maximum matchings of planar graphs, Carnegie Mellon Tech. Report, 1977.

[5] J. O'Rourke, *Art gallery theorems and algorithms*, Oxford U. Press, 1987.

[6] W. Tutte, A theorem on planar graphs, *Trans. Amer. Math. Soc.* **82** (1956), 99-116.

Voronoi Diagrams of Moving Points in the Plane

Leonidas J. Guibas
DEC Systems Research Center and
Computer Science Department
Stanford University, USA

Joseph S.B. Mitchell *
Operations Research
Cornell University, USA

Thomas Roos **
Computer Science Department
University of Würzburg, F.R.G.

Abstract

Consider a set of n points in the Euclidean plane each of which is continuously moving along a given trajectory. At each instant in time, the points define a Voronoi diagram. As the points move, the Voronoi diagram changes continuously, but at certain critical instants in time, topological *events* occur that cause a change in the Delaunay diagram. In this paper, we present a method of *maintaining* the Voronoi diagram over time, while showing that the number of topological events has a nearly cubic upper bound of $O(n^2\lambda_s(n))$, where $\lambda_s(n)$ is the maximum length of an (n, s)-Davenport-Schinzel sequence and s is a constant depending on the motions of the point sites. In the special case of points moving at constant speed along straight lines, we get $s = 4$, implying an upper bound of $O(n^3 2^{\alpha(n)})$, where $\alpha(n)$ is the extremely slowly-growing inverse of Ackermann's function. Our results are a linear-factor improvement over the naive quartic bound on the number of topological events.

In addition, we show that if only k points are moving (while leaving the other $n - k$ points fixed), there is an upper bound of $O(k\, n\, \lambda_s(n) + (n - k)^2\, \lambda_s(k))$ on the number of topological events, which is nearly quadratic if k is constant.

We give a numerically stable algorithm for the update of the topological structure of the Voronoi diagram, using only $O(\log n)$ time per event (which is worst-case optimal per event).

1 Introduction

One of the most fundamental data structures in computational geometry is the *Voronoi diagram*. In its most general form, the Voronoi diagram $VD(S)$ of a set S of n objects in a space E is a subdivision of this space into maximal regions, so that all points within a given region have the same nearest neighbor in S with respect to a given distance measure d.

Shamos and Hoey [ShHo 75] introduced the Voronoi diagram for a finite set of points in the Euclidean plane $\mathbb{E}^2$ into to the field of computational geometry, providing the first efficient

---

*Partially supported by a grant from Hughes Research Laboratories, Malibu, CA, and by NSF Grant ECSE-8857642.

**Work on this paper by Thomas Roos was supported by the Deutsche Forschungsgemeinschaft (DFG) under contract (No 88/10 - 1).

algorithm for its computation. Since then, Voronoi diagrams in many variations have appeared throughout the algorithmic literature; see, for example, [Le 82], [Ed 86], [ChEd 87], [DrLe 78], [Fo 86], [Ya 87], and [Ro 89].

A problem of recent interest has been that of allowing the set of objects S to vary over time. This "dynamic" version of the problem has been studied by [ImSuIm 89], [AuImTo 90] and [Ro 90].[1]

The problem we consider here is the following: We are given a set S of n points in the Euclidean plane each of which is continuously moving along a given trajectory. At each instant in time, the points define a Voronoi diagram. As the points move, the Voronoi diagram changes continuously, but at certain critical instants in time, topological *events* occur that cause a change in the Delaunay diagram. Our goal is to *maintain* the Voronoi diagram over time in some useful data structure.

Our main result is to prove a nearly cubic upper bound of $O(n^2 \lambda_s(n))$ on the number of topological events, where $\lambda_s(n)$ is the maximum length of an (n, s)-Davenport-Schinzel sequence and s is a constant depending on the motions of the point sites. In the special case of points moving at constant speed along straight lines, we get $s = 4$, implying an upper bound of $O(n^3 2^{\alpha(n)})$, where $\alpha(n)$ is the extremely slowly-growing inverse of Ackermann's function. Our results are a linear-factor improvement over the naive quartic bound on the number of topological events.

In the case that only k of the n points of S are moving (while the remaining $n - k$ stay fixed), our bound on the number of events becomes $O(k n \lambda_s(n) + (n - k)^2 \lambda_s(k))$, which is nearly quadratic for fixed k.

We also present a numerically stable algorithm for the update over time of the topological structure of the Voronoi diagram, using only $O(\log n)$ time for each topological change. It is known (cf. [Ro 91]) that this update time is worst-case optimal.

2 The Topological Structure of Voronoi Diagrams

This section summarizes the elementary definitions and properties of Euclidean Voronoi diagrams of point sets. We let $d(\cdot, \cdot)$ denote Euclidean distance.

We are given a finite set $S := \{P_1, \ldots, P_n\}$ of $n \geq 3$ points in the Euclidean plane $\mathbb{E}^2$. The perpendicular *bisector of P_i and P_j* is defined by

$$B_{ij} := \{x \in \mathbb{E}^2 \mid d(x, P_i) = d(x, P_j)\},$$

and the *Voronoi polygon of P_i* by

$$v(P_i) := \{x \in \mathbb{E}^2 \mid \forall_{j \neq i} \ d(x, P_i) \leq d(x, P_j)\}.$$

The vertices of the Voronoi polygons are called *Voronoi points* and the bisector portions on the boundary are called *Voronoi edges*. Finally the *Voronoi diagram of S* is defined by

$$VD(S) := \{v(P_i) \mid P_i \in S\}.$$

The embedding of the Voronoi diagram provides a planar straight line graph that we call the *geometrical structure* of the underlying Voronoi diagram.

---

[1]The term *dynamic* in this paper is taken to mean that the sites S are moving along trajectories, not that data can be inserted or deleted from the problem.

Now we turn our attention to the dual graph of the Voronoi diagram, the so-called *Delaunay triangulation* $DT(S)$. If S is in general position – i.e. no four points of S are cocircular and no three points of S are collinear – every bisector part in $VD(S)$ corresponds to an edge and every Voronoi point in $VD(S)$ to a triple in $DT(S)$. The use of the dual graph not only has numerically advantages, but also allows a clearer separation between geometrical and topological aspects.

We use a *one-point-compactification* to simplify our discussion. We augment set S by adding the "point at infinity", yielding a new set of sites $S' := S \cup \{\infty\}$. The extended Delaunay triangulation is then given by

$$DT(S') = DT(S) \cup \{(P_i, \infty) \mid P_i \in S \cap \partial CH(S)\};$$

i.e., in addition to the Delaunay triangulation $DT(S)$, every point on the boundary of the convex hull $\partial CH(S)$ is connected to ∞. We call the underlying graph of the extended Delaunay triangulation $DT(S')$ the *topological structure* of the Voronoi diagram. In contrast with $DT(S)$, $DT(S')$ has the nice property that there are exactly three triples adjacent to each triple in $DT(S')$. Let $v(P_i, P_j, P_k)$ denote the center of the circumcircle $C(P_i, P_j, P_k)$ of the three points $P_i, P_j, P_k \in S$. We can characterize the triples in $DT(S')$ as follows:

$$\{P_i, P_j, P_k\} \in DT(S') \iff v(P_i, P_j, P_k) \text{ is a Voronoi point in } VD(S).$$
$$\iff C(P_i, P_j, P_k) \text{ contains no point of } S \text{ in its interior.}$$
$$\{P_i, P_j, \infty\} \in DT(S') \iff P_i \text{ and } P_j \text{ are neighboring points of } S \text{ on the boundary of the convex hull } \partial CH(S).$$

Since $DT(S')$ is a complete triangulation of the extended plane $\overline{\mathbb{E}^2}$ – i.e. every triple is bounded by exactly three edges and every edge belongs to exactly two triples – Euler's formula implies that the number of edges and triples of the topological structure $DT(S')$ of the Voronoi diagram $VD(S)$ is linear in n.

Also, note that the hardest part of constructing a Voronoi diagram is to determine its topological structure, since the geometrical structure of a Voronoi diagram can be derived in linear time from the Delaunay triangulation $DT(S')$. Furthermore, the geometrical structure is determined only locally by its topological structure, namely in the neighborhood of the corresponding Voronoi point. This implies the possibility of a local update of the Voronoi diagram after a local change of one or more points in S.

3 Topological Events

We turn now to the problem of describing the changes in the topological structure of a set of continuously moving points. We state several results without proof; the formal proofs can be found in [Ro 91].

We are given a finite set of $n \geq 3$ continuous trajectory curves in the Euclidean plane $\mathbb{E}^2$, $S := S(t) := \{P_1(t), \ldots, P_n(t)\}$. We make the following assumptions about the trajectories:

(A)   *The points move without collisions, i.e.* $\forall_{i \neq j}\ \forall_{t \in \mathbb{R}}\ P_i(t) \neq P_j(t)$

(B)   *There exists a moment $t_0 \in \mathbb{R}$ when $S(t_0)$ is in general position.*[2]

Let us consider the situation at a moment $t \in \mathbb{R}$ when all points in $S(t)$ are in general position. By investigating the continuity of a suitable product of determinants, it is easy to see, that a sufficiently small continuous motion of the points does not change the fact that the points are in general position.

The topological structure $DT(S')$ is completely determined on the one hand by the *active Voronoi points*[3] in $VD(S)$ which correspond to the triangles of $DT(S)$, and on the other hand by those pairs of points in S that lie on the boundary of the convex hull $\partial CH(S)$ (or by the extended triples in $DT(S')$). Therefore we only have to distinguish two different situations in which the topological structure changes:

Case (1)   The appearance (disappearance) of an inactive (active) Voronoi point.

Case (2)   The appearance (disappearance) of a point on the boundary of the convex hull.

In accordance with the characterization for triples in $DT(S')$, the first case occurs when four points of $S(t + \varepsilon)$ become cocircular, while the second case occurs when three points of $S(t + \varepsilon)$ become collinear, for some small $\varepsilon > 0$.

In both cases the loss of the general position of the points in $S(t + \varepsilon)$ (for a short time) is necessary for a change of the topological structure of the Voronoi diagram. This leads to our first basic theorem that describes the local stability of the topological structure $DT(S')$.

Theorem 1   For a finite set S of points in general position, the topological structure of the Voronoi diagram is *locally stable* under sufficiently small continuous motions of the sites. However, the geometrical structure changes in the neighborhood of the moving points.

Thus, the loss of general position of the points $S(t)$ is necessary for changing the topological structure $DT(S'(t))$. In order to address the question of sufficient conditions, we proceed with an investigation of the *elementary changes* of the topological structure of a Voronoi diagram. One can show that such changes can be characterized as "SWAP"s of adjacent triangles in $DT(S')$, except in degenerate cases; see the figure below.

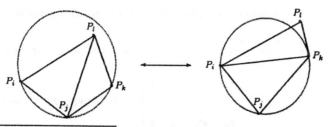

---

[2] In other words, there exists an instant in which no four points in $S'(t_0)$ are cocircular.
[3] i.e. those Voronoi points that appear in the first order Voronoi diagram.

The transition is equivalent to a *fusion* of two existing Voronoi points, which come together and disappear at the instant of cocircularity, and the creation of two new Voronoi points. This observation leads to the following theorem:

Theorem 2 Elementary changes in the topological structure of the Voronoi diagram $VD(S)$ are characterized by SWAPs of adjacent triples in $DT(S')$, except in degenerate cases.

Note that the one-point-compactification allows us the convenience of treating both cases similarly: as simple SWAPs in the extended dual graph $DT(S')$.

In the following we call a pair of adjacent triples in $DT(S')$ a *quadrilateral*. It is easy to see that every quadrilateral corresponds to an edge of the topological structure $DT(S')$. Therefore the number of quadrilaterals is also linear.

We have been ignoring a technicality caused by degeneracies: it may be that more than four points in $S(t)$ are cocircular at the same instant or that more than three points in $S(t)$ are collinear at the same instant. In other words, it may be that two or more adjacent quadrilaterals swap at the same time.[4] In this case, we can use the linear-time algorithm presented in [Ag 87] to retriangulate the interior of the convex polygon described by the cocircular points at a moment $t + \varepsilon$. However, it is necessary to select $\varepsilon > 0$ in such a way, that the moment of retriangulation precedes the next topological event.

4 Upper Bounds

In the previous section topological events are characterized by moments of cocircularity or collinearity of neighboring points. Therefore it is necessary, that the zeros of the functions $INCIRCLE(\ldots)$ and $CCW(\ldots)$ introduced by [GuSt 85] can be calculated.[5] For that we demand the following additional assumption that is achieved, for example, in the case of *polynomial curves of bounded degree*.

(C) The functions $INCIRCLE(P_i, P_j, P_k, P_l)$ and $CCW(P_i, P_j, P_k)$ have at most $s \in O(1)$ zeros.

Assumption (C) implies that each quadrilateral in $DT(S')$ generates at most a constant number of topological events. With this assumption, we obtain a trivial $s\binom{n+1}{4} \in O(n^4)$ upper bound on the maximum number of topological events.

By a Davenport-Schinzel sequence argument, we improve the naive upper bound by (roughly) a linear factor.

---

[4]Quadrilaterals that are not adjacent may swap in any order, since they do not affect each other.
[5]The functions are defined as follows

$$INCIRCLE(P_i, P_j, P_k, P_l) := \begin{vmatrix} x_{P_i} & y_{P_i} & x_{P_i}^2 + y_{P_i}^2 & 1 \\ x_{P_j} & y_{P_j} & x_{P_j}^2 + y_{P_j}^2 & 1 \\ x_{P_k} & y_{P_k} & x_{P_k}^2 + y_{P_k}^2 & 1 \\ x_{P_l} & y_{P_l} & x_{P_l}^2 + y_{P_l}^2 & 1 \end{vmatrix} \quad \text{and} \quad CCW(P_i, P_j, P_k) := \begin{vmatrix} x_{P_i} & y_{P_i} & 1 \\ x_{P_j} & y_{P_j} & 1 \\ x_{P_k} & y_{P_k} & 1 \end{vmatrix}$$

See [GuSt 85] and [Ro 91] for several basic properties of these functions.

First, due to assumption (C), it is clear that the maximum number of extended SWAPs is at most $s \binom{n}{3} \in O(n^3)$, since this is the maximum number of moments at which three points of S can become collinear. Therefore we only have to deal with such topological events where four points of S become cocircular. The basic observation is that every topological event is related to one quadrilateral leaving the four bounding Delaunay edges of this quadrilateral unchanged.

With that, we are able to determine the total number of topological events by adding for every imaginable Delaunay edge (P_i, P_j) the number of adjacent topological events that do not destroy this edge.

With this intention, we consider an arbitrary pair $(P_i(t), P_j(t))$ of different points and the corresponding bisector $B_{ij}(t, \mu)$ which is given by the formulation below:

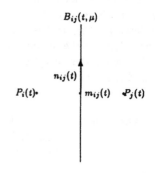

$$B_{ij}(t, \mu) \quad := \quad m_{ij}(t) + \mu \, n_{ij}(t) \quad \text{where} \quad \mu \in \mathbb{R},$$

$$m_{ij}(t) \quad := \quad \frac{P_i(t) + P_j(t)}{2} \quad \text{and}$$

$$n_{ij}(t) \quad := \quad \begin{pmatrix} P_{i2}(t) - P_{j2}(t) \\ P_{j1}(t) - P_{i1}(t) \end{pmatrix}$$

In addition, $P_i(t)$ and $P_j(t)$ define two open half-planes

$$h_{ij}^{>}(t) \quad := \quad \{x \in \mathbb{E}^2 \mid n_{ij}(t)^T [x - P_i(t)] > 0\} \quad \text{and}$$

$$h_{ij}^{<}(t) \quad := \quad \{x \in \mathbb{E}^2 \mid n_{ij}(t)^T [x - P_i(t)] < 0\}$$

which are bounded by the straight line through these two points.

Now, whenever the Delaunay edge (P_i, P_j) exists, there are exactly two triples $\{P_i, P_j, P_k\}$, $\{P_i, P_j, P_l\} \in DT(S')$ adjacent to the Delaunay edge (P_i, P_j), where

$$P_k \in \left(h_{ij}^{>} \cap S\right) \cup \{\infty\} \quad \text{and} \quad P_l \in \left(h_{ij}^{<} \cap S\right) \cup \{\infty\}$$

If we look at the μ-values $\mu_x(t)$ of the centers of the circumcircles $C(P_i(t), P_j(t), P_x(t))$ on the bisector $B_{ij}(t, \mu)$, we obtain the following characterization for the upper triple $\{P_i, P_j, P_k\}$:

$$\mu_k(t) \quad := \quad \min_{P_x \in S' \setminus \{P_i, P_j\}} \mu_x(t) \quad \text{where}$$

$$\mu_x(t) \quad := \quad \begin{cases} \infty & \text{if } P_x = \infty \text{ or } P_x \notin h_{ij}^{>} \cap S \\ \dfrac{[P_x(t) - P_j(t)]^T [P_x(t) - P_i(t)]}{2 \, n_{ij}(t)^T [P_x(t) - P_i(t)]} & \text{otherwise} \end{cases}$$

Naturally, an analogous construction can be done for the lower triple. If we investigate those moments when the upper triple changes[6] we can restrict ourselves to those intervals in which $h_{ij}^{>} \cap S \neq \emptyset$.

Next, we look closer at the functions $\mu_x(t)$ and their pairwise points of intersection:

---

[6]Notice, that P_k can only be replaced by another point of $(h_{ij}^{>} \cap S) \cup \infty$, because the Delaunay edge (P_i, P_j) is not destroyed during the topological event.

Case (1) $\qquad \mu_x(t) = \mu_y(t) < \infty$

Both circumcircles $C(P_i, P_j, P_x)$ and $C(P_i, P_j, P_y)$ are identical, which implies that all four points are cocircular. By assumption (C), this can happen only s times.

Case (2) $\qquad \mu_x(t) = \mu_y(t) = \infty$

These moments have no influence on the complexity of the minimum function $\mu_k(t)$, since we have restricted ourselves to intervals where $\mu_k(t) < \infty$.

Finally, we can summarize both cases with the statement that two different functions $\mu_x(t)$ and $\mu_y(t)$ have at most s relevant intersections. Thus, the theory of Davenport-Schinzel sequences ([Sh 88]) implies that the minimum function $\mu_k(t)$ has worst case complexity $O(\lambda_s(n))$, where $\lambda_s(n)$ is the maximum length of a Davenport-Schinzel sequence of length n and order s. Summing over all $\binom{n}{2}$ pairs of points (P_i, P_j), we obtain the following theorem. (The lower bound is shown later.)

Theorem 3 Given a finite set $S(t)$ of n continuous trajectories satisfying the assumptions (A), (B) and (C), the maximum number of topological events over time is $O(n^2 \lambda_s(n))$. $\Omega(n^2)$ is a lower bound on the worst-case number of topological events.

We now briefly examine the special case in which only k of the points S are moving (while the other $n - k$ points remain fixed in the plane). We partition the set S' in a natural way:

$$S' = S_m \cup S_f,$$

where S_m and S_f denote the set of moving and fixed points, respectively. Thereby we use the same technique for the estimation of the maximum number of topological events, but with one difference: We consider the fixed and the moving pairs of points (P_i, P_j) separately.

First, we look at the $\binom{n-k}{2}$ fixed pairs (P_i, P_j). We show that every such pair will generate only $O(\lambda_s(k))$ topological events (rather than $O(\lambda_s(n))$). This can be seen easily if we investigate the μ_x-functions defined above. Any fixed point $P_x \in S_f \setminus \{P_i, P_j\}$ leads to a constant μ_x function. From this it follows that

$$\mu_k(t) \quad := \quad \min_{P_x \in S' \setminus \{P_i, P_j\}} \mu_x(t)$$

$$= \quad \min \left\{ \min_{P_x \in S_m} \mu_x(t) \, , \, \mu_{min} \right\},$$

where μ_{min} is the minimum function of the constant functions μ_x with $P_x \in S_f \setminus \{P_i, P_j\}$. This proves that the function $\mu_k(t)$ has at most $O(\lambda_s(k + 1))$ pieces.

On the other hand, each of the remaining $\binom{n}{2} - \binom{n-k}{2} \in O(kn)$ moving pairs (P_i, P_j) can generate at most $O(\lambda_s(n))$ topological events – as we have seen above.

Now we add both cases and obtain a maximum number of $O(k n \lambda_s(n) + (n - k)^2 \lambda_s(k))$ moments of cocircularity of neighboring points. Also, it is easy to see that the number of extended topological events is bounded by $O(k n^2)$. In summary, we state the following generalization of the last theorem.

Theorem 4   Given a finite set $S(t)$ of k continuous curves and $n - k$ fixed points under the assumptions (A), (B) and (C). Then, during the entire flow of the points, there appear at most $O(k\,n\,\lambda_s(n) + (n - k)^2\,\lambda_s(k))$ topological events. $\Omega(k\,n)$ is a lower bound on the worst-case number of topological events.

To prove the lower bound claim, we present a configuration of $k + n \in \Theta(n)$ points where $\Omega(k\,n)$ topological events appear. We fix the set S_f in such a way that all circumcircles defined by triples in $DT(S_f)$ have a nonempty intersection (see [HaDe 59]). We construct the set of k curves S_m of the remaining points in the following way: we let one point after the other cross the nonempty intersection. If we leave sufficient time between these crossings, the topological sub-structure $DT(S_f)$ is destroyed only by one point of each curve of S_m. Thus, each point in S_m generates $\Omega(n)$ topological events in $DT((S_f \cup S_m)')$. (This also shows the $\Omega(n^2)$ lower bound for the motion of all n points.)

Finally, we look at the special case of moving only one point $P_i \in S$, while leaving the other points fixed. By a simple argument we can improve the last theorem in this case. As already stated, topological events are characterized by SWAPs of adjacent pairs of triples, in which the original pair disappears and the two dual triples appear. Now, if we move only one point of S, while the others remain fixed, we note that one of the four triples contains points of $S' \setminus \{P_i\}$ exclusively. Since the points of this set are fixed, topological events can only occur when the moving point P_i passes a (possibly degenerate) circumcircle defined by triples in $DT(S' \setminus \{P_i\})$. Since there are only a linear number of triples in $DT(S' \setminus \{P_i\})$, and since the collision-free curve of P_i can cross each circumcircle at most s times, we obtain:

Theorem 5   Given a finite set $S(t)$ of one continuous curve and $n - 1$ fixed points under the assumptions (A), (B) and (C). Then, during the entire flow of the moving point, there appear at most $O(n)$ topological events.

Constant-speed straight-line trajectories:

In the special case that $S(t)$ represents a set of trajectories for points that move at constant speed along straight trajectories, we can specialize our results. Let the trajectories be given by $P_i(t) = \vec{a}_i t + \vec{b}_i$, where $\vec{a}_i, \vec{b}_i \in \mathbb{R}^2$ are (given) constant vectors, for every $i \in \{1, \ldots, n\}$. In this case (and also in the case of general polynomial curves) the previously defined functions $INCIRCLE(\ldots)$ and $CCW(\ldots)$ are also polynomials. By a simple investigation of the degree of these polynomials we see that assumption (C) is satisfied with $s = 4$. (Assumption (B) prevents these polynomials from degeneration.) In this case we know that $\lambda_4(n) = O(n\,2^{\alpha(n)})$, where $\alpha(n)$ is the extremely slowly-growing inverse of Ackermann's function (see [AgShSh 89]).

Remarks:

(1). If, in addition to the points moving along known trajectories $S(t)$, each point P_i has an associated (additive) weight $w_i(t)$, varying in time according to a bounded-degree polynomial, then our methods generalize to show (essentially) cubic upper bounds on the number of topological events in the resulting *weighted* Voronoi diagram defined by the Voronoi polygons

$$v(P_i) := \{x \in \mathbb{E}^2 \,|\, \forall_{j \neq i} \ d(x, P_i) + w_i \leq d(x, P_j) + w_j\}.$$

(2). B. Aronov has pointed out that, in the special case of points moving at unit speeds along straight-line trajectories, the degree of $INCIRCLE(\ldots)$ is 3 (rather than 4), so that the above argument leads to a bound of $O(n^3 \alpha(n))$.

(3). Using the familiar technique of "linearization" (see, e.g., [ImSuIm 89]), one can easily get a bound of $O(n^3)$ on the number of topological events in the case of points moving along straight-line trajectories: The Voronoi diagram corresponds to the lower envelope of a set of parabolic surfaces $(\min_i\{(x - x_i(t))^2 + (y - y_i(t))^2\}$, for moving points $(x_i(t), y_i(t)) = (a_i t + b_i, c_i t + d_i))$, which, when expanded (upon removing the common additive term of $x^2 + y^2$ from the minimization over i), can be written as the lower envelope of a set of n hyperplanes in a higher-dimensional space — namely, six-dimensional space in the new variables t, t^2, x, y, xt, yt. Such a lower envelope of hyperplanes has complexity $O(n^{\lfloor d/2 \rfloor}) = O(n^3)$ [Ed 87]. Note, however, that the linearization method leads to higher degree (super-cubic) polynomials in n when the algebraic degree of the trajectories $S(t)$ grows; in contrast, the Davenport-Schinzel arguments continue to show nearly-cubic bounds, for any fixed algebraic degree of $S(t)$.

(4). In the time since we originally derived our combinatorial result, several other researchers have independently arrived at the same or very similar result. In particular, [ImIm 90] and [FuLe 91] have obtained similar (nearly-cubic) bounds. Also, at least for the case of points moving along straight lines with identical speeds, it has been pointed out to us that P. Agarwal, B. Aronov, J. O'Rourke, and M. Sharir (and, likely, others) have independently arrived at cubic upper bounds. M. Sharir has also recently shown, using linearization, that, for some other special cases of moving points, the number of topological events is at most $O(n^2)$.

(5). The challenging open problem is to obtain subcubic bounds on the number of topological events, or to obtain super-quadratic lower bounds. See the discussion in the recent Computational Geometry Column [O'Ro 91].

5 Dynamic Scenes

The topological structure of a Voronoi diagram under continuous motions of the points in S can be maintained by the following algorithm:

Algorithm : **Preprocessing :**

1. Compute the topological structure $DT(S'(t_0))$ of the starting position.

2. For every existing quadrilateral in $DT(S'(t_0))$ calculate the potential topological events.

3. For the set of the potential topological events create an event queue (priority queue).

Iteration :

1. Determine the next topological event and decide whether it is a SWAP or a RETRIANGULATION.

2. Process the topological event and do an update of the event queue.

We look closer to the individual steps of the algorithm and their time and storage requirements. In the first preprocessing step, we compute the initial Delaunay triangulation $DT(S(t_0))$ and augment it with extended dual edges, obtaining $DT(S'(t_0))$ in time $O(n \log n)$ and space $O(n)$ (e.g., see [GuSt 85]).

In the second preprocessing step, we continue with a flow of the quadrilaterals in $DT(S'(t_0))$. For quadrilaterals $\{P_i, P_j, P_k, P_l\} \in DT(S'(t_0))$ we calculate the zeros of the function $INCIRCLE(P_i, P_j, P_k, P_l)$ and for those of the form $\{P_i, P_j, P_k, \infty\} \in DT(S'(t_0))$ we compute the zeros of the function $CCW(P_i, P_j, P_k)$. Under assumption (C), this step can be done in linear time and space.

In the third preprocessing step, we build up the event queue for the set of potential topological events. The topological events are stored in a priority queue according to their temporal appearance, with the corresponding quadrilateral stored with each event. Using assumption (C), this step and therefore the entire preprocessing step requires $O(n \log n)$ time and $O(n)$ space.

To determine the next topological event, we simply pop the event queue (time $O(\log n)$). Assuming that the number of cocircular and collinear points in the degenerate cases remains constant, then one can decide in constant time if the event is a SWAP or a RETRIANGULATION.

Now, each SWAP destroys only four quadrilaterals while creating only four new quadrilaterals. Thus, in order to update the event queue, all we have to do is to delete the four destroyed quadrilaterals and their corresponding topological events in the event queue and to insert the four new ones. (In the degenerate case, we may need to destroy and create several quadrilaterals.) Thus, we spend amortized time $O(\log n)$ per event (which [Ro 91] shows is worst-case optimal under linear motions of the points). In summary, we have:

Theorem 6 Given a finite set $S(t)$ of k continuous curves and $n - k$ fixed points under the assumptions (A), (B) and (C). After preprocessing requiring (optimal) time $O(n \log n)$ (and space $O(n)$), we can maintain the topological structure in time $O(\log n)$ per event (which is worst-case optimal).

6   Concluding Remarks and Open Problems

We have presented an algorithm for maintaining Voronoi diagrams of moving points over time.

Our upper bound on the number of events is nearly cubic, based on a Davenport-Schinzel argument. The lower bound examples we can exhibit show only $\Omega(n^2)$ events. The major open question remaining is to determine tighter bounds on the number of events. This question is one of a large class of questions relating to the complexity of the lower envelope of surfaces in three dimensions — a class of questions that remain largely unanswered.

The algorithm presented here has been implemented on a SUN workstation, using [SuIr 89] methods for numerically stable evaluation of the function $INCIRCLE(\dddot{.}.)$. Extensive tests suggest that the number of topological events grows like $\Theta(n\sqrt{n})$ in the *average case* under linear motions chosen at random.

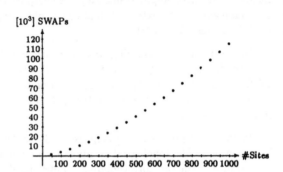

Acknowledgement

Thomas Roos wishes to thank Professor Hartmut Noltemeier for his helpful comments and Gerhard Albers for the nice implementation. J. Mitchell thanks the DEC Systems Research Center for its generous support during his visit (1988) to work with L. Guibas on the research discussed here.

References

[AbMü 88]   S. Abramowski and H. Müller, *Collision Avoidance for Nonrigid Objects*, in H. Noltemeier (ed.): ZOR - Zeitschrift für Operations Research, Vol. 32, 1988, pp 165 - 186

[Ag 87]   A. Aggarwal, L. Guibas, J. Saxe and P. Shor, *A Linear Time Algorithm for Computing the Voronoi Diagram of a Convex Polygon*, Proc. of the 19th Annual ACM Symposium on Theory of Computing, New York City, 1987, pp 39 - 45

[AgShSh 89]   A. Aggarwal, M. Sharir and P. Shor, *Sharp Upper and Lower Bounds on the Length of General Davenport-Schinzel Sequences*, Journal of Combinatorial Theory, Series A, Vol. 52, 1989, pp 228 - 274

[At 85]   M.J. Atallah, *Some Dynamic Computational Geometry Problems*, Computers and Mathematics with Applications, Vol. 11, 1985, pp 1171 - 1181

[AuImTo 90] H. Aunuma, H. Imai, K. Imai and T. Tokuyama, *Maximin Locations of Convex Objects and Related Dynamic Voronoi Diagrams*, Proc. of the 6th ACM Symposium on Computational Geometry, Berkeley, 1990, pp 225 - 234

[Au 88] F. Aurenhammer, *Voronoi Diagrams – A Survey*, Report 263, Nov. 1988, Institute für Informationsverarbeitung, Techn. Univ. Graz

[ChEd 87] B. Chazelle and H. Edelsbrunner, *An Improved Algorithm for Constructing kth-Order Voronoi Diagrams*, IEEE Transactions on Computers, Vol. C-36, Nov. 1987, No. 11, pp 1349 - 1354

[DrLe 78] R.L. Drysdale III and D.T. Lee, *Generalized Voronoi Diagrams in the Plane*, Proc. 16th Annual Allerton Conference on Communications, Control and Computing, Oct. 1978, pp 833 – 842

[Ed 86] H. Edelsbrunner, *Edge-Skeletons in Arrangements with Applications*, Algorithmica 1986, Vol. 1, pp 93 - 109

[Ed 87] H. Edelsbrunner, *Algorithms in Combinatorial Geometry*, EATCS Monographs in Computer Science, Springer - Verlag, Berlin - Heidelberg, 1987

[EdO'RSe 86] H. Edelsbrunner, J. O'Rourke and R. Seidel, *Constructing Arrangements of Lines and Hyperplanes with Applications*, SIAM J. Comput., Vol. 15, No. 2, May 1986, pp 341 – 363

[Fo 86] S. Fortune, *A Sweep-line Algorithm for Voronoi Diagrams*, Proc. 2nd Annual ACM Symp. Computational Geometry, Yorktown Heights, 1986, pp 313 – 322

[FuLe 91] J-J. Fu and R.C.T. Lee, *Voronoi Diagrams of Moving Points in the Plane*, Int. Journal of Computational Geometry & Applications, Vol. 1, No. 1, 1991, pp 23 – 32

[ImIm 90] H. Imai and K. Imai, *Voronoi Diagrams of Moving Points*, Proc. Int. Computer Symp., Taiwan, 1990, pp 600 – 606

[Go 83] I.G. Gowda, D.G. Kirkpatrick, D.T. Lee and A. Naamad, *Dynamic Voronoi Diagrams*, IEEE Trans. on Information Theory, Vol. IT-29, No. 5, Sept. 1983, pp 724 - 731

[GuKnSh 90] L. Guibas, D.E. Knuth and M. Sharir, *Randomized Incremental Construction of Delaunay and Voronoi Diagrams*, Proc. 17th Intern. Colloquium on Automata, Languages and Programming ICALP 90, LNCS 443, Springer, 1990, pp 414 - 431

[GuSt 85] L. Guibas and J. Stolfi, *Primitives for the Manipulation of General Subdivisions and the Computation of Voronoi Diagrams*, ACM Transactions on Graphics, Vol. 4, No. 2, April 1985, pp 74 - 123

[HaDe 59] H. Hadwiger und H. Debrunner, *Kombinatorische Geometrie in der Ebene*, Monographies de L'Enseignement Mathématique, No. 2, Université Genève, 1959

[HaSh 86] S. Hart and M. Sharir, *Nonlinearity of Davenport-Schinzel Sequences and of Generalized Path Compression Schemes*, Combinatorica, 1986, Vol. 6, pp 151-177

[ImSuIm 89]   K. Imai, S. Sumino and H. Imai, *Geometric Fitting of Two Corresponding Sets of Points*, Proc. 5th ACM Symp. on Computational Geometry, 1989, pp 266 - 275

[Le 82]   D.T. Lee, *On k-Nearest Neighbor Voronoi Diagrams in the Plane*, IEEE Transactions on Computers, Vol. C-31, No. 6, June 1982, pp 478 - 487

[LeSh 87]   D. Leven and M. Sharir, *Planning a Purely Translational Motion for Convex Objects in Two-Dimensional Space Using Generalized Voronoi Diagrams*, Discrete & Computational Geometry, 1987, Vol. 2, pp 9 – 31

[No 88]   H. Noltemeier, *Computational Geometry and its Applications*, Proceedings Workshop CG '88, Universität Würzburg, März 1988, LNCS 333, Springer Verlag, 1988

[O'Ro 91]   J. O'Rourke, *Computational Geometry Column 12*, SIGACT News, Vol. 22, No. 2, Spring 1991, pp 26 – 29

[PrSh 85]   F.P. Preparata and M.I. Shamos, *Computational Geometry – An Introduction*, Springer - Verlag, New York, 1985

[Ro 88]   T. Roos, *Voronoi Diagramme*, Diplomarbeit, Universität Würzburg, 1988

[Ro 89]   T. Roos, *k - Nearest - Neighbor Voronoi Diagrams for Sets of Convex Polygons, Line Segments and Points*, Proceedings 15th Intern. Workshop on Graph-Theoretic Concepts in Computer Science WG89, LNCS 411, Springer - Verlag, Berlin - Heidelberg - New York

[Ro 90]   T. Roos, *Voronoi Diagrams over Dynamic Scenes (Extended Abstract)*, Proceedings 2nd Canadian Conference on Computational Geometry, Ottawa, 1990

[Ro 91]   T. Roos, *Voronoi Diagrams over Dynamic Scenes*, to appear in Discrete Applied Mathematics, 1991

[RoNo 91]   T. Roos and H. Noltemeier, *Dynamic Voronoi Diagrams in Motion Planning*, to appear on 15th IFIP Conference on System Modeling and Optimization, Zurich, 1991

[ShHo 75]   M.I. Shamos and D. Hoey, *Closest - Point Problems*, Proc. 16 th Annual Symp. on FOCS, 1975, pp 151 – 162

[Sh 88]   M. Sharir, *Davenport-Schinzel Sequences and their Geometric Applications*, pp 253-278, NATO ASI Series, Vol. F40, Theoretical Foundations of Computer Graphics and CAD, R.A. Earnshaw (Ed.), Springer-Verlag Berlin Heidelberg, 1988

[SuIr 89]   K. Sugihara and M. Iri, *Construction of the Voronoi Diagram for One Million Generators in Single-Precision Arithmetic*, private communications, 1989, to appear in Proc. of IEEE

[Ya 87]   C.K. Yap, *An $O(n \log n)$ Algorithm for the Voronoi Diagram of a Set of Simple Curve Segments*, Discrete & Computational Geometry, 1987, Vol. 2, pp 365 – 393

Using Maximal Independent Sets to Solve Problems in Parallel

Takayoshi Shoudai

Department of Control Engineering and Science

Kyushu Institute of Technology, Iizuka 820, Japan

E-mail:shoudai@ces.kyutech.ac.jp

Satoru Miyano

Research Institute of Fundamental Information Science

Kyushu University 33, Fukuoka 812, Japan

E-mail:miyano@rifis.sci.kyushu-u.ac.jp

Abstract

By using an $O((\log n)^2)$ time EREW PRAM algorithm for a maximal independent set problem (MIS), we show the following two results: (1) Given a graph, the maximal vertex-induced subgraph satisfying a hereditary graph property π can be found in time $O(\Delta^{\lambda(\pi)}T_\pi(n)(\log n)^2)$ using a polynomial number of processors, where $\lambda(\pi)$ is the maximum of diameters of minimal graphs violating π and $T_\pi(n)$ is the time needed to decide whether a graph with n vertices satisfies π. (2) Given a family $C = \{c_1, \ldots, c_m\}$ of subsets of a finite set $S = \{1, \ldots, n\}$ with $S = \bigcup_{i=1}^m c_i$, a minimal set cover for S can be computed on an EREW PRAM in time $O(\alpha\beta(\log(n+m))^2)$ using a polynomial number of processors, where $\alpha = \max\{|c_i| \mid i = 1, \ldots, m\}$ and $\beta = \max\{|d_j| \mid j = 1, \ldots, n\}$.

1 Introduction

We show a way of employing the parallel algorithms for the maximal independent set problem (MIS) [3, 4, 5] to solve problems in which maximal or minimal solutions are searched. For the bounded degree maximal subgraph problems, we have constructed an NC algorithm by employing the NC algorithms for MIS [8]. This paper extends the technique developed in [8] and gives NC algorithms for two kinds of such problems.

The first problem is to find a maximal set of vertices which induces a subgraph satisfying a given graph property π. The other is the minimal set cover problem that is, given a collection $C = \{c_1, ..., c_m\}$ with $c_i \subset S = \{1, ..., n\}$, to find a collection $C' \subseteq C$ such that every element in S is contained in some $c \in C'$ but no proper subcollection $C'' \subset C'$ does not have this property.

These problems are easily solved in polynomial time by straightforward greedy sequential algorithms. However, these algorithms are hardly parallelizable since they are P-complete [7]: It is shown in [6] that the lexicographically first maximal subgraph problem for a given property π is P-complete if π is hereditary, nontrivial and polynomial-time testable. The same fact also holds for the greedy minimal set cover algorithm.

For the maximal subgraph problem, we need some restrictions on the property to solve the problem in NC. A graph property π is called *local* if the diameter of any minimal graph violating π is bounded by some constant. For such local property π, we consider the problem of finding a maximal vertex-induced subgraph which satisfies π and, simultaneously, whose maximum vertex degree is at most Δ, where Δ is a given constant. We prove that this problem can be solved in NC by using MIS if π is testable in NC.

For the minimal set cover problem, we also show an algorithm which employs an MIS algorithm. This algorithm can be implemented on an EREW PRAM in time $O(\alpha\beta(\log(n + m)))^2)$ using a polynomial number of processors, where $\alpha = \max\{|c_i| \mid i = 1, ..., m\}$ and $\beta = \max\{|d_j| \mid j = 1, ..., n\}$ with $d_j = \{c_i \mid j \in c_i\}$. This implies that if $\alpha\beta = O((\log(n+m))^k)$ then the problem is solvable in NC.

The algorithms for these problems are described by a scheme which applies MIS repeatedly. Thus we do not directly deal with parallelization of the problems. Our concern is how to employ an MIS algorithm to solve problems in parallel.

2 Maximal subgraph problem for a local property

Let $G = (V, E)$ be a graph. For a subset U of vertices, the *induced subgraph of U* is the graph defined by $G[U] = (U, E[U])$, where $E[U]$ consists of edges whose endpoints are both in U.

Let π be a property on graphs. We say that a graph $G = (V, E)$ is a *minimal graph violating π* with respect to vertices if G violates π and the vertex-induced subgraph $G[U]$ of U satisfies π for every subset U of V with $U \neq V$. The property π is called *local* with respect to vertices if $\lambda(\pi) = \sup\{\text{diameter}(G) \mid G \text{ is a minimal graph violating } \pi \text{ with respect to vertices}\}$ is finite.

Remark 1 A minimal graph violating a property π with respect to vertices must be connected if π is local.

A property π on graphs is called *hereditary* with respect to vertices if for every graph $G = (V, E)$ satisfying π, the vertex-induced subgraph $G[U]$ also satisfies π for every subset $U \subseteq V$.

Theorem 1 Let π be a graph property which is local and hereditary with respect to vertices. Then a maximal subgraph of a graph $G = (V, E)$ which satisfies π and whose maximum degree is at most Δ can be computed on an EREW PRAM in time $O(\Delta^{\lambda(\pi)}T_\pi(n)(\log n)^2)$ using a polynomial number of processors, where $T_\pi(n)$ is the time needed to decide whether a graph with n vertices satisfies π.

Proof. For subsets W and U of vertices with $W \cap U = \emptyset$, let $E_U^W = \{\{v, w\} \subseteq W \mid dist_{G[U \cup \{v,w\}]}(v, w) \leq \lambda(\pi) \text{ with } v \neq w\}$ and $N_U(w) = \{u \in U \mid dist_{G[U]}(u, w) \leq \lambda(\pi) - 1\}$, where $dist_G(\{v, w\})$ is the length of the shortest path between v and w in G. Then let $H_U^W = (W, E[W] \cup E_U^W)$. The required set U of vertices is computed together with a set W of vertices such that $W \cap U = \emptyset$. Initially let $W = V$ and $U = \emptyset$. At each iteration of the algorithm, a maximal independent set I of H_U^W is computed and added to U while vertices which induce a graph violating π or make the degree of some vertex greater than Δ are deleted from W together with I. This is iterated $\Delta^{\lambda(\pi)}$ times. Formally the algorithm is described as follows:

```
 1  begin /* G = (V, E) is an input */
 2      W ← V; U ← ∅;
 3      while W ≠ ∅ do
 4          begin
 5              Find a maximal independent set I of H_U^W;
 6              U ← U ∪ I;
 7              W ← W - I;
 8              W ← W - {w ∈ W | G[U ∪ {w}] violates π or deg(G[U ∪ {w}]) > Δ}
 9          end
10  end
```

E_U^W consists of unordered pairs of vertices such that if the vertices are added to U at the same iteration the induced subgraph of U may violate the property. We show that this algorithm computes a maximal subset U whose induced subgraph satisfies π and maximum degree is at most Δ.

Let $W_0 = V$ and $U_0 = \emptyset$. Then the graph $H_{U_0}^{W_0}$ is the same as $G = (V, E)$. Therefore in the first iteration, a maximal independent set of G is computed at line 5. For $i = 1, ..., \Delta^{\lambda(\pi)}$, let U_i, I_i and W_i be the contents of variables U, I and W at the end of ith iteration, respectively.

Obviously, $W_i \cap U_i = \emptyset$ for $i = 0, ..., \Delta^{\lambda(\pi)}$. We assume that the induced subgraph $G[U_{i-1}]$ satisfies π and the maximum degree of $G[U_{i-1}]$ is at most Δ.

Let $\{w, u\}$ be an edge in E with $w \in W_i$ and $u \in U_i$. Line 8 deletes every vertex which is adjacent to more than Δ vertices in U_i or adjacent to a vertex v in U_i with $deg_{G[U_i]}(v) = \Delta$. Therefore u is adjacent to at most Δ vertices in U_i and $deg_{G[U_i \cup \{w\}]}(u) \leq \Delta$. Moreover, $|N_{U_i}(w)| \leq \Delta^{\lambda(\pi)-1}$. Hence, for each w in W_i, we see that

$$A_i(w) = \sum_{u \in N_{U_i}(w)} deg_{G[U_i \cup \{w\}]}(u) \leq \Delta^{\lambda(\pi)}.$$

To show that W becomes empty within $\Delta^{\lambda(\pi)}$ iterations of the while-loop, it suffices to prove that

$$A_i(w) > A_{i-1}(w)$$

for each w in W_i. Since w is not in the maximal independent set I_i of $H_{U_{i-1}}^{W_{i-1}}$ computed by line 5, w is adjacent to a vertex v in $I_i \subseteq W_{i-1}$ via an edge $\{w, v\}$ in $E[W_{i-1}]$ or $E_{U_{i-1}}^{W_{i-1}}$.

Case 1. $\{w, v\} \in E[W_{i-1}]$: Then $\{w, v\}$ is an edge in $G[U_i \cup \{w\}]$. Hence $deg_{G[U_i \cup \{w\}]}(v) \geq 1$. Since $v \in N_{U_i}(w)$ and $v \notin N_{U_{i-1}}(w)$, we see that $A_i(w) \geq A_{i-1}(w) + deg_{G[U_i \cup \{w\}]}(v) > A_{i-1}(w)$.

Case 2. $\{w, v\} \in E_{U_{i-1}}^{W_{i-1}}$: Then there is a path $w, u_1, ..., u_{k-1}, v$ with $k \leq \lambda(\pi)$ and $u_j \in U_{i-1}$ $(j = 1, ..., k-1)$ in $G[U_{i-1} \cup \{w, v\}]$. Since $v \in W_{i-1}$, $W_{i-1} \cap U_{i-1} = \emptyset$ and $w \neq v$, we see $v \notin U_{i-1} \cup \{w\}$. Hence $\{v, u_{k-1}\}$ is not an edge in $G[U_{i-1} \cup \{w\}]$. On the other hand, v is in U_i and u_{k-1} is in $U_{i-1} \subseteq U_i$. Hence $\{v, u_{k-1}\}$ is an edge in $G[U_i \cup \{w\}]$. Therefore $deg_{G[U_i \cup \{w\}]}(u_{k-1}) > deg_{G[U_{i-1} \cup \{w\}]}(u_{k-1})$. Since $u_{k-1} \in N_{U_{i-1}}(w) \subset N_{U_i}(w)$, we see that $A_i(w) > A_{i-1}(w)$.

We now show that $deg(G[U_i]) \leq \Delta$ and $G[U_i]$ satisfies π.

Claim 1. $deg(G[U_i]) \leq \Delta$.

Proof. For a vertex u in U_{i-1}, if u is adjacent to a vertex w in I_i via an edge in E, then no other vertex in I_i is adjacent to u since I_i is also an independent set with respect to $E_{U_{i-1}}^{W_{i-1}}$. Therefore the degree of u in $G[U_{i-1} \cup I_i]$ remains to be at most Δ since $deg(G[U_{i-1} \cup \{w\}]) \leq \Delta$ by the algorithm. For a vertex u in I_i, $deg_{G[U_{i-1} \cup I_i]}(u)$ is at most Δ since u is adjacent to at most k vertices in U_{i-1} and since I_i is an independent set with respect to $E[W_{i-1}]$. Hence $deg_{G[U_{i-1} \cup I_i]}(u) \leq \Delta$.

Claim 2. $G[U_i]$ satisfies π.

Proof. We assume that $G[U_i]$ does not satisfy π. Then, there is a minimal subset $S \subseteq U_i$ such that $G[S]$ violates π. Since $S \subseteq U_i$ and $U_i = U_{i-1} \cup I_i$, we see that $S = (S \cap U_{i-1}) \cup (S \cap I_i)$. The set $S \cap I_i$ contains at least two vertices since if $S \cap I_i$ consists of only one vertex then line 8 deletes the vertex at the last iteration. Therefore there are two distinct vertices v, w such that $\{v, w\} \in E$ or there are at most $\lambda(\pi) - 1$ vertices in $S \cap U_{i-1}$ which construct a path between v and w since $\text{diameter}(G[S]) \leq \lambda(\pi)$. For each case, $\{v, w\}$ are in $E[W_{i-1}]$ or $E_{U_{i-1}}^{W_{i-1}}$ since $v, w \in I_i \subset W_{i-1}$. It contradicts the fact that $v, w \in S \cap I_i \subset I_i$ and I_i is a maximal independent set with respect to $E[W_i] \cup E_{U_{i-1}}^{W_{i-1}}$. Hence $G[U_i]$ satisfies π.

Since only vertices which violate the property π or the condition of maximum degree Δ are deleted from W and since π is hereditary, the resulting set U is a maximal subset which induces a subgraph satisfying π when W becomes empty.

MIS can be solved on an EREW PRAM in $O((\log n)^2)$ time using a polynomial number of processors [5]. It is not hard to see that the steps other than MIS can also be implemented on an EREW PRAM in $O((\log n)^2)$ time using a polynomial number of processors. Hence the total algorithm can be implemented using the same amount of time and processors. $\square$

Remark 2 At line 8 of the algorithm, for each $w \in W$, it is sufficient to decide whether $G[N_U(w) \cup \{w\}]$ satisfies π and $deg(G[N_U(w) \cup \{w\}]) \leq \Delta$. Therefore, the time needed to compute line 8 depends only on Δ and $\lambda(\pi)$.

Finding a maximal subgraph of maximum degree k takes $O(k^2 (\log n)^2)$ time using a polynomial number of processors [8]. This is a special case of Theorem 1 for $\pi =$ "maximum degree k", $\lambda(\pi) = 2$ and $\Delta = k$. For a graph of maximum degree Δ and $\pi =$ "k cycle free", it takes $O(\Delta^{\lfloor k/2 \rfloor} (\log n)^2)$ time to find a maximal subgraph satisfying π of maximum degree Δ since $\lambda(\pi) = \lfloor k/2 \rfloor$.

3   Solving the minimal set cover problem using MIS

Let $C = \{c_1, ..., c_m\}$ be a family of subsets of a finite set $S = \{1, ..., n\}$. A subset S' of S is called a *hitting set* for C if $c_i \cap S' \neq \emptyset$ for all $i = 1, ..., m$. A subset S'' of S is called a *co-hitting set* if $c_i \not\subseteq S'$ for all $i = 1, ..., m$. We say that C is a *set cover* if $\bigcup_{i=1}^n c_i = S$.

It should be noticed that S' is a hitting set for C and only if $S - S'$ is a co-hitting set for C. Therefore, S' is a minimal hitting set for C if and only if $S - S'$ is a maximal co-hitting set for C.

The problem of finding a hitting set is closely related to the set cover problem. For a family $C = \{c_1, ..., c_m\}$ with $\bigcup_{i=1}^n c_i = \{1, ..., n\}$, let

$$d_j = \{c_i \mid j \in c_i \in C\}$$

for $j = 1, ..., n$. Then each d_j is not empty. Let $D = \{d_1, ..., d_n\}$ and $C' \subseteq C$ be a minimal hitting set for D. Then $d_j \cap C' \neq \emptyset$ for each $j = 1, ..., n$. Therefore there is some $c_i \in d_j \cap C'$. Thus $j \in c_i$. Hence C' is a set cover of $\{1, ..., n\}$ and also can be seen that C' is minimal.

Theorem 2 Let $C = \{c_1, ..., c_m\}$ be a family of distinct subsets of a finite set $S = \{1, ..., n\}$. Let $\alpha = \max\{|c_i| \mid i = 1, ..., m\}$ and $\beta = \max\{|d_j| \mid j = 1, ..., n\}$, where $d_j = \{c_i \mid j \in c_i\}$. Then a minimal hitting set for C can be computed on an EREW PRAM in time $O(\alpha\beta(\log(n + m))^2)$ using a polynomial number of processors with respect to n and m.

Hence, if $\alpha\beta = O((\log(n + m))^k)$, then a minimal hitting set can be computed in NC.

Proof. We consider the following algorithm that finds a maximal co-hitting set for C_0:

```
/* A family C_0 = {c_1, ..., c_m} with c_i ⊆ S_0 = {1, ..., n} for i = 1, ..., m is given. */
/* We assume that S_0 = ⋃_{c∈C_0} c and |c_i| ≥ 2 for i = 1, ..., m. */
 1  begin
 2      S ← S_0; C ← C_0;
 3      W ← ∅; /* W gets a maximal co-hitting set */
 4      while S ≠ ∅ do
 5          begin
 6              E ← ∅;
 7              par c ∈ C do
 8                  begin
 9                      Choose two distinct vertices v, w from c ∩ S;
10                      Add the edge {v, w} to E
11                  end;
12              Find a maximal independent set I of the graph G = (S, E);
13              W ← W ∪ I;
14              S ← S − I;
15              U ← {u ∈ S | c ∩ S ⊆ W ∪ {u} for some c ∈ C};
16              par c ∈ C do if c ∩ U ≠ ∅ then delete c from C;
17              S ← S − U;
18              V ← S − ⋃_{c∈C} c
19              W ← W ∪ V
20              S ← S − V;
21          end
22  end
```

The variable W gets a maximal co-hitting set. Let I_i, C_i, U_i, W_i and S_i be the contents of the variables I, C, U, W and S just after the ith iteration of the while-loop, respectively. For convenience, let $W_0 = \emptyset$ and $U_0 = \emptyset$. Let $U_i^* = U_0 \cup \cdots \cup U_i$. We also let E_i be the set of edges constructed during lines 7-11. Then from the algorithm we can easily see that S_0, S_{i-1} and W_i are represented as the following disjoint unions (Figure 1):

(1) $S_i \cup W_i \cup U_i^* = S_0$.

(2) $S_{i-1} = I_i \cup U_i \cup V_i \cup S_i$.

(3) $W_i = W_{i-1} \cup I_i \cup V_i$.

Claim 1. For $c \in C_i$, $c \cap S_i$ contains at least two elements.

Proof. By the assumption on the input, Claim 1 obviously holds for $i = 0$. Assume that the claim holds for i and $S_{i+1} \neq \emptyset$. Let c be in C_i. Then $c \cap U_i = \emptyset$ from line 16 and $c \cap V_i = \emptyset$ from line 18. Therefore from (2) we see that $c \cap S_i = c \cap (S_{i-1} - I_i)$. If $c \cap S_i = \emptyset$, then $U_i = S_{i-1} - I_i$ from line 15. This yields $S_i = \emptyset$ from line 17. This is a contradiction since S_i is assumed not empty. On the other hand, if $c \cap S_i = \{u\}$, then $c \cap S_i \subseteq W_{i-1} \cup I_i \cup \{u\}$. This means that u is in U_i and, therefore, $c \cap U_i \neq \emptyset$, a contradiction. Thus $|c \cap S_i| \geq 2$.

Claim 2. W_i is a co-hitting set for C_0.

Proof. We assume that $S_{i-1} \neq \emptyset$. Obviously, $W_0 = \emptyset$ is a co-hitting set for C_0. Assume that W_{i-1} is a co-hitting set for C_0. Let c be in C_0.

Case 1. $c \notin C_i$: c was deleted during the jth iteration for some $1 \leq j \leq i$. Then $c \cap U_j \neq \emptyset$. Hence there is u in $c \cap U_j \subseteq U_i^*$. By (1) u is not in W_i. Therefore we have $c \not\subseteq W_i$.

Case 2. $c \in C_i$: c is also in C_{i-1}. Then by Claim 1 there are v, w in $c \cap S_{i-1}$ with $v \neq w$ and $\{v, w\} \in E_i$. Since I_i is an independent set, $v \notin I_i$ or $w \notin I_i$. Since W_{i-1} is a co-hitting set for C_0, we have $c \not\subseteq W_{i-1}$. Since no element in S_{i-1}, hence no element in I_i, is in W_{i-1}, v or w is not in $W_{i-1} \cup I_i$. Therefore $c \not\subseteq W_{i-1} \cup I_i$. On the other hand, $c \cap V_i = \emptyset$ by line 18. Therefore $c \not\subseteq W_{i-1} \cup I_i \cup V_i = W_i$.

Claim 3. For any $u \in U_i$, there is $c \in C_{i-1}$ such that $c \subseteq W_i$.

Proof. By line 15, for $u \in U_i$ there is $c \in C_{i-1}$ such that $c \cap (S_{i-1} - I_i) \subseteq W_{i-1} \cup I_i \cup \{u\}$. Then $c \cap S_{i-1} \subseteq W_{i-1} \cup I_i \cup \{u\}$. Note that for $c \in C_{i-1}$ we have $c \cap U_{i-1}^* = \emptyset$ by line 16. Then

$$
\begin{aligned}
c &= c \cap (S_{i-1} \cup W_{i-1} \cup U_{i-1}^*) && \text{(by (1))} \\
&= (c \cap S_{i-1}) \cup (c \cap W_{i-1}) \cup (c \cap U_{i-1}^*) \\
&\subseteq W_{i-1} \cup I_i \cup \{u\}. && \text{(by } c \cap U_{i-1}^* = \emptyset)
\end{aligned}
$$

Let t be the integer such that $S_t = \emptyset$. Then by (1) $S_0 = W_t \cup U_t^*$. From Claim 2 W_t is a co-hitting set. Claim 3 asserts that for any $u \in U_t^*$ there is some c with $c \subseteq W_t \cup \{u\}$. Therefore W_t is a maximal co-hitting set for C_0.

Claim 4. $t \leq \alpha\beta$.

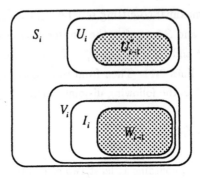

Figure 1: Relation between S_i, I_i, V_i, W_{i-1} and U_i^*

Proof. For $u \in S_i$, we define

$$B_i(u) = \{v \mid u \neq v \text{ and } \{u, v\} \subseteq c \cap S_i \text{ for some } c \in C_i\}.$$

It is easy to see that $|B_i(u)| \leq \alpha\beta$. Then it suffices to show that

$$|B_i(u)| < |B_{i-1}(u)|$$

for each $u \in S_i$. If $u \in S_i$, then u is not in I_i from line 14. Since I_i is a maximal independent set, there is v with $\{u, v\} \in E_i$. Therefore $\{u, v\} \subseteq c \cap S_{i-1}$ for some $c \in C_{i-1}$. Hence v is in $B_{i-1}(u)$. However, v is not in S_i since v is in I_i. Therefore v is not in $B_i(u)$.

As in the proof of Theorem 1, the part of finding a maximal independent set can be implemented on an EREW PRAM in $O((\log(n + m))^2))$ time using polynomially many processors with respect to n and m. The other steps can also be implemented with at most the same amount of time and processors.□

The following corollary is obtained in a straightforward way from Theorem 2:

Corollary 1 Let $C = \{c_1, ..., c_m\}$ be a family of subsets of a finite set $S = \{1, ..., n\}$ such that $S = \bigcup_{i=1}^{m} c_i$. Let $\alpha = \max\{|c_i| \mid i = 1, ..., m\}$ and $\beta = \max\{|d_j| \mid j = 1, ..., n\}$, where $d_j = \{c_i \mid j \in c_i\}$. Then a minimal set cover for S can be computed on an EREW PRAM in time $O(\alpha\beta(\log(n + m))^2)$ using a polynomial number of processors with respect to n and m.

Hence, if $\alpha\beta = O((\log(n + m))^k)$, then a minimal set cover can be computed in NC.

Remark 3 An NC approximation algorithm for the set cover problem is shown in [1]. But it should be noted here that their algorithm does not produce a minimal set cover.

4 Conclusion

We have shown that parallel MIS algorithms are useful to solve the minimal set cover problem and the maximal subgraph problem for a property "local and of degree at most Δ". However, the idea of using MIS does not seem to work for other properties, for example, "acyclic", "planar", which are not local. MIS locates at an interesting position in the NC hierarchy. It is in NC^2 but unlikely to belong to classes such as AC^1 and DET shown in [2]. It is not difficult to see that the algorithms shown in this paper can be transformed to NC^1-reductions to MIS. Hence the results in this paper give some new problems NC^1-reducible to MIS.

References

[1] B. Berger, J. Rompel and P.W. Shor, Efficient NC algorithm for set cover with application to learning and geometry, *Proc. 30th IEEE FOCS* (1989) 54-59.

[2] S.A. Cook, Taxonomy of problems with fast parallel algorithms, *Inform. Contr.* **64** (1985) 2-22.

[3] M. Goldberg and T. Spencer, A new parallel algorithm for the maximal independent set problem, *SIAM J. comput.* **18** (1989) 419-427.

[4] R.M. Karp and A. Wigderson, A fast parallel algorithm for the maximal independent set problem, *J. Assoc. Comput. Mach.* **32** (1985) 762-773.

[5] M. Luby, A simple parallel algorithm for the maximal independent set problem, *Proc. 17th ACM STOC* (1985) 1-10.

[6] S. Miyano, The lexicographically first maximal subgraph problems: P-completeness and NC algorithms, *Math. Systems Theory* **22** (1989) 47-73.

[7] S. Miyano, S. Shiraishi and T. Shoudai, A list of P-complete problems, RIFIS-TR-CS-17, Research Institute of Fundamental Information Science, Kyushu University, 1989 (revised in December, 1990).

[8] T. Shoudai and S. Miyano, Bounded degree maximal subgraph problems are in NC, *Proc. Toyohashi Symposium on Theoretical Computer Science* (1990) 97-101.

Fast Parallel Algorithms for Coloring
Random Graphs[*]

Zvi M. Kedem [1]   *Krishna V. Palem* [2]   *Grammati E. Pantziou* [3]
Paul G. Spirakis [1,3]   *Christos D. Zaroliagis* [3]

(1) Dept of Computer Science, Courant Institute of Math. Sciences
NYU, 251 Mercer St, New York, NY 10012, USA

(2) IBM Research Division, T.J. Watson Research Center,
P.O. Box 704, Yorktown Heights, NY 10598, USA

(3) Computer Technology Institute, P.O. Box 1122, 26110 Patras, Greece
Computer Sc and Eng Dept, University of Patras, Greece

Abstract

We improve here the *expected* performance of parallel algorithms for graph coloring. This is achieved through new *adaptive* techniques that may be useful for the average-case analysis of many graph algorithms. We apply our techniques to:
(a) the class $G_{n,p}$ of random graphs. We present a parallel algorithm which colors the graph with a number of colors at most twice its chromatic number and runs in time $O(\log^4 n/\log\log n)$ almost surely, for $p = \Omega((\log^{(3)} n)^2/\log^{(2)} n)$. The number of processors used is $O(m)$ where m is the number of edges of the graph.
(b) the class of all k-colorable graphs, uniformly chosen. We present a parallel algorithm which actually *constructs* the coloring in *expected* parallel time $O(\log^2 n)$, for constant k, by using $O(m)$ processors on the average. This problem is not known to have a polynomial time algorithm in the worst case.

1 Introduction

We examine here the *average-case parallel* complexity of graph coloring problems (which are known to be NP-hard in the worst case). By average case we mean that the inputs are selected randomly from some natural family of distributions parametrized by problem size. We consider two families of random graphs: (a) The class $G_{n,p}$ of graphs of n vertices where each edge may be independently present with probability p (see [6]). For this class of graphs we modify an algorithm of [4] and do a tight analysis, through which we show that our algorithm colors the input with a number of colors at most twice its chromatic number, in parallel time $O(\log^4 n/\log\log n)$ by using $O(n^2 p)$ processors for $p = \Omega((\log^{(3)} n)^2/\log^{(2)} n)$, on a CRCW PRAM, with probability at least $1 - n^{-d}$ $(d > 1)$. The algorithm of [4] finds the same number of colors, uses the same number of processors on a CRCW PRAM and runs in $O(\log^5 n/\log\log n)$ time for $p = \Omega(\log n/n)$, with probability at least $1 - O(1/\log n)$. If $p < O(\log n/n)$ both algorithms have the same performance on a CRCW PRAM.
(b) The class of all k-colorable graphs, for k constant and edge probability $p = \Omega(\sqrt{\log n/n})$, where each graph is uniformly chosen. Here the instance is known to have the property we are seeking and

---

[*]This work was partially supported by the EEC ESPRIT Basic Research Action No. 3075 (ALCOM), by the Ministry of Industry, Energy and Technology of Greece and by the NSF grant CCR-89-6949.

our task is to exhibit a proof of this. From the worst-case point of view such problems are just as hard as those in which we do not know whether the instance has the property. Dyer and Frieze in [5] showed how to color such graphs in expected *polynomial* time for constant k, p. Here we show how to color such graphs with k colors (thus demonstrating that they are k-colorable) in *expected parallel* time $O(\log^2 n)$ and by using an *expected* number of $O(m)$ processors, where m is the number of edges in the graph.

Both our results are achieved through interesting and quite general techniques for analyzing the expected performance of parallel algorithms (see also [7, 10]). We call them *adaptive* techniques. The analysis stems from the fact that, up to now, such analyses are quite ad-hoc and complicated. The uniform randomness of the input is quickly *conditioned* by the algorithm's progress. Thus, at later steps the statistical properties of the data structure depend on the computation history. However, two crucial facts may act on the algorithm's benefit:

1. In each parallel step, *many* processors may act *independently*. Thus, a lower bound on the probability of satisfactory progress may be calculated.

2. The algorithm can tune its parameters *adaptively* to allow for competing events to be tolerated.

The analysis of the algorithm can also be an "adaptive" one, where we may wish not to always prove that a magnitude stays "close to the mean", just to leave room for competing events to happen.

Another interesting remark is the fact that parallelizing the sequential *greedy* technique actually works in random graphs! This is evident in both of our results. For example, Kucera in [8] used a sequential "greedy minimum degree" coloring method to achieve average polynomial time for coloring k-colorable graphs. Although a *block-execution* of such a greedy method (where many sequential steps are now done in one parallel step) *does not* produce the same graph at its end, however the statistical properties of the produced structure seem to be quite similar to that of the sequential technique.

2 The Techniques

2.1 General Remarks

An algorithm A, working on a random input I, may go through a sequence of *phases* $\Phi_1, \Phi_2, \ldots$. During each phase Φ_i, the progress of the algorithm can be (in many cases) measured by the number S_i of "successes" of a number N_i of independent trials $(t_1^i, t_2^i, \ldots, t_{N_i}^i)$, each succeeding with corresponding probability $p_1^i, \ldots, p_{N_i}^i$. These elementary experiments may alter a crucial object which characterizes the algorithm's progress. Let us call it *remaining work* and assume it to be W_i at the beginning of phase Φ_i.

Such a scenario as above appears (for example) in parallel algorithms working on random graphs, or probabilistic parallel algorithms with unreliable PRAMs (or gates), etc. There, even each elementary parallel *step* can be considered as a phase since all the processors may (independently) attempt to do something (i.e. trials) at each step.

2.2 What happens in each phase

In order to calculate the statistics of the sequence $W_0, W_1, W_2, \ldots$ and the phase number z for which W_z becomes small (e.g. 0 or less than $K(|I|)$, e.g. $\log|I|$) we must first estimate the progress in each phase. The following may be then used, possibly in combination:

(a) Chernoff bounds

Let us first assume that $p_1^i = \cdots = p_{N_i}^i = p(i)$. Then from [2] we have for any $\beta(i) : 0 < \beta(i) < 1$,

$$\Pr[S_i > (1 + \beta(i))N_i p(i)] \leq \exp\left(-\frac{\beta^2(i)}{2}N_i p(i)\right)$$

and also,

$$\Pr[S_i < (1 - \beta(i))N_i p(i)] \leq \exp\left(-\frac{\beta^2(i)}{2} N_i p(i)\right)$$

Let us remark here that *we have freedom in choosing a suitable* $\beta(i)$. For example, a $\beta(i) = 1/\sqrt{N_i}$ may still provide an excellent probability of "staying close to the mean $N_i p(i)$" and/or in other cases this may be achieved by a constant $\beta(i) = \beta$ (independent of the phase). Bollobas (in his book [1]) and also Littlewood in [9] provide *even tighter* bounds for the Bernoulli tails. But the above will be enough for our purposes.

(b) Large Deviation Theorems

The number S_i of successes can be measured by an "indicator" Y_j^i such that $Y_j^i = 1$ iff $t_j^i = success$ else $Y_j^i = 0$. Then if $p(i) = (p_1^i + \cdots + p_{N_i}^i)/N_i$ we have from [7] for any $a > 0$

$$\Pr[S_i > a + N_i p(i)] < \exp(-2a^2/N_i)$$

Note that this is a very "tight" bound in the sense that the right-hand member goes to zero as soon as $a = \Omega(\sqrt{N_i})$.

Clearly, the larger the S_i the less W_{i+1} is going to be (in fact, in many applications, $W_{i+1} \leq W_i - S_i$).

2.3 How many phases will finish the work

After we have calculated the progress of phase Φ_i, we can argue about the relation of W_{i+1} and W_i.

(a) Generalized Martingales

Assume here that we are only able to prove that $E[W_{i+1} \mid W_i] \leq g_1^{(i)} W_i$ (i.e. the conditional on the size of W_i mean value of W_{i+1} is less than or equal to a fraction of W_i), where $0 < g_1^{(i)} < 1$. From Markov's inequality we have

$$\Pr[W_{i+1} \geq a] \leq \frac{E[W_{i+1}]}{a}$$

thus, conditioning on W_i's size we have

$$\Pr[W_{i+1} < a \mid W_i] \geq 1 - \frac{E[W_{i+1} \mid W_i]}{a} \geq 1 - \frac{g_1^{(i)} W_i}{a}$$

Hence, by picking $a = g_2^{(i)} W_i$ where $1 > g_2^{(i)} > g_1^{(i)}$, we get

Lemma 1 *In a generalized martingale* $\Pr[W_{i+1} < g_2^{(i)} W_i] \geq 1 - g_1^{(i)}/g_2^{(i)}$ *where* $0 < g_1^{(i)} < g_2^{(i)} < 1$.

Let $g_3^{(i)} = 1 - g_1^{(i)}/g_2^{(i)}$. Notice that $0 < g_3^{(i)} < 1$. Assume that somehow we are able to prove that $\exists g_3 : g_3^{(i)} > g_3 > 0, \forall i$, and also $\exists g_2 : g_2^{(i)} < g_2 < 1, \forall i$. Then, after an *expected number* of at most $1/g_3$ phases, the remaining work will drop down by a factor, i.e. the event $W_{nextphase} < g_2 W_{currentphase}$ will indeed happen (and in all cases $W_{i+1} < W_i$ anyway!). (To prove this, just consider the geometric process with success probability g_3.) But then we conclude that

Lemma 2 *With* g_2 *and* g_3 *as above, the expected number* $E[z]$ *of phases for* $W_z < 1$ *is bounded above by*

$$\frac{\log W_1}{g_3 \log(1/g_2)}$$

Note that lemma 2 implies a *logarithmic* expected number of phases! (All logarithms in this paper are to the base e.)

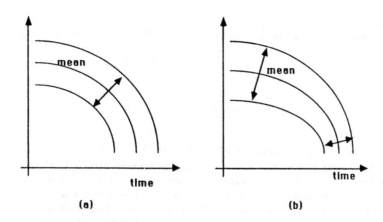

Figure 1: (a) Uniform narrow bounds. (Total probability may go to zero!) (b) Adaptive bounds, allow for competing events to be tolerated.

(b) Fast Progress with high probability

Instead of starting by bounds on the conditional mean of W_{i+1} we may be able to use the fact that S_i stays *close* to the mean progress (mean number of successes) in phase Φ_i. Again here we stress the remark that "how close to the mean" we want to be is a *parameter* and it can be turned *adaptively* to allow for competing events to be tolerated (see fig.1).

For the sake of the argument, assume here that $W_{i+1} \leq W_i - S_i$ and we have that "$S_i \geq (1 - \beta(i))N_ip(i)$" with probability at least $1 - \exp(-\beta^2(i)N_ip(i)/2)$. Let $\mathcal{E}_i$ denote the event "$S_i \geq (1 - \beta(i))N_ip(i)$". It is usually the case that $E[S_i] = N_ip(i)$ is also equal to some constant times W_i, i.e. $E[S_i] = c_iW_i$. Then, we get $W_{i+1} \leq W_i(1 - (1 - \beta(i))c_i)$ if $\mathcal{E}_i$ holds. We want $\mathcal{E}_i$ to hold continuously for all $i < z$ such that $W_z < \xi \log W_1$ for the first time. (Here $\xi > 1$ is a suitable constant.) But

$$W_z \leq W_1 \prod_{i=1}^{z-1}(1 - (1 - \beta(i))c_i)$$

The probability of this is

$$\Pr\left[\bigcap_{i=1}^{z-1} \mathcal{E}_i\right] = 1 - \Pr[\exists i : \overline{\mathcal{E}_i}]$$

But

$$\Pr[\exists i : \overline{\mathcal{E}_i}] \leq \sum_{i=1}^{z-1} \Pr[\overline{\mathcal{E}_i}] \leq \sum_{i=1}^{z-1} \exp\left(-\frac{\beta^2(i)}{2}N_ip(i)\right)$$

$$\leq \sum_{i=1}^{z-1} \exp\left(-\frac{\beta^2(i)}{2}c_iW_i\right) \leq \sum_{i=1}^{z-1} W_1^{-c_i\xi\frac{\beta^2(i)}{2}}$$

Thus fast progress with good probability would imply that there is a small z (e.g. a polylog $z = (\log W_1)^{O(1)}$) satisfying the following two constraints:

$$(A) \quad \prod_{i=1}^{z-1}(1 - (1 - \beta(i))c_i) < \xi\frac{\log W_1}{W_1}$$

and the total error probability is at most

$$(B) \quad \sum_{i=1}^{z-1} W_1^{-c_i\xi\frac{\beta^2(i)}{2}}$$

(i.e. it would be behaving as $W_1^{-\xi}, \xi' > 1$).

In many cases such a z can be determined. For example, when $\beta(i) = 1/2, \forall i$ and $c_i = c$, then (A) becomes

$$\left(1 - \frac{c}{2}\right)^{z-1} < \xi \frac{\log W_1}{W_1}$$

implying

$$z \geq 1 + \frac{\log W_1 - \log \log W_1 - \log \xi}{\log\left(\frac{1}{1-c/2}\right)}$$

Then (B) gives total error probability $\leq O(\log W_1) \cdot W_1^{-c\xi/8} \longrightarrow 0$. Note that in this case non-adaptive constants work just because no competing events are present. Note also that this method may indicate faster-than-logarithmic progress, especially when $p(i)$ is very close to 1.

Note: Although the "generalized martingale" method gives logarithmic mean number of phases, it does not say much about the probability that the *actual* number of phases is small. Thus, the "fast progress" technique actually provides tighter bounds (more statistical information).

We claim the following generalization of a martingale theorem.

Theorem 1 *Let $W_0, W_1, ..., W_n$ be a generalized martingale with $|W_i - W_{i-1}| \leq k, \forall i$. Then*

$$\Pr[W_n > \lambda] < e^{-(\lambda^2 k/2n)g_1}$$

Proof: Similar to the proof of the corresponding theorem in [11], page 55. ∎

3 Coloring the $G_{n,p}$ Class

Let $q = 1 - p$ and $\Psi_{n,p} = n \log(1/q)/\log(1 + np)$. We know from [1] that the sequential greedy coloring algorithm almost surely colors the graph $G_{n,p}$ with at most $\Psi_{n,p}(1 + o(1))$ colors (by removing succesively independent sets and make each of them a color class). It is also known that the chromatic number $\chi(G_{n,p}) \geq \Psi_{n,p}/2$. The sequential algorithm can be easily parallellized if we use a parallel greedy independent set algorithm (see e.g. [3]) for discovering each color class. Let us call it the *PARALLEL GREEDY COLORING* algorithm. Obviously, this algorithm may have a number of iterations proportional to the number of color classes ($\Omega(n/\log n)$ for dense random graphs), therefore it does not constitute an efficient parallelization. In [4] a parallel algorithm was presented, called *PARALLEL COLORING*, which colors $G_{n,p}$ almost surely with $\Psi_{n,p}(1 + o(1))$ colors in $O(\log^5 n/\log \log n)$ time and one processor per edge on a CRCW PRAM. We modified this algorithm in order to get tighter results and a tighter analysis.

Let $G(V, E)$ be an instance of $G_{n,p}$ with m edges. Our algorithm proceed in stages. Let n_i be the number of uncolored vertices at the beginning of stage i. A high-level description of the algorithm follows.

ALGORITHM *MODIFIED PARALLEL COLORING*
(1) if $p < (\log n)/n$ then goto (2)
    else while $n_i > c \cdot \omega(n) \cdot \log n$ do
            run the *STAGE* algorithm (* see 3.1 below for details *)
        od
(2) run the *PARALLEL GREEDY COLORING* algorithm

Here c is a constant ≥ 1 and $\omega(n)$ is an increasing function $\leq \log n$. (Note that $\omega(n) = O(1)$ in [4].)

In this section we will do a tight analysis of the *STAGE* algorithm. Although the algorithm is similar to that of [4], its parameters are also modified. The end result is a *new* algorithm which runs in expected parallel time $O(\log^4 n/\log \log n)$ and one processor per edge in the CRCW PRAM model, uses as many colors as the algorithm of [4] and has *smaller* failure probability. Our analysis works for a large class of dense random graphs, i.e. $p \geq (\log \log \log n)^2/\log \log n$. This is a clear improvement over the algorithm of [4] (for dense random graphs), which demonstrates the value of the adaptive probabilistic analysis.

3.1 The algorithm *STAGE*

The algorithm proceeds in stages, *until* only at most $c \cdot \omega(n) \cdot \log n$ vertices remain uncolored, where $\omega(n)$ is a function $\leq \log n$ (to be chosen) and $\omega(n) \longrightarrow +\infty$ as $n \longrightarrow +\infty$, and c is a suitable constant, $c > 1$. When we arrive at this small number of uncolored vertices, we then run the *PARALLEL GREEDY COLORING* algorithm.

We now discuss the ith stage of the *STAGE* algorithm: suppose n_i vertices remain uncolored at the beginning of the stage ($n_0 = n$). We use the parameters D_i, τ_i, F_i of the stage (values will be chosen shortly). At any time we have τ_i "active" color classes to which we try to add vertices. An active class that acquires D_i vertices is called "full" and stops being active. At this point, a new (empty) color class replaces it, so that exactly τ_i classes are active at a time.

Stage i has n_i/F_i phases, each processing F_i vertices. Each vertex u of the F_i vertices of a phase checks whether there are edges joining it to at least one vertex in each active class. If so, u is *incompatible* with all active classes and has its coloring *deffered* to a future stage. A vertex compatible with one or more classes "attempts" to enter in all of them (and if it succeeds in more than one then a color is randomly chosen). As in [4] we "padd" an active class of $d < D_i$ vertices with $D_i - d$ "dummy" vertices adjacent to each of the F_i vertices with probability p. (This only makes things worse, but now a vertex u of the F_i has the same probability, $(1-p)^{D_i}$, to be compatible with any particular class.)

The vertices that are compatible with a class are input to a parallel greedy independent set algorithm (thus τ_i such algorithms are running in parallel here). Vertices rejected by any of these τ_i parallel greedy independent set algorithms have their coloring deffered to a later stage.

3.2 Definitions and basic remarks

Definition 1 *Let $p_s(i)$ be the probability that a particular vertex, not yet colored, gets colored in stage i.*

Definition 2 *Let $p_f(i) = 1 - p_s(i)$ be the probability that a particular vertex is deffered for a later stage during stage i.*

Lemma 3 $\forall \beta_i \in (0,1), n_{i+1} \leq n_i - (1 - \beta_i)p_s(i)n_i$ *with probability at least* $1 - \exp(-\beta_i^2 p_s(i)n_i/2)$.

Proof: By Chernoff's bound, $\Pr[\# \text{ vertices get colored during stage } i \geq (1 - \beta_i)p_s(i)n_i] \geq 1 - \exp(-\beta_i^2 p_s(i)n_i/2)$. ∎

Corollary 1 $\forall \beta_i \in (0,1), n_{i+1} \leq n_i(\beta_i + p_f(i))$ *with probability at least* $1 - \exp(-\beta_i^2 p_s(i)n_i/2)$.

Corollary 2 $\forall \beta_i \in (0,1)$ *and* $n_i \geq c \cdot \omega(n) \cdot \log n$,

$$\Pr[n_{i+1} \leq n_i(\beta_i + p_f(i))] \geq 1 - n^{-c\frac{\beta_i^2}{2}p_s(i)\omega(n)}$$

Definition 3 *Let x be the number of stages until $n_i \leq c \cdot \omega(n) \cdot \log n$.*

Lemma 4 *The parallel time of the STAGE algorithm is*

$$O\left(\max\{\log^2 n, \omega(n)\log n \log\log n\} + \left(\sum_{i=0}^{x} \frac{n_i}{F_i}\right) \cdot TG(i) \cdot TC(i)\right)$$

where $TG(i) =$ time to run the parallel greedy MIS on F_i's vertices that go to one color class, and $TC(i) =$ time to check incompatibilities between vertices in F_i and color classes.

Proof: Clear from the *STAGE* algorithm and from [3]. ∎

Lemma 5 *The parallel time of the STAGE algorithm is*

$$O\left(\max\{\log^2 n, \omega(n)\log n \log\log n\} + \left(\sum_{i=0}^{x} \frac{n_i}{F_i}\right) \cdot \log n\right)$$

Proof: $TG(i) = O(\log n)$ by [3] and $n_i \leq n$. Also $TC(i) = O(1)$ for CRCW PRAMs. ∎

3.3 The estimation of $p_s(i)$

A vertex u (of the F_i vertices) is compatible with color class j, with probability $(1-p)^{D_i}$ for every $j = 1, 2, ..., \tau_i$. Let $q = 1 - p$.

Definition 4 *Let C_j^i be the set of vertices (of a particular group of the F_i vertices) that are compatible with active color class j.*

Lemma 6 *The $|C_j^i|$ has an expected value $E[|C_j^i|] = q^{D_i} F_i, \forall i$.*

Proof: Obvious. ∎

We now note that (1) the parallel greedy MIS algorithm achieves the same result (maximal lexicographic independent set) with the sequential greedy MIS (see e.g. [4]), and (2) the cardinality of the MIS thus achieved is $\Theta(\log |C_j^i|)$ with very high probability (see e.g. [11]). (Here "very high probability" means probability at least $1 - |C_j^i|^{-a}$ for $a > 1$.)

Now, since the greedy MIS is created in a random graph, for any particular vertex u that belongs to $|C_j^i|$, the probability that u *is chosen into the MIS set (for j) is* $\Theta((\log |C_j^i|)/|C_j^i|)$. Now, we choose

$$F_i = \frac{n_i}{\log^2(1 + n_i p)} \quad , \quad \tau_i = \frac{2 n_i p}{\log^2(1 + n_i p)}$$

and we also choose D_i so that

$$q^{D_i} = \frac{\log^k n_i \cdot \log^2(1 + n_i p)}{1 + n_i p}$$

for some $k > 1$, and therefore

$$D_i = \frac{\log(1 + n_i p) - k \log \log n_i - 2 \log \log(1 + n_i p)}{\log(1/q)}$$

Then,

$$F_i q^{D_i} \approx \frac{1}{p} \log^k n_i \quad (*)$$

(actually $F_i q^{D_i} = n_i \log^k n_i / (1 + n_i p)$).

Since q^{D_i} must be less than or equal to 1 and k must be the same for each stage i we select $k = \log(p \log n)/c \log \log \log n$ where c is a suitable constant.

Lemma 7 *Let $\gamma \in (0, 1)$. Then $\forall j = 1, ..., \tau_i$*

$$\Pr\left[|C_j^i| > (1 - \gamma) E[|C_j^i|]\right] \geq 1 - \exp\left(-\frac{\gamma^2}{2} E[|C_j^i|]\right) \geq 1 - \exp\left(-\frac{\gamma^2}{2p} \log^k n_i\right)$$

Proof: By Chernoff bound, lemma 6 and equation $(*)$. ∎

Definition 5 *Let E_i be the number of active color classes that are compatible with a vertex in F_i.*

Lemma 8 *Let $\gamma' \in (0, 1)$. Then for any vertex $u \in F_i$*

$$\Pr\left[(1 - \gamma') \tau_i q^{D_i} \leq E_i \leq (1 + \gamma') \tau_i q^{D_i}\right] \geq 1 - 2 \exp\left(-\frac{(\gamma')^2}{2} \tau_i q^{D_i}\right)$$

Proof: Note that $\Pr[u$ is compatible with a particular color class$] = q^{D_i}$ and thus $E[E_i] = \tau_i q^{D_i}$. The rest follows from Chernoff bound. ∎

Remark: Since $\tau_i q^{D_i} = 2 \log^k n_i$ and $n_i \geq c \cdot \omega(n) \cdot \log n$, then the event described by lemma 8 holds with very high probability.

Now, clearly for a vertex u

$$\Pr[u \text{ goes to some class with which it is compatible}] = 1 - \Pr[u \text{ fails to go to any class}]$$

$$= 1 - \left(1 - \Theta\left(\frac{\log|C_j^i|}{|C_j^i|}\right)\right)^{E_i} \geq 1 - \exp\left(-E_i \cdot \Theta\left(\frac{\log|C_j^i|}{|C_j^i|}\right)\right)$$

Conditioned on the events of lemmas 7 and 8 this is

$$\geq 1 - \exp\left(-(\tau_i q^{D_i})\Theta\left(\frac{\log(F_i q^{D_i})}{F_i q^{D_i}}\right)\right) \geq 1 - \exp(-\Theta(2kp\log\log n_i)) \geq 1 - \frac{1}{\log^\lambda n_i}$$

Since we want to get a $\lambda > 1$, kp must be greater than or equal to 1. From the value of k we have that p must be at least

$$\frac{(\log\log\log n)^2}{\log\log n}$$

Note that all conditioning holds with probability $\geq 1 - n_i^{-a}$ thus, it can be removed and the above result is preserved.

Lemma 9 *For any vertex u in stage i we have*

$$p_s(i) \geq 1 - \frac{2}{\log^\lambda n_i}, \quad \lambda > 1$$

Proof: We have that

$$p_s(i) \geq \Pr[u \text{ is compatible with at least one class}] \cdot \Pr[u \text{ goes to some of its compatible classes}]$$

$$\geq \left(1 - \left(1 - q^{D_i}\right)^{\tau_i}\right) \cdot \left(1 - \frac{1}{\log^\lambda n_i}\right) \geq \left(1 - \exp\left(-q^{D_i}\tau_i\right)\right) \cdot \left(1 - \frac{1}{\log^\lambda n_i}\right)$$

$$\geq \left(1 - \exp(-\log^k n_i)\right) \cdot \left(1 - \frac{1}{\log^\lambda n_i}\right) \geq 1 - \frac{2}{\log^\lambda n_i}$$

∎

Corollary 3 $p_f(i) < 2/\log^\lambda n_i$, $\lambda > 1$.

3.4 The time of $STAGE$

From corollary 1 we now get $n_{i+1} \leq n_i(\beta_i + 2\log^{-\lambda} n_i)$ with probability at least

$$1 - \exp\left(-\frac{\beta_i^2}{2}\left(1 - \frac{2}{\log^\lambda n_i}\right)n_i\right)$$

By selecting $\beta_i = \sqrt{\log n_i}$ (adaptive Chernoff bound) we get $n_{i+1} \leq n_i(2/\sqrt{\log n_i})$ with probability at least

$$1 - n^{-c\omega(n)\frac{1}{\log n_i}\left(1 - \frac{2}{\log^\lambda n_i}\right)}$$

and finally we get, by choosing $\omega(n) = \log n$

$$n_{i+1} \leq n_i\frac{2}{\sqrt{\log n_i}} \quad (**)$$

with probability at least $1 - n^{-d}$ (for some $d > 1$ that can be controlled). The recursive relation $(**)$ proves that $n_x \leq c \cdot \omega(n) \cdot \log n$ when

$$x = \Theta\left(\frac{\log n}{\log\log n}\right)$$

By applying the above to lemma 5 we get

Theorem 2 *The parallel time of the algorithm STAGE is*

$$O\left(\log^2 n \log\log n + \frac{\log^2(1+np)\log^2 n}{\log\log n}\right) = O\left(\frac{\log^4 n}{\log\log n}\right)$$

with probability at least $1 - n^{-d'}$ *($d' > 1$ appropriate constant).*

3.5 The number of colors used

The number of colors used is L,

$$L \leq \sum_{i\geq 0}\left(\frac{n_i - n_{i+1}}{D_i} + \tau_i\right) \leq \sum_{i\geq 0}\left(\frac{(n_i - n_{i+1})\log(1/q)}{\log(1+n_ip) - k\log\log n_i - 2\log\log(1+n_ip)} + \frac{2n_ip}{\log^2(1+n_ip)}\right)$$

$$\leq \log(1/q)\left(\frac{n}{\log(1+np)} - \sum_{i\geq 1}\frac{n_i}{\log(1+n_{i-1}p)} + \sum_{i\geq 1}\frac{n_i}{\log(1+n_ip)} + O\left(k^2\sum_{i\geq 0}\frac{n_i\log\log n_i}{\log^2(1+n_ip)}\right)\right)$$

since $k = \log(p\log n)/c\log\log\log n$. In [4] it is proved that

$$\sum_{i\geq 1}\frac{n_i}{\log(1+n_ip)} - \sum_{i\geq 1}\frac{n_i}{\log(1+n_{i-1}p)} = o\left(\frac{n}{\log(1+np)}\right)$$

We are able to prove that (see [10])

$$\sum_{i\geq 0}\frac{k^2 n_i\log\log(1+n_ip)}{\log^2(1+n_ip)} = o\left(\frac{n}{\log(1+np)}\right)$$

Thus, the number of colors used is $\Psi_{n,p}(1 + o(1))$.

Our main result is the following.

Theorem 3 *The MODIFIED PARALLEL COLORING algorithm colors a random graph $G_{n,p}$, with $p \in (0,1) - [(\log n)/n, (\log\log\log n)^2/\log\log n)$, almost surely with $\Psi_{n,p}(1 + o(1))$ colors in parallel time $O(\log^4 n/\log\log n)$ by using one processor for each edge of the graph on a CRCW PRAM.*

Note that the algorithm should know p in advance, for the computation of D_i, τ_i, F_i. This can be done in the same way as in [4] by counting the degrees of a few vertices in the graph.

4 Coloring k-colorable Graphs

4.1 Preliminaries

In this section we parallelize the greedy (sequential) algorithm of [8] (which is also appeared as a procedure in [5]) for coloring k-colorable graphs, with edge probability $\geq d\sqrt{\log n/n}$, d constant, $d > 1$. The algorithm takes as input a k-colorable graph (drawn equally likely from the uniform distribution over all k-colorable graphs with N vertices) and colors it with exactly k colors.

We prove here that such a coloring can be achieved in $O(\log^2 N)$ expected time, using a linear (to the edges) number of processors on an EREW PRAM. Thus we have an efficient (randomized) parallel solution of an NP-complete problem. In the sequel, let $G = (V(G), E(G))$ denote an arbitrary graph and let $N = |V(G)|$, $M = |E(G)|$. Similarly to [5], we have the following models for k-colorable graphs.

1. There are k color classes, called *blocks*, each having the same number of $n = N/k$ vertices. An edge between different classes has probability $p(n)$ of being present (choices made independently for each edge). Here, as in [5] we assume that k is constant.

2. There are fixed blocks but we select M inter-block edges at random for some M.

3. It has two variants:

    (a) Consider all the ways of choosing k blocks as a partition of $V(G)$ and then select M inter-block edges. Select a graph uniformly from the above sample space. Note that the same k-colorable graph may occur more than once.

    (b) It is the uniform distribution obtained by model 3a, by removing all duplicates of the same graph.

4. It has also two variants (models 4a and 4b) corresponding to the models 3a and 3b, if we allow M to vary.

Thus the model we are interested in is clearly model 4b, i.e. the uniform distribution over all k-colorable graphs with N vertices.

We prove our result for model 1. In the special case where p is constant the result carries out to the other models by simply following the steps of the "translation" from one model to the other presented in [5].

Before proceeding to the algorithm we shall give some notation. For $v \in V(G)$, let $\Gamma(v) = \{w | (v, w) \in E(G)\}$. Similarly, for $S \subseteq V(G), \Gamma(S) = \cup_{v \in S} \Gamma(v) - S$. Also we define the degree of a vertex v in relation with a set $X \subseteq V(G)$ as follows: $\delta_X(v) = |\Gamma(v) \cap X|$.

4.2 The algorithm and its analysis

We shall give now a "parallel greedy" algorithm to parallelize the sequential greedy technique of [8].

PROCEDURE *PARCOLOR1*
for $i = 1$ to $k - 1$ do
begin
$\quad X_i = \emptyset$;
$\quad Y_i = V(G) - \cup_{j<i} X_j$;
$\quad j = 1$;
$\quad$ **repeat**
$\qquad\qquad$ select set $A_j = \{v_1, v_2, ..., v_{2^{j-1}}\} \subseteq Y_i$ such that
$\qquad\qquad$ $v_1, v_2, ..., v_{2^{j-1}}$ are the 2^{j-1} minimal degree vertices in Y_i;
$\qquad\qquad$ **for all** $v, w \in A_j$ **do**
$\qquad\qquad\qquad\qquad$ **if** $\{v, w\} \in E(G)$ **then** *PARCOLOR1* has failed
$\qquad\qquad\qquad\qquad$ **else** $X_i = X_i \cup A_j$; $Y_i = Y_i - A_j - \Gamma(A_j)$;
$\qquad\qquad$ **od**
$\qquad\qquad$ $j = j + 1$;
$\quad$ **until** $Y_i = \emptyset$
end
if $X_k = V(G) - \cup_{j=1}^{k-1} X_j$ is stable then $X_1, ..., X_k$ is a k-coloring
else *PARCOLOR1* has failed

Lemma 10 *If k is constant and $p \geq d\sqrt{\log n / n}$ (d constant, $d > 1$) then PARCOLOR1 fails with probability $O(e^{-n^a})$, $0 < a < 1$, when G is a random graph selected under model 1.*

Proof: It suffices to show that each iteration of the for-loop terminates with a block of G in one color class with failure probability $\leq e^{-n^\gamma}, 0 < \gamma < 1$. If this is the case and $S_i, i = 1, 2, ..., k$, is the event that the ith iteration of the for-loop succeeds, the probability for success of *PARCOLOR1* is $\Pr[S_1 \cap ... \cap S_k] = 1 - \Pr[\exists i : \overline{S_i}] \geq 1 - k \Pr[\overline{S_i}] \geq 1 - k e^{-n^\gamma}$.

Suppose without loss of generality (wlog), that during the first iteration of the for-loop of *PARCOLOR1* the first v selected ($v \in A_0$) belongs to the block B_1. In the sequel we use the term "very high probability" for probability at least $1 - e^{-n^\alpha}$, $0 < a < 1$. We will prove the following: if all vertices of sets $A_1, A_2, ..., A_j$ (selected during the first j iterations of the repeat-until loop) belong to B_1, then the vertices of A_{j+1} belong also to B_1, with very high probability. We first prove the following claim.

Claim. Suppose that $v \in B_1 \cap Y_1$ and $v' \in B_i \cap Y_1$, for some $i \neq 1$, after the jth iteration of the repeat-until loop of *PARCOLOR1*. Then $\Pr[\delta_{Y_1}(v') < \delta_{Y_1}(v)] \leq e^{-n^{a'}}$, for $0 < a' < 1$.

Proof of Claim. Let $r_0 = \lceil (\log k - \log p + \log(1 + c) - \log(1 - c))/ - \log(1 - p) \rceil$ for some constant $0 < c < 1$ which will be defined later.

Case 1. Suppose that $|X_1| = |\cup_{i=1}^j A_i| = r$ and $r < r_0$. Then for all blocks $B_j, j \neq 1$ we have from Chernoff bounds that,

$$\Pr[|B_j - \Gamma(X_1)| > (1 + \beta(r))n(1 - p)^r] \leq \exp\left(-\frac{\beta^2(r)}{2}n(1 - p)^r\right)$$

for any $\beta(r) : 0 < \beta(r) < 1$. Also if v belongs to some $B_i, i \neq j$, then

$$\Pr\left[|B_j - \Gamma(X_1 \cup \{v\})| > (1 + \beta(r))n(1 - p)^{r+1}\right] \leq \exp\left(-\frac{\beta^2(r)}{2}n(1 - p)^{r+1}\right)$$

Thus, if a vertex v belongs to $B_1 \cap Y_1$ it will have at most

$$(1 + \beta(r))(k - 1)[n(1 - p)^r - n(1 - p)^{r+1}] = (1 + \beta(r))(k - 1)np(1 - p)^r$$

neighbors with very high probability. If a vertex $v' \in B_i \cap Y_1$, for some $i \neq 1$ then it will have at least $(1 - \beta(r))[np + (k - 2)np(1 - p)^r]$ neighbors with very high probability (since a vertex $v' \in B_i, i \neq 1$, it will have at least $(1 - \beta(r))np$ neighbors in B_1). But if we choose

$$\beta(r) = \frac{1 - (1 - p)^r}{1 + 2(k - 1)(1 - p)^r}$$

we have that $(1 - \beta(r))[np + (k - 2)np(1 - p)^r] \geq (1 + \beta(r))(k - 1)np(1 - p)^r$ with probability at least $1 - e^{-n^a}$, $0 < a < 1$, for all $p \geq d\sqrt{\log n/n}$.

Case 2. If $|X_1| = |\cup_{i=1}^j A_i| = r$ and $r \geq r_0$ we have for each $2 \leq i \leq k$ (by Chernoff bound) that

$$\Pr[|B_i \cap Y_1| > (1 + c)n(1 - p)^{r_0}] \leq \exp\left(-\frac{c^2}{2}n(1 - p)^{r_0}\right)$$

for some $0 < c < 1$. If a vertex v belongs to $B_1 \cap Y_1$, it will have at most $(k - 1)n(1 - p)^{r_0}$ neighbors with probability at least $1 - \exp(-\frac{c^2}{2}n(1 - p)^{r_0})$. If a vertex v' belongs to $B_i \cap Y_1$, for some $i \neq 1$, then (again by Chernoff bound) it will have at least $(1 - c)np$ neighbors in B_1 with probability $\geq 1 - \exp(-c^2np/2)$. From the value of r_0 we have that $(1 - c)np > (1 + c)(k - 1)n(1 - p)^{r_0}$ with very high probability. This ends the proof of the claim. ∎

Suppose now wlog, that $\delta_{Y_1}(v_1) \leq \delta_{Y_1}(v_2) \leq ... \leq \delta_{Y_1}(v_{2|A_j|})$. Let E_i be the event "v_i indeed belongs to B_1". Then

$$\Pr[\text{ all vertices of } A_{j+1} \text{ belong to } B_1] = \Pr[E_{2|A_j|} \cap E_{2|A_j|-1} \cap ... \cap E_1]$$

$$= \Pr[E_{2|A_j|} \mid E_{2|A_j|-1} \cap ... \cap E_1] \cdots \Pr[E_2 \mid E_1] \Pr[E_1]$$

Let us call the above quantity Q. But,

$$\Pr[E_1] = \Pr[v_1 \in B_1] = \Pr[\forall v \in B_j, j \neq 1 : \delta_{Y_1}(v) \geq \delta_{Y_1}(v_1)]$$

$$= 1 - \Pr[\exists v \in B_j, j \neq 1 : \delta_{Y_1}(v) < \delta_{Y_1}(v_1)]$$

$$= 1 - \sum_{v \in B_j, j \neq 1} \Pr[\delta_{Y_1}(v) < \delta_{Y_1}(v_1)] \geq 1 - (k - 1)ne^{-n^{a_1}} \quad \text{(by claim)}$$

By using an argument similar to that of case 1 of the claim, we can prove that $\Pr(E_2 \mid E_1) \geq 1 - (k-1)ne^{-n^{a_2}}$ if $r < r_0$. (The case in which $r \geq r_0$ is straightforward.) By induction on the size of A_{j+1} we are able to prove that for all $k = 2, ..., 2|A_j|$ we have: $\Pr[E_k \mid E_{k-1} \cap ... \cap E_1] \geq 1 - (k-1)ne^{-n^{a_k}}$, if $r < r_0$. (Again, the case $r \geq r_0$ is straightforward.) Thus

$$Q \geq \left(1 - (k-1)ne^{-n^\delta}\right)^{2^j} \geq 1 - (k-1)n2^j e^{-n^\delta}$$

for some $\delta = \min_k \{a_k\}$. Let R_j be the event "all the vertices of A_j belong to B_1", $j = 1, ..., \lceil \log n \rceil$. Then

$$\Pr[\text{ the first iteration of the for-loop ends with a block of } G \text{ in } X_1] =$$

$$= \Pr[R_1 \cap ... \cap R_{\lceil \log n \rceil}] = 1 - \Pr[\exists j : \overline{R_j}] \geq 1 - \sum_{j=1}^{\lceil \log n \rceil} n(k-1)2^j e^{-n^\delta}$$

$$\geq 1 - (k-1)n^2 e^{-n^\delta} \geq 1 - e^{-n^\gamma}$$

for some $0 < \gamma < 1$. ∎

Lemma 11 *Procedure PARCOLOR1 can be implemented to run on an EREW PRAM in time $O(\log^2 n)$ by using $O(M)$ processors, with probability of success $1 - e^{-n^a}$, $0 < a < 1$.*

Proof: Clearly, we need $O(\log n)$ time in order to compute the degree of each vertex. Also the repeat-until loop takes $O(\log n)$ iterations. This gives a total of $O(\log^2 n)$ time. It is also clear that we need one processor for every edge of the graph. The success probability comes from lemma 10. ∎
Note: We have shown that *PARCOLOR1* gives us an "almost sure" solution algorithm. However, its failure probability is not small enough to get *expected* polylogarithmic time and polynomial number of processors. We shall show now how to get efficient *expected* behaviour in the special case where p is constant.

4.3   Efficient expected behaviour

From [5] we know of an algorithm (called there *COLOR2*) which is a brute-force one. It *guesses* large k-colored subgraphs of G and uses an "only available color" rule to color most of G; then it uses complete enumeration of the remainder, provided this is small enough. Fortunately *COLOR2* can be parallelized by trying all its major steps in parallel in a straightforward way.

Lemma 12 *COLOR2 can be implemented on an EREW PRAM to run in $O(\log n)$ time by using $O(n^{1+8k^3 \log k} \cdot \log^k n)$ processors. Furthermore, it has failure probability $O(e^{-n \log n})$ for large n and G selected under model 1 with k, p constants.*

Proof: See [10]. ∎
   Now, if we use *PARCOLOR1* and combine it with the parallel version of procedure *COLOR2* (say, *PARCOLOR2*), we have a parallel algorithm for coloring a k-colorable graph (selected under model 1). Let us call the above algorithm *PARCOLOR*.

Theorem 4 *The algorithm PARCOLOR finds a k-coloring of a random graph selected under model 1 (with constant k, p), in $O(\log^2 n)$ expected time by using an expected number of $O(M)$ processors.*

Proof: Let T, P be the time and number of processors respectively, used by *PARCOLOR*. Similarly, let T_1, P_1 be the resource bounds of *PARCOLOR1* and T_2, P_2 those of *PARCOLOR2*. Also, let $Z_1(Z_2)$ be the event that *PARCOLOR1* (*PARCOLOR2*) succeeds. Then we have (by using also lemmas 11,12)

$$E[T] = T_1 \Pr[Z_1] + T_2 \Pr[\overline{Z_1} \cap Z_2] + O(\log n) \Pr[\overline{Z_1} \cap \overline{Z_2}]$$
$$\leq O(\log^2 n) + O(\log n \cdot e^{-n^a}) + O(\log n \cdot e^{-n \log n}) = O(\log^2 n)$$
$$E[P] = P_1 \Pr[Z_1] + P_2 \Pr[\overline{Z_1} \cap Z_2] + k^{kn} \Pr[\overline{Z_1} \cap \overline{Z_2}]$$
$$\leq O(M) + O(n^{1+8k^3 \log k} \cdot \log^k n \cdot e^{-n^a}) + O(k^{kn} \cdot e^{-n \log n}) = O(M)$$

■

As we noted in section 4.1, the above result can be "translated" model by model in exactly the same way as in [5]. (Note that also here, we want to know p in advance. But a good lower bound on p can be obtained using the same technique as in [5].) Thus we have the following.

Corollary 4 *The algorithm PARCOLOR finds a k-coloring of a random graph drawn equally likely from the uniform distribution over all k-colorable graphs with N vertices, with constant k, p, in $O(\log^2 N)$ expected time by using $O(M)$ expected number of processors.*

5 Acknowledgements

We are grateful to Ludek Kucera for his valuable remarks and suggestions on improving the paper, and to George Moustakides for his valuable help.

References

[1] B. Bollobas, "Random Graphs", Academic Press, London, 1985.

[2] H. Chernoff, "A measure of asymptotic efficiency for tests based on the sum of observations", *Ann. Math. Statist.* 23 (1952), 493-509.

[3] A. Calkin, A. Frieze, "Probabilistic Analysis of a Parallel Algorithm for Finding Maximal Independent Sets", *Random Structures & Algorithms*, Vol.1, No.1, 39-50, 1990.

[4] D. Coppersmith, P. Raghavan, M. Tompa, "Parallel Graph Algorithms that are Efficient on Average", *Proc. of the 28th Annual IEEE FOCS*, 1987, pp.260-269.

[5] M. E. Dyer, A. M. Frieze, "The Solution of Some Random NP-hard Problems in Polynomial Expected Time", *Journal of Algorithms*,10, 451-489, 1989.

[6] P. Erdos, A. Renyi, "On random graphs I", *Publ. Math. Debrecen*, 6(1959), 290-297.

[7] Z. Kedem, K. Palem, P. Spirakis, "Adaptive average case analysis", unpublished manuscript, 1990.

[8] L. Kucera, "Expected behaviour of graph coloring algorithms", *Proc. of Fundamentals in Computation Theory* LNCS, Vol.56, pp.447-451, Springer-Verlag, 1977.

[9] J.E. Littlewood, "On the probability in the tail of a binomial distribution", *Adv. Appl. Probab.*, 1(1969), 43-72.

[10] G. Pantziou, P. Spirakis, C. Zaroliagis, "Coloring Random Graphs Efficiently in Parallel, through Adaptive Techniques", CTI TR-90.10.25, Computer Technology Institute, Patras. Also presented in the ALCOM Workshop on Graph Algorithms, Data Structures and Computational Geometry, Berlin 3-5 October, 1990.

[11] J. Spencer, "Ten Lectures on the Probabilistic Method", SIAM, 1987.

Optimal Vertex Ordering of a Graph and its Application to Symmetry Detection

X. Y. Jiang,* H. Bunke
Institut für Informatik und angewandte Mathematik
Universität Bern, Länggass-Strasse 51, 3012 Bern, Switzerland

Abstract

We consider in this paper the problem of optimal vertex ordering of a graph. The vertex ordering and an optimality measure are defined. It is proved that the optimal ordering problem can be transformed into the well-known minimum-weight spanning tree problem. Some properties of optimal vertex orderings are investigated. And finally, the application of the optimal vertex ordering technique to an algorithm for detecting the symmetry of polyhedra is discussed in some detail.

1 Introduction

A geometrical entity, say a point set, a polygon, or a polyhedron, is symmetrical if its shape remains unchanged under an affine transformation. Because of many applications of symmetry information in pattern recognition, robotics, and computer graphics, several algorithms for symmetry detection have appeared in the literature. In particular, optimal algorithms for detecting the symmetry of polyhedra are known. While the asymptotic behavior of them is good, these optimal algorithms are rather intricate and an actual implementation is quite difficult. Thus, they are primarily of theoretical interest. In [9, 10] we have proposed a new algorithm for detecting the symmetry of polyhedra. Our algorithm is simple and can be easily implemented. In designing the algorithm we encountered the need for ordering the vertices of the vertex-edge graph of a polyhedron. As the overall time complexity of the symmetry detection algorithm depends on the quality of the vertex ordering, we are faced with the problem of finding an optimal vertex ordering of a graph. This paper deals with this problem. The main results are an algorithm for determining the optimal vertex ordering and some properties of the optimal ordering. As a consequence of this result, a reduction of the time complexity in the algorithm proposed in [9, 10] from cubic to quadratic has been achieved.

The rest of this paper is organized as follows. The next section gives the definition of a vertex ordering and an optimality measure, and shows some basic results. In section 3 we propose an algorithm for determining the optimal vertex ordering and prove its correctness. In section 4 we consider the upper bound of the optimality measure. The use of the optimal vertex ordering technique in our symmetry detection algorithm mentioned above is described in some detail in section 5. And finally, a concluding section highlights the most significant results of this paper.

---

*The support of the Swiss National Science Foundation under the NFP-23 program, grant no. 4023-027026, is gratefully acknowledged.

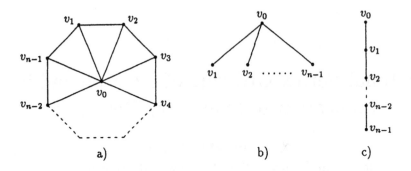

Figure 1: a) The graph of a $(n-1)$-pyramid. b) A non-optimal ordering. c) An optimal ordering.

2 Definitions and basic results

By definition, a graph is a pair $G = (V, E)$, where $V = \{v_0, v_1, \cdots, v_{n-1}\}$ is the set of vertices and $E \subseteq V \times V$ is the set of edges. In this work we assume that G is simple and connected. Given a graph G and a vertex $v_0 \in V$, an *ordering* of the vertices of G beginning with v_0 is a sequence

$$v_0 v_1' v_2' \cdots v_{n-1}', \quad v_i' \in V - \{v_0\}, \quad i = 1, 2, \cdots, n-1, \quad v_i' \neq v_j' \text{ for } i \neq j$$

together with a function $f : V - \{v_0\} \to V$ such that each vertex $v_i' \in V - \{v_0\}$ is mapped to a vertex u in the subsequence $v_0 v_1' v_2' \cdots v_{i-1}'$ by the function f such that u is adjacent in G to v_i', i.e.

$$\forall v_i' (\exists u (u \in \{v_0, v_1', v_2', \cdots, v_{i-1}'\} \wedge (u, v_i') \in E \wedge u = f(v_i')))$$

We call the vertex $u = f(v_i')$ the *father* of v_i'. For a given graph G and a vertex v_0, there are many possible orderings, in general. Let $d_G(v)$ denote the degree of the vertex $v \in V$ in G, i.e., the number of edges incident to v. An ordering is called *optimal* in G if the sum of the fathers' degrees of all vertices except v_0, i.e.

$$\sum_{i=1}^{n-1} d_G(f(v_i'))$$

is minimal among all possible orderings beginning with v_0.

Example 1. The vertex-edge graph G of a $(n-1)$-pyramid is shown in Fig. 1a). A possible ordering beginning with v_0 is $v_0 v_1 v_2 \cdots v_{n-1}$ with $f(v_i) = v_0$, $i = 1, 2, \cdots, n-1$. Another ordering beginning with v_0 is $v_0 v_1 v_2 \cdots v_{n-1}$ with $f(v_i) = v_{i-1}$, $i = 1, 2, \cdots, n-1$. □

The vertex ordering of a graph can also be represented by a rooted tree. A tree is called *rooted* if one vertex, the root, is distinguished. If (u, v) is an edge of a rooted tree such that u lies on the path from the root to v, then u is said to be the *father* of v and v is the *son* of u. In representing a vertex ordering of a graph beginning with v_0 by a rooted tree, we choose v_0 as the root. A vertex v_i is the father of another vertex v_j in the tree if v_i is the father of v_j in the ordering, i.e. $v_i = f(v_j)$. The tree representations for the two orderings given in example 1 are, for instance, shown in Fig. 1b) and 1c). Note that while the mapping from an ordering into a tree representation is unambiguous, the reverse is not true. The tree in Fig. 1b) may in fact correspond to any of the $(n-1)!$ orderings $v_0 v_1' v_2' \cdots v_{n-1}'$ where $v_i' \in \{v_1, v_2, \cdots, v_{n-1}\}$ and $f(v_i') = v_0$. All these orderings, however, are equivalent in the sense that, for all of them, the sum of the fathers' degrees is identical. Thus, this ambiguity is not essential.

Each of the trees shown in Fig. 1b) and 1c), respectively, is a rooted spanning tree of the graph G. Obviously, every spanning tree rooted at a vertex v_0 is an ordering of G. Thus our initial problem of finding an optimal vertex ordering can be stated in terms of tree representation as follows. Given a graph $G = (V, E)$ and a vertex $v_0 \in V$, find a spanning tree T rooted at v_0 such that the sum

$$SFD_T(v_0) = \sum_{u \in V - \{v_0\}} d_G(f(u))$$

of the fathers' degrees of all vertices in G, except v_0, is minimal where $f(u) = v$ if v is the father of u in T. Note that, for the calculation of SFD, not the degrees of the father vertices in T but in the original graph G are summed up. In the following we use $SFD_{opt}(v_0)$ to denote $SFD_T(v_0)$ for some optimal spanning tree T of G rooted at v_0.

Example 2. For the vertex-edge graph of the $(n-1)$-pyramid in example 1, a non-optimal spanning tree T is shown in Fig. 1b). For this tree we have

$$
\begin{aligned}
SFD_T(v_0) &= d_G(f(v_1)) + d_G(f(v_2)) + \cdots + d_G(f(v_{n-1})) \\
&= d_G(v_0) + d_G(v_0) + \cdots + d_G(v_0) \\
&= (n-1) + (n-1) + \cdots + (n-1) \\
&= (n-1)^2
\end{aligned}
$$

As another example, an optimal spanning tree T^* is shown in Fig. 1c). For this optimal spanning tree we get

$$
\begin{aligned}
SFD_{opt}(v_0) &= d_G(f(v_1)) + d_G(f(v_2)) + \cdots + d_G(f(v_{n-1})) \\
&= d_G(v_0) + d_G(v_1) + \cdots + d_G(v_{n-2}) \\
&= (n-1) + 3 + \cdots + 3 \\
&= 4n - 7
\end{aligned}
$$

While the SFD of the non-optimal spanning tree is quadratically bounded by the number of vertices, $SFD_{opt}(v_0)$ is only linearly bounded by the number of vertices. □

Now we establish a lemma which represents $SFD_T(v_0)$ without using the function f.

Lemma 1 *Let $G = (V, E)$ be a simple connected graph and T be a spanning tree rooted at v_0. Then the equation*

$$SFD_T(v_0) = \sum_{v_i \in V} d_G(v_i) \cdot d_T(v_i) + d_G(v_0) - 2|E|$$

holds.

Proof. Clearly, $SFD_T(v_0)$ can be written as

$$SFD_T(v_0) = \sum_{v \in V} d_G(v) \cdot s(v)$$

where $s(v)$ is the number of vertices u such that $f(u) = v$, i.e. the number of sons of v in T. As v_0 has $d_T(v_0)$ sons and all other vertices v_i have $d_T(v_i) - 1$ sons in T, we get

$$SFD_T(v_0) = d_G(v_0) \cdot d_T(v_0) + \sum_{v_i \in V - \{v_0\}} d_G(v_i) \cdot (d_T(v_i) - 1)$$

$$= \sum_{v_i \in V} d_G(v_i) \cdot d_T(v_i) + d_G(v_0) - \sum_{v_i \in V} d_G(v_i)$$

As the sum of the degrees of all the vertices of a graph is twice the number of edges, we have

$$SFD_T(v_0) = \sum_{v_i \in V} d_G(v_i) \cdot d_T(v_i) + d_G(v_0) - 2|E|$$

This concludes our proof. $\square$

Based on this lemma, our problem of finding an optimal vertex ordering can be reformulated as follows. Given a graph $G = (V, E)$, find a spanning tree T such that the sum

$$S_T = \sum_{v_i \in V} d_G(v_i) \cdot d_T(v_i)$$

is minimal. Then the optimal spanning tree T, if rooted at $v_0 \in V$, gives us the optimal vertex ordering beginning with v_0, and we have

$$SFD_{opt}(v_0) = S_T + d_G(v_0) - 2|E|$$

Lemma 2 *Let $G = (V, E)$ be a simple connected graph. For two vertices $u, w \in V$,*

$$SFD_{opt}(u) - SFD_{opt}(w) = d_G(u) - d_G(w)$$

holds.

Proof. Let T^* (T') be a spanning tree of G which, if rooted at u (w), gives us the optimal vertex ordering beginning with u (w). Note that T^* and T' may be identical. Anyway $S_{T^*} = S_{T'}$ holds as S_{T^*} and $S_{T'}$ are both minimal among the S_T's for all spanning trees T of G. Thus, we have

$$SFD_{opt}(u) - SFD_{opt}(w) = (S_{T^*} + d_G(u) - 2|E|) - (S_{T'} + d_G(w) - 2|E|)$$

$$= d_G(u) - d_G(w)$$

This concludes our proof. $\square$

3   The algorithm

Using the basic results proved in the last section, our initial problem of finding the optimal vertex ordering of a graph G has been reduced to finding a spanning tree T such that S_T is minimal among all spanning trees of G. The main result of this section is that this problem can be transformed into the well-known minimum-weight spanning tree problem. Thus any of the known algorithms for that purpose can be used to solve our vertex ordering problem.

Given a simple connected graph $G = (V, E)$. To each edge (u, v) of G, we assign a weight

$$w((u, v)) = d_G(u) + d_G(v) \tag{1}$$

We can prove

Theorem 1 *A minimum-weight spanning tree T^* of G minimizes S_T among all spanning trees T of G.*

Proof. Let T be a spanning tree of G. We count the sum of weights of all edges in T,

$$W_T = \sum_{(u,v)\in T} w((u,v)) = \sum_{(u,v)\in T} (d_G(u) + d_G(v))$$

As this sum is simply a summation of degrees $d_G(v_i)$ for $v_i \in V$, we can write it as

$$W_T = \sum_{v_i \in V} d_G(v_i) \cdot c(v_i)$$

where $c(v_i)$ represents how many times $d_G(v_i)$ is counted in the sum. As $d_G(v_i)$ occurs exactly once in the sum for each neighbor of v_i in T, and v_i has $d_T(v_i)$ neighbors in T, we get $c(v_i) = d_T(v_i)$. Thus,

$$W_T = \sum_{v_i \in V} d_G(v_i) \cdot d_T(v_i) = S_T$$

As T^* is a minimum-weight spanning tree, we have

$$W_{T^*} = \min\{W_T \mid T \text{ is a spanning tree of } G\}$$

This leads to

$$S_{T^*} = \min\{S_T \mid T \text{ is a spanning tree of } G\}$$

Thus, T^* minimizes S_T among all spanning trees T of G. $\square$

From theorem 1 and the discussions in the last section we propose the following algorithm for finding an optimal vertex ordering of a graph $G = (V, E)$, given a vertex $v_0 \in V$.

1. Construct the weighted graph G using the weighting function in (1).

2. Find a minimum-weight spanning tree T of the weighted graph G.

3. Choose v_0 as the root of T. The resulting rooted tree gives an optimal vertex ordering beginning with v_0.

The main part of our algorithm is to find a minimum-weight spanning tree. Many algorithms are known for doing this. Most famous are Prim's algorithm [16] and Kruskal's algorithm [15]. Both have a quadratic time complexity. More efficient algorithms are known. In [18], for example, an alternative method is given which has a lower time bound of $O(|E| \log \log |V|)$ but which is more intricate. For other improved algorithms see [3, 4, 5, 12, 13, 14]. There is also a number of parallel algorithms for finding minimum-weight spanning trees. For a survey, see [1, 17].

4 Upper bounds for SFD

Given a graph $G = (V, E)$ and a vertex $u \in V$, we investigate the upper bound of $SFD_{opt}(u)$. The main result is that, if G is 3-connected and planar, $SFD_{opt}(u)$ is linearly bounded by the number of vertices and the number of edges of G. For graphs which are not 3-connected or not planar the upper bound can be quadratic.

For proving our results we need a theorem due to D. W. Barnette [2, 7].

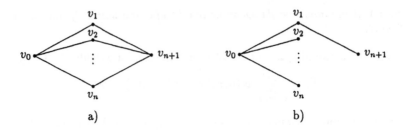

Figure 2: a) The complete bipartite graph $K_{2,n}$. b) An optimal spanning tree of $K_{2,n}$.

Theorem 2 *Every 3-connected and planar graph contains a spanning tree of maximal degree 3.*

By means of this theorem we can prove the following theorem.

Theorem 3 *For a 3-connected and planar graph $G = (V, E)$ and a vertex $u \in V$, the inequality*

$$SFD_{opt}(u) \leq 4|E| + max\{d_G(v)|v \in V\}$$

holds.

Proof. Because of Barnette's theorem, there exists, for G, a spanning tree T of maximal degree 3. As $SFD_{opt}(u) = SFD_{T^*}(u)$ for some optimal spanning tree T^*, we have

$$
\begin{aligned}
SFD_{opt}(u) &\leq SFD_T(u) \\
&= \sum_{v_i \in V} d_G(v_i) \cdot d_T(v_i) + d_G(u) - 2|E| \\
&\leq 3 \sum_{v_i \in V} d_G(v_i) + d_G(u) - 2|E|
\end{aligned}
$$

As the sum of the degrees of all the vertices of a graph is twice the number of edges, we get

$$
\begin{aligned}
SFD_{opt}(u) &\leq 4|E| + d_G(u) \\
&\leq 4|E| + max\{d_G(v)|v \in V\}
\end{aligned}
$$

This concludes our proof. □

For each vertex u of a 3-connected planar graph, $SFD_{opt}(u)$ is therefore linearly bounded by the number of edges. As, in a connected planar graph, the number of vertices and that of edges are of the same order, $SFD_{opt}(u)$ is also linearly bounded by the number of vertices of the graph.

For a graph which is not 3-connected or not planar, the linear upper bound is no more guaranteed to be true. As an example, we first consider the complete bipartite graph $K_{2,n}$ in Fig. 2a) which is planar but not 3-connected. For the vertex v_0, one optimal spanning tree is shown in Fig. 2b). Thus

$$SFD_{opt}(v_0) = \sum_{i=1}^{n} d_G(v_0) + d_G(v_1) = n^2 + 2$$

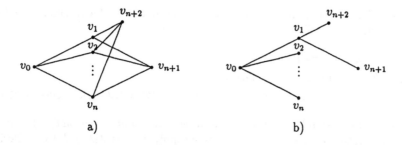

Figure 3: a) The complete bipartite graph $K_{3,n}$. b) An optimal spanning tree of $K_{3,n}$.

In this case $SFD_{opt}(v_0)$ is quadratically bounded by the number of vertices and edges. Another example $K_{3,n}$, which is 3-connected but not planar, is shown in Fig. 3a). For the vertex v_0, one optimal spanning tree is given in Fig. 3b). We have

$$SFD_{opt}(v_0) = \sum_{i=1}^{n} d_G(v_0) + 2 \cdot d_G(v_1) = n^2 + 6$$

which is also quadratically bounded by the number of verties and edges. Finally, we consider the complete graph $K_n, n \geq 5$. K_n is 3-connected but not planar. For each vertex v_0 of K_n,

$$SFD_{opt}(v_0) = (n-1) \cdot (n-1) = (n-1)^2$$

This expression is linearly bounded by the number of edges but quadratically bounded by the number of vertices. Note that, as the degree of a vertex u is in the worst case linear in the number of vertices, the $SFD_{opt}(u)$ of a vertex can be at worst quadratically bounded by the number of vertices.

5   An application to symmetry detection

As mentioned in section 1, our motivation of investigating the optimal vertex ordering problem lies in the need for designing an algorithm for detecting the symmetry of polyhedra. In this section we sketch the symmetry detection algorithm to highlight the application of the optimal vertex ordering technique developed so far.

5.1   Symmetry analysis

A polyhedron comprising n vertices, m edges and h faces can be defined as a graph $G = (V, E)$ and a set F with

$$V = \{v_1, v_2, \ldots, v_n\}$$
$$E = \{e_k = (v_{k1}, v_{k2}) \mid k = 1, 2, \ldots, m\}$$
$$F = \{f_k = (v_{k1}, v_{k2}, \ldots, v_{kl_k}) \mid k = 1, 2, \ldots, h\}$$

where V represents the set of vertices, E is the set of edges, and F is the set of oriented faces where each face is represented by the closed chain of its vertices.

Our algorithm proposed in [9, 10] detects all rotational symmetries of a polyhedron. A rotational symmetry $Sym(\theta, R)$ is defined as a bijective mapping $\theta : V \to V$ together with a spatial rotation, represented by a 3×3 rotation matrix R. Furthermore, the following conditions must be fulfilled.

1. $v_k \in V \Longrightarrow R \cdot v_k = \theta(v_k)$.
2. $(v_k, v_j) \in E \Longrightarrow (\theta(v_k), \theta(v_j)) \in E$. (2)
3. $f_k = (v_{k1}, v_{k2}, \ldots, v_{kl}) \in F \Longrightarrow \theta(f_k) = (\theta(v_{k1}), \theta(v_{k2}), \ldots, \theta(v_{kl})) \in F$.

The first condition simply says that the location of the rotated version $\theta(v_k)$ of any vertex v_k must be identical to the location obtained by applying the rotation R to v_k. Besides this geometric condition two additional topological constraints must be satisfied. Condition 2 means that the rotated versions $\theta(v_k)$ and $\theta(v_j)$ of any two vertices v_k and v_j must be connected by an edge if v_k and v_j are connected by an edge. Finally, if a chain of vertices forms a face, the rotated version of that chain must form a face, too.

The task of the symmetry detection is to determine all bijective mappings θ satisfying the conditions 1-3. We tackle this problem by a hypothesis-and-verification method.

5.2 Hypothesis generation

Because of the fact that three vertices and their rotated versions under the bijective mapping θ determine the underlying rotation unambiguously, the strategy of hypothesis generation and verification can be stated as follows. We generate all possible mappings of three initial vertices v_{i1}, v_{i2}, v_{i3} and subsequently calculate the rotation matrix R (hypothesis generation). Then the mappings of all remaining vertices $v_{i4}, v_{i5}, \ldots, v_{in}$ under θ can be uniquely determined because the mapping of v_{ij}, $\theta(v_{ij})$, must have the coordinates $R \cdot v_{ij}$ for any $j = 4, 5, \ldots, n$. After the completion of θ the conditions 2 and 3 in (2) will be checked (verification).

For the sake of efficiency, we select three neighboring vertices of the vertex chain of some face as the three initial vertices v_{i1}, v_{i2}, v_{i3}. In this case their possible mappings subject to a rotational symmetry must also be three neighboring vertices of some vertex chain. Thus, there are totally $2m$ different hypotheses.

Given three initial vertices v_{i1}, v_{i2} and v_{i3}, we build a hypothesis $\theta(v_{i1}) = v_{j1}, \theta(v_{i2}) = v_{j2}, \theta(v_{i3}) = v_{j3}$ for all $2m$ possibilities of three neighboring vertices v_{j1}, v_{j2}, v_{j3} of some vertex chain. The transformation matrix R from v_{i1}, v_{i2}, v_{i3} to v_{j1}, v_{j2}, v_{j3} can be calculated by

$$R = [v_{j1} \ v_{j2} \ v_{j3}] \cdot [v_{i1} \ v_{i2} \ v_{i3}]^{-1}$$

Then it is checked whether R is a rotation matrix, that is, whether $RR^T = I_3$ holds. If R is really a rotation matrix the verification is started.

5.3 Verification

For each hypothesis the mapping θ must be completed. This means that we check for each vertex $v_{ik}, k = 4, 5, \ldots, n$, whether there exists a vertex v_{jk} with the coordinates $R \cdot v_{ik}$. For each mapping found in this way, the conditions 2 and 3 in (2) will be verified. Assume that the completion of θ needs t time. In [9, 10] we have shown that the verification of the conditions 2 and 3 runs in $O(m)$ time. As there are totally $2m$ hypotheses, the overall complexity of the symmetry detection algorithm amounts to $t^* = O(m(t + m))$ which crucially depends on t.

With an arbitrary verification order $v_{i4}, v_{i5}, \ldots, v_{in}$ we must examine, for each vertex v_{ik}, all remaining vertices. In this case $t = O(n^2)$ and t^* is of cubic order. In the following we will define a special verification order by the optimal vertex ordering technique which leads to $t = O(m)$.

If v_{ik} is a neighbor of v_{ij}, $\theta(v_{ik})$ must be in the neighborhood of $\theta(v_{ij})$ because of the condition 2. We look for a special verification order satisfying

$$\forall v_{ik}(\exists v_{ij}(1 \leq j < k \wedge (v_{ik}, v_{ij}) \in E)), \text{ for } k \geq 4 \tag{3}$$

In other words, when adding a new vertex to the verification order we consider only such vertices as candidate that are a neighbor of at least one vertex already selected. We call v_{ij} the *father* of v_{ik}, denoted $v_{ij} = f(v_{ik})$. With this special verification order the mapping of v_{ik} under θ can be restricted to the neighbors of $\theta(f(v_{ik}))$. Thus,

$$t = \sum_{k=4}^{n} d_G(\theta(f(v_{ik}))) = \sum_{k=4}^{n} d_G(f(v_{ik}))$$

Clearly, $v_{i1}, v_{i2}, v_{i3}, v_{i4}, \ldots, v_{in}$ is a vertex ordering of G, and T is of the same order as the SFD of this ordering. To reduce T, we use the algorithm described in section 3 to find an optimal vertex ordering beginning with v_{i1}. First, we assume that the optimal ordering looks like

$$v_{i1}, v_{i2}, v_{i3}, v'_{i4}, v'_{i5}, \ldots, v'_{in} \tag{4}$$

with some function f and its SFD is

$$SFD_{opt}(v_{i1}) = d_G(f(v_{i2})) + d_G(f(v_{i3})) + \sum_{k=4}^{n} d_G(f(v'_{ik}))$$

We recall that Steinitz' theorem [6, 7] says that the vertex-edge graph of any convex polyhedron is 3-connected and planar. Furthermore, as any non-convex polyhedron (without holes) is homeomorphic to some convex polyhedron, its graph is also 3-connected and planar. Thus, G is 3-connected and planar. According to Theorem 3,

$$SFD_{opt}(v_{i1}) \leq 4|E| + \max\{d_G(v)|v \in V\}$$

Therefore, if we use the verification order $v'_{i4}, v'_{i5}, \ldots, v'_{in}$,

$$\begin{aligned}
t &= \sum_{k=4}^{n} d_G(f(v'_{ik})) \\
&= SFD_{opt}(v_{i1}) - d_G(f(v_{i2})) - d_G(f(v_{i3})) \\
&\leq 4|E| + \max\{d_G(v)|v \in V\} - d_G(f(v_{i2})) - d_G(f(v_{i3}))
\end{aligned}$$

Thus, $t = O(m)$ and t^* is of quadratic order. If the optiml vertex ordering, however, doesn't have the form (4), i.e. v_{i2} (or v_{i3}) is not the second (or third) vertex of the optimal ordering, we can always construct an ordering satisfying (3) by deleting v_{i2}, v_{i3} in the optimal ordering and subsequently inserting v_{i2}, v_{i3} between v_{i1} and the second vertex. If the optimal ordering has a father function f, we define the father function of the resulting (non-optimal) ordering as f':

$$f'(v_{i2}) = v_{i1}, \ f'(v_{i3}) = v_{i2}, \ f'(v) = f(v), \text{ for } v \in V - \{v_{i1}, v_{i2}, v_{i3}\}$$

Notice that the SFD of the new ordering is of the same order as $SFD_{opt}(v_{i1})$. Thus, $t = O(m)$ holds, too. In summary, the verification order derived from the optimal vertex ordering beginning with v_{i1} guarantees that t^* is quadratic.

The symmetry detection algorithm discussed above is only sketched to highlight the application of the optimal vertex ordering technique. For more details, see [9, 10]. The symmetry information acquired by the algorithm has actually been used in an experimental machine vision system for speeding up the process of 3-D object recognition [8, 11].

6 Conclusion

In this paper we have investigated the problem of optimal vertex ordering of a graph. The measure of optimality was defined as the sum of fathers' degrees (SFD). We have demonstrated that this problem can be transformed into the well-known minimum-weight spanning tree problem. Thus, any of the known algorithms for that purpose can be used to solve our optimal vertex ordering problem. We have proved that, for a 3-connected planar graph, the SFD of the optimal ordering is linearly bounded by the number of vertices and edges. For graphs which are not 3-connected or not planar, examples were given to illustrate that the linear upper bound is no longer guaranteed to be true. Finally, we have discussed the application of our optimal vertex ordering technique to an algorithm for detecting the symmetry of polyhedra.

Acknowledgements

The authors would like to express their sincere thanks to B. Jaggi of the Mathematical Institute of the University of Bern for many valuable discussions. In particular, the proof of Theorem 1 is due to B. Jaggi which greatly simplifies our original proof by induction.

References

[1] S. G. Akl, The design and analysis of parallel algorithms, Prentice Hall, 1989.

[2] D. W. Barnette, Trees in polyhedral graphs, *Canad. J. Math,*, **18**, 731–736, 1966.

[3] D. Cheriton, R. E. Tarjan, Finding minimum spanning-trees, *SIAM J. on Comput.*, **5** (4), 724–742, 1976.

[4] E. W. Dijkstra, An note on two problems in connection with graphs, *Numerische Mathematik*, **1** (5), 269–271, 1959.

[5] H. N. Gabow, Two algorithms for generating weighted spanning trees in order, *SIAM J. on Comput.*, **6**, 139–150, 1977.

[6] B. Grünbaum, Convex polytopes, Interscience Publishers, 1967.

[7] B. Grünbaum, Polytopal graphs, in D. R. Fulkerson (Ed.), *Studies in graph theory*, The Mathematical Association of America, 201–224, 1975.

[8] X. Y. Jiang, H. Bunke, Recognition of overlapping convex objects using interpretation tree search and EGI matching, *Proc. of SPIE Conf. on Applications of Digital Image Processing XII*, Vol. 1153, San Diego, 611–620, 1989.

[9] X. Y. Jiang, H. Bunke, Determining symmetry of polyhedra, Proc. of Int. Workshop on Visual Form, Capri, Italy, 1991.

[10] X. Y. Jiang, H. Bunke, Detektion von Symmetrien polyedrischer Objekte, in R. E. Grosskopf (Ed.), *Mustererkennung 1990*, Proc. of 12th German Pattern Recognition Symposium (DAGM), Informatik Fachberichte 254, Springer-Verlag, 225-231, 1990.

[11] X. Y. Jiang, H. Bunke, Recognizing 3-D objects in needle maps, *Proc. of 10th Int. Conf. on Pattern Recognition*, Atlantic City, New Jersey, 237–239, 1990.

[12] D. B. Johnson, Priority queues with update and minimum spanning trees, *Inf. Process. Lett.*, **4**, 53–57, 1975.

[13] A. Kershenbaum, R. Van Slyke, Computing minimum spanning trees efficiently, *Proc. Ass. Comput. Math. Conf.*, 517–528, 1972.

[14] V. Kevin, M. Whitney, Algorithm 422 — Minimal spanning tree, *CACM*, **15**, 273–274, 1972.

[15] J. B. Kruskal Jr., On the shortest spanning sub-tree and the traveling salesman problem, *Proceedings of Annual ACM Conference*, 518–527, 1956.

[16] R. C. Prim, Shortest connection networks and some generalisations, *Bell System Tech. J.*, **36**, 1389–1401, 1957.

[17] M. J. Quinn, N. Deo, Parallel graph algorithms, *Computing Surveys*, **16** (3), 319–348, 1984.

[18] A. C. Yao, An $O(|E| \log \log |V|)$ algorithm for finding minimum spanning tree, *Inf. Process. Lett.*, **4**, 21–23, 1975.

EDGE SEPARATORS FOR GRAPHS OF BOUNDED GENUS WITH APPLICATIONS

Ondrej Sýkora and Imrich Vrťo *
Computing Centre, Slovak Academy of Sciences
Dúbravská 9, 84235 Bratislava, Czecho-Slovakia

Abstract

We prove that every n-vertex graph of genus g and maximal degree k has an edge separator of size $O(\sqrt{gkn})$. The upper bound is best possible to within a constant factor . This extends known results on planar graphs and similar results about vertex separators. We apply the edge separator to the isoperimetric number problem, graph embeddings and lower bounds for crossing numbers.

1 Introduction

Many divide and conquer algorithms on graphs are based on finding a small set of vertices or edges whose removal divides the graph roughly in half. Applications include VLSI layouts [14], Gaussian elimination [15] and graph embeddings [16].

Formally, a class of graphs has $f(n)$ vertex (edge) separator if every n-vertex graph in the class has a vertex (edge) cutset of size $f(n)$ that divides the graph in two parts having no more than $2n/3$ vertices. Lipton and Tarjan [17] proved that planar graphs have $O(\sqrt{n})$ vertex separator. The genus of a graph is the minimum number of handles that must be added to a sphere so that the graph can be imbedded in the resulting sphere with no crossing edges. Djidjev [6] and Gilbert, Hutchinson and Tarjan [9] proposed algorithms that find an $O(\sqrt{gn})$ vertex separator for graphs of genus g. Further generalization was done in [19,22]. Miller [18] and Diks at al. [4] showed that every n-vertex planar graph of maximal degree k has an $O(\sqrt{kn})$ edge separator. Extensions of these results can be found in [8].

In this paper we prove that any n-vertex graph of positive genus g and maximal degree k has an $O(\sqrt{gkn})$ edge separator. This bound is best possible to within a constant factor. The separator can be found in $O(g + n)$ time provided that we start with an imbedding of the graph in its genus surface. We apply the edge separator to the isoperimetric problem, to efficient embedding of graphs of genus g into various classes of graphs including trees, meshes and hypercubes and to finding lower bounds on crossing numbers of $K_n, K_{m,n}$ and Q_n drawn on surfaces of genus g.

*Both authors were supported by a research grant from Humboldt Foundation, Bonn, Germany

2   Separation of graphs of genus g

2.1   Upper bound

We prove a stronger "weighted" version of the edge separator theorem mentioned in introduction. Before proving it we state some notions and an important lemma.

Let $G = (V, E)$ be an n-vertex graph of genus $g > 0$ and maximal degree k whose vertices have nonnegative weights summing to 1 such that no weight exceed 2/3. Let us denote the sum of weights of vertices belonging to a set X as weight(X). Let $\mid E \mid = m$.

Lemma 2.1 [7] *If G has a breadth first search spanning tree of radius r and rooted in a vertex t then there is a partition of V into three sets A, B, C such that no edge joins a vertex in A with a vertex in B, weight$(A) \leq 2/3$, weight$(B) \leq 2/3$ and $\mid C \mid \leq (4g + 2)r + 1$, where $t \in C$.*

Theorem 2.1 *There is a partition of V into sets A, B and a set of edges D such that weight$(A) \leq 2/3$, weight$(B) \leq 2/3$, $\mid D \mid \leq 5\sqrt{3gkn}$ and every edge between A and B belongs to D.*

Proof: Let G be connected and t be an arbitrary vertex of G. Suppose that G has a breadth first search spanning tree of radius r rooted in t. According to Lemma 2.1 there is a set of vertices C separating G such that $\mid C \mid \leq (4g + 2)r + 1$.

Assume that $r < \sqrt{m/((4g + 2)k)}$. Let D be the set of all edges incident to the set C. Hence

$$\mid D \mid < \left(\sqrt{\frac{m}{(4g + 2)k}}(4g + 2) + 1 \right) k \leq (\sqrt{6} + 1)\sqrt{gkm}.$$

By distributing properly the set C between A and B we receive the desired partition.

If $r \geq \sqrt{m/((4g + 2)k)}$ then we set $s = \sqrt{m/((4g + 2)k)}$. We can assume $s > 1$, otherwise $m < (4g + 2)k$ and we have a trivial edge separator of size $m = \sqrt{m}\sqrt{m} \leq \sqrt{6gkm}$. Partition the vertices of G into levels $U_0, U_1, ..., U_r$ according to their distance from t. Define $L_j = \{(u, v) \mid u \in U_{i-1}, v \in U_i, i = j \bmod s, i = 1, 2, 3, ..., r\}$. As $\mid \bigcup_{j=0}^{s-1} L_j \mid \leq m$ and $\mid L_i \cap L_j \mid = 0$, for $i \neq j$, there exists s_0 such that $\mid L_{s_0} \mid \leq m/s$. By removing the edges of L_{s_0} the graph G is partitioned into connected components $G_i = (V_i, E_i), i = 1, 2, 3, ...$

If weight$(V_i) \leq 2/3$ for all i then we easily combine the desired partition A, B. Set $D = L_{s_0}$. Thus

$$\mid D \mid \leq \frac{m}{s} \leq \frac{m}{\sqrt{m/((4g + 2)k)}} \leq \sqrt{(4g + 2)km} \leq \sqrt{6gkm}.$$

Let weight$(V_i) > 2/3$ for some i. Let $h(l)$ be the highest (lowest) level of vertices in V_i. Delete the vertices of G at levels $> h$. Shrink the vertices of G at levels $< l$ into the root t. The result is a graph H_i of genus $\leq g$, $\mid V_i \mid + 1$ vertices and radius s. Apply Lemma 2.1 to the graph H_i. We obtain a partition of H_i into sets A_i, B_i and C_i such that

$$\text{weight}(A_i) \leq \frac{2}{3}(\text{weight}(V_i) + \text{weight}(t)), \quad \text{weight}(B_i) \leq \frac{2}{3}(\text{weight}(V_i) + \text{weight}(t))$$

and

$$|\,C_i\,| \le (4g+2)s + 1.$$

Delete t from H_i. Let D_i denote the set of edges incident to C_i. Removing D_i from G_i we partition V_i into two sets having weights $\le 2/3$. Thus we have divided G into components whose weights do not exceed $2/3$. One can easily combine the components into the desired partition A and B. The total number of deleted edges is

$$|\,D\,| = |\,L_{s_0}\,| + |\,D_i\,| \le \frac{m}{s} + (4g+2)ks = 2\sqrt{(4g+2)km} \le 2\sqrt{6gkm}.$$

Finnaly, suppose $g \le n/48$. Then

$$|\,D\,| \le 2\sqrt{6gkm} \le 2\sqrt{6gk(3n+6g)} \le \sqrt{75gkn}.$$

If $g \ge n/48$ then we have a trivial separator of size

$$m = \sqrt{m}\sqrt{m} \le \sqrt{3n+6g}\sqrt{\frac{kn}{2}} \le \sqrt{75gkn}.$$

In case that G is not connected we apply the above proof to the component with the greatest weight. □

Our proof can be directly transformed to an algorithm for finding the edge separator. Provided that we start with an imbedding of G in its genus surface the time complexity is $O(m) = O(g+n)$ because finding both the set L_{s_0} and the vertex cut from Lemma 2.1 requires $O(g+n)$ time.

For some applications it can be useful to have an edge cut that divides the graph into two parts whose numbers of vertices differ at most by 1. Such edge cuts are called bisectors.

Corollary 2.1 *Any n-vertex graph of genus g and maximal degree k has a bisector of size $48\sqrt{gkn}$.*

The proof follows the method used in Corollary 3 of [17].

2.2   Lower bound

In this section we prove that the bound in Theorem 2.1 is tight to within a constant factor whenever $gk = O(n)$. We show this for the unweighted version of Theorem 2.1, i.e. all vertices have the same weight. We essentially use the following lemma.

Lemma 2.2 [9] *There exists a constant α such that for infinitely many $g, n_0, g < n_0$, there is a regular graph G_0 with n_0 vertices, genus g and of degree 6 whose every vertex cut dividing G_0 into parts having $\le 3/4$ vertices has size $\ge \alpha\sqrt{gn_0}$.*

Corollary 2.2 *Every edge cut that divides G_0 into parts having $\le 3n/4$ vertices has size at least $\alpha\sqrt{gn_0}$.*

Proof: Let the claim be false. Hence there is an edge cut D of G_0 of size $|\,D\,| < \alpha\sqrt{gn_0}$ that partitions the vertices of G_0 in A and B, $|\,A\,|, |\,B\,| \le 3n/4$. Let $|\,A\,| \le |\,B\,|$. Delete all vertices that are incident to edges from D and belong to B. We have constructed a vertex cut of size $< \alpha\sqrt{gn_0}$ that divides G_0 into parts having $\le 3n/4$ vertices. □

Theorem 2.2 *For $k = 0 \bmod 6$ and infinitely many $g, n, gk < 2n$ there is a graph G of n vertices, genus g and maximal degree k such that every edge separator of G has size $\Omega(\sqrt{gkn})$.*

Proof: Consider the graph G_0 from Lemma 2.2. Replace each edge of G_0 by $k/6$ new parallel edges. Put one new vertex on each new edge. We get a graph $G = (V, E)$ of genus g, maximal degree k and with n vertices where $n = n_0 k/2 + n_0$. Consider a minimal edge separator of G, i.e. we have a partition of V in A, B such that $| A | \leq 2n/3, | B | \leq 2n/3$ and every edge between A and B belongs to a set D. Our aim is to prove that $| D | = \Omega(\sqrt{gkn})$. Let $V_0 \subset V$ denote the set of vertices that correspond to the vertices of G_0. Denote $A_1 = A \cap V_0, B_1 = B \cap V_0$. It holds $| A_1 | + | B_1 | = n_0$. Assume $| A_1 | \leq | B_1 |$. We distinguish two cases:

1. $| A_1 | \geq n_0/4$. Then $| B_1 | \leq 3n_0/4n$. According to Corollary 2.2 , G contains at least $\alpha\sqrt{gn_0}$ tuples (u, v) such that $u \in A, v \in B$ and u and v are joined by $k/6$ paths of length 2. From each such path at least one edge must belong to D. Hence

$$| D | \geq \alpha\frac{k}{6}\sqrt{gn_0} = \alpha\frac{k}{6}\sqrt{\frac{2gn}{k+2}} = \frac{\alpha\sqrt{6}}{12}\sqrt{gkn}.$$

2. $| A_1 | < n_0/4$. We estimate the number of vertices in $A - A_1$ that have a neighbour in B_1. Each such vertex together with the neighbour defines an edge that must belong to D. Hence

$$| D | \geq | A - A_1 | - | \{\text{vertices of } A - A_1 \text{ that have both neighbours in } A_1\} | \geq$$

$$\geq \frac{n}{3} - | A_1 | - | A_1 | \frac{k}{2} \geq \frac{n}{3} - \frac{n_0}{4}\left(1 + \frac{k}{2}\right) \geq \frac{n}{12} \geq \frac{\sqrt{6}}{24}\sqrt{gkn}. \quad \square$$

3 Applications

In this section we apply the edge separator to the isoperimetric problem, to graph embeddings and to finding lower bounds for crossing numbers of complete, bipartite and hypercube graphs drawn on a surface of genus g.

3.1 Isoperimetric number

The isoperimetric number $i(G)$ of a graph $G = (V, E)$ is defined as

$$i(G) = \min\{\frac{| \delta(X) |}{| X |} : X \subset V, 1 \leq |X| \leq \frac{|V|}{2}\},$$

where $\delta(X)$ is a set of edges having one edge in X and the other in $V - X$. The quantity $i(G)$ is a discrete analog of the well-known Cheeger [2] isoperimetric constant measuring the minimal possible ratio between the size of the surface and the volume of a geometric figure. Isoperimetric numbers for important graphs are computed in [20]. Boshier [1] proved that if G is an n-vertex graph of genus g and maximal degree k then

$$i(G) \leq \frac{3(g+2)k}{\sqrt{n/2 - 3(g+2)}},$$

for $n > 18(g+2)^2$. Our edge separator immediately implies the following improvement.

Theorem 3.1

$$i(G) \leq 15\sqrt{\frac{3gk}{n}}.$$

Proof: Let A, B and D be the sets from the unweighted version of Theorem 2.1. Clearly

$$i(G) = \min_{X \subset V} \frac{|\delta(X)|}{|X|} \leq \frac{|D|}{\min\{|A|, |B|\}} \leq 15\sqrt{\frac{3gk}{n}}. \qquad \Box$$

3.2   Graph embeddings

Many computational problems can be mathematically formulated as the graph embedding, e.g. representing some kind of data structure by another data structure [16], simulation of interconnection networks of parallel computers [21] and laying out circuits in standard format [14].

Let $G_1 = (V_1, E_1), G_2 = (V_2, E_2)$ be graphs such that $|V_1| \leq |V_2|$. An embedding of G_1 into G_2 is a couple of mappings (ϕ, ψ) satisfying

$$\phi : V_1 \to V_2 \quad \text{is an injection}$$

$$\psi : E_1 \to \{\text{set of all simple paths in } G_2\}$$

such that if $(u, v) \in E_1$ then $\psi((u, v))$ is a path between $\phi(u)$ and $\phi(v)$.

We shall study the following measures of the quality of the embedding

$$\mathrm{adil}(\phi, \psi) = \sum_{e \in E_1} \frac{|\psi(e)|}{|E_1|},$$

where $|\psi(e)|$ denotes the length of the path $\psi(e)$.

$$\exp(\phi, \psi) = \frac{|V_2|}{|V_1|}$$

$$\mathrm{cg}(\phi, \psi) = \max_{e \in E_2} |\{f \in E_1 : e \text{ belongs to } \psi(f)\}|.$$

The above measures are called the average dilation, expansion and congestion.

Lipton and Tarjan [16] pointed out that edge separators can be used to embed graphs with a small average dilation. Applying their method Diks et al. [4,5] proved the following results on graph embeddings.

Theorem 3.2 *Every n-vertex planar graph of maximal degree k can be embedded in the path, 2-dimensional mesh, d-dimensional mesh $(d \geq 3)$, complete binary tree and hypercube with average dilations $\Theta(\sqrt{kn}), O(\sqrt{k}\log(n/k)), \Theta(d\sqrt[d]{k}), \Theta(\log k), O(\log k)$ and expansion $O(1)$.*

Using the same method we can extend this result as follows.

Theorem 3.3 *Every n-vertex graph of genus g and maximal degree k can be embedded in the path, 2-dimensional mesh, d-dimensional mesh $(d > 2)$, complete binary tree and hypercube with average dilations: $O(\sqrt{gkn}), O(\sqrt{gk}\log(n/k)), O(d\sqrt{g}\sqrt[d]{k}), O(\sqrt{g}\log k)$ and $O(\sqrt{g}\log k)$, and expansion $O(1)$.*

For the first average dilation we show an optimal lower bound. We use a method of [3].

Lemma 3.1 *Let $\mid V_1 \mid = n$ and G_2 be an n-vertex path. Let $f_p(n)$ be the size of the minimal edge cut that divides G_1 into two parts having exactly p and $n - p$ vertices. Then*

$$\mathrm{adil}(\phi, \psi) \geq \frac{1}{\mid E_1 \mid} \sum_{i=1}^{n/2} f_i(n).$$

Theorem 3.4 *Any embedding (ϕ, ψ) of the graph G constructed in Theorem 2.2 into the n-vertex path requires*

$$\mathrm{adil}(\phi, \psi) = \Omega(\sqrt{gkn}).$$

Proof: Let β be the constant behind the Ω in Theorem 2.2. Then setting $G_1 = G$ in Lema 3.1 we get

$$\mathrm{adil}(\phi, \psi) \geq \frac{1}{\mid E \mid} \sum_{i=1}^{n/2} f_i(n) \geq \frac{1}{\mid E \mid} \sum_{i=n/3}^{n/2} f_i(n) \geq \left(\frac{n}{2} - \frac{n}{3}\right) \beta \sqrt{gkn} \frac{k+2}{2kn} = \Omega(\sqrt{gkn}). \quad \square$$

When G_2 is a path then the minimal congestion is usually called cutwidth. This notion has applications in VLSI design. Yannakakis [23] stated an open problem to find a good approximation to the cutwidth of planar graphs. This was partially solved in [4]. We extend the result for graphs of positive geni.

Theorem 3.5 *Any n-vertex graph of positive genus g and maximal degree k has a cutwidth of $O(\sqrt{kgn})$. This bound can not be improved in general.*

Proof: By breaking the graph recursively into roughly equal parts using the edge separator and embedding the parts into subpaths one can easily estimate the cutwidth by

$$\sqrt{75kg}(\sqrt{n} + \sqrt{\frac{2}{3}n} + \sqrt{\left(\frac{2}{3}\right)^2 n} + ...) = O(\sqrt{gkn}).$$

Because the cutwidth of a graph is not smaller than the size of its minimal edge separator the graph from Theorem 2.2 has the cutwidth at least $\Omega(\sqrt{gkn})$. $\quad \square$

3.3   Lower bounds on crossing numbers

In this subsection we apply Theorem 2.1 to finding lower bounds for the crossing numbers of complete, bipartite and hypercube graphs drawn in an orientable surface of genus g.

The orientable surface S_g of genus g is obtained from a sphere by adding g handles. The crossing number $\mathrm{cr}_g(G)$ of a graph G is defined as the least number of crossings when G is drawn in S_g. Very little is known on $\mathrm{cr}_g(G)$. In [10] and [11] it is proved that

$$\mathrm{cr}_1(K_n) = \Theta(n^4), \quad \mathrm{cr}_1(K_{m,n}) = \Theta(m^2 n^2).$$

Kainen[12] showed that

$$\mathrm{cr}_g(Q_n) = \Theta(\gamma - g),$$

for $g \geq \gamma - 2^{n-4}$, where Q_n denotes the n-dimensional hypercube graph and γ its genus.

The following theorem which describes a lower bound methods for finding crossing numbers, was originally proved for planar graphs [13]. Our extension to graphs of genus g is straightforward.

Theorem 3.6 *Let $G = (V, E)$ be a graph. Let $\mathrm{mec}(G)$ denote the size of the minimal edge cut that divides G in two parts having $\leq 2 \mid V \mid /3$ vertices. Suppose that the class of graphs of genus g and of maximal degree k has an $f_{g,k}(n)$ edge separator. Then*

$$\mathrm{cr}_g(G) \geq f_{g,k}^{-1}(\mathrm{mec}(G)) - \mid V \mid .$$

Corollary 3.1

$$\mathrm{cr}_g(G) \geq \frac{\mathrm{mec}^2(G)}{75gk} - \mid V \mid . \tag{1}$$

Proof: From Theorem 2.1 we have $f_{g,k}(n) = \sqrt{75gkn}$.

Theorem 3.7

$$\mathrm{cr}_g(K_n) > \frac{n^4}{6075g} - \frac{n^3}{2} .$$

Proof: Define a new graph H_n as follows. Consider a drawing of K_n in S_g with minimal number of crossings. Let v be an arbitrary vertex of K_n. Let $u_0, u_1, u_2, ..., u_{n-2}$ be its neighbours. In S_g find a region hoheomorfic to an open disc that contains v and no crossings. Denote $p = \lfloor n/2 \rfloor$. Place new vertices $u_{i1}, u_{i2}, ..., u_{ip}$ on the edge (u_i, v) so that they lie in the region, for $i = 0, 1, 2, ..., n - 2$. Add edges $(u_{i,j}, u_{(i+1)\mathrm{mod}(n-1),j})$ for $i = 0, 1, 2, ..., n - 2$ and $j = 1, 2, 3, ..., p$. Delete the vertex v. The resulting graph H_n has $n(n-1)p$ vertices, genus g and degree 4. Clearly it holds

$$\mathrm{cr}_g(K_n) \geq \mathrm{cr}_g(H_n) \tag{2}$$

It remains to find $\mathrm{mec}(H_n)$. In what follows, we show that

$$\mathrm{mec}(H_n) \geq \frac{2}{9}n^2. \tag{3}$$

Setting $G = H_n$ and substituting 2, 3 into 1 we obtain the desired result.

Lemma 3.2

$$\mathrm{mec}(H_n) \geq \frac{2}{9}n^2.$$

Proof: We use a method of Leighton [13]. Recall the definition of the embedding and the congestion. Leighton proved that

$$\mathrm{mec}(G_2) \geq \frac{\mathrm{mec}(G_1)}{\mathrm{cg}(\phi, \psi)}.$$

Let $2K_{n(n-1)p}$ denote the complete graph on $n(n-1)p$ vertices whose each edge is replaced by two new parallel edges. Set $G_2 = H_2, G_1 = 2K_{n(n-1)p}$. We construct an embedding (ϕ, ψ) of G_1 in G_2 such that

$$\mathrm{cg}(\phi, \psi) \leq 2(n-1)^2 p^2.$$

Noting that

$$\mathrm{mec}(2K_{n(n-1)p}) \geq \frac{4}{9}n^2(n-1)^2p^2$$

we immediately receive the lower bound for $\mathrm{mec}(H_n)$.

We shall construct 2 paths between any two vertices of H_n so that the congestion be as small as possible. Let us call the graph induced by the vertices $u_{ij}, i = 0, 1, 2, ..., n-2, j = 1, 2, 3, ..., p$, a cobweb.

Let u_{ij} and u_{rs} belong to the same cobweb. Suppose $j \geq s$.

If $i = r$ then we join u_{ij} and u_{rs} by two shortest (identical) path.

If $i \neq r$ then we join u_{ij} with u_{rs} by two paths $u_{ij}, u_{i+1,j}, ..., u_{rj}$ and $u_{ij}, u_{i-1,j}, ..., u_{rj}$ and prolong these paths to u_{rs} as above.

Let two vertices of H_n belong to different cobwebs. Let (x, y) be an edge joining these cobwebs. Join $x(y)$ to the vertex belonging to the same cobweb as $x(y)$ by two path as above. Connect the paths by adding twice the edge (x, y). Detailed counting analysis shows that

$$\mathrm{cg}(\phi, \psi) \leq 2(n-1)^2p^2. \quad \square$$

Using the same approach as above we can prove:

Corollary 3.2

$$\mathrm{cr}_g(K_{m,n}) > \frac{m^2n^2}{1200g} - \frac{mn(m+n)}{2}, \quad \mathrm{cr}_g(Q_n) > \frac{4^n}{1500g} - n^2 2^{n-1}.$$

4 Conclusions

Our paper leaves several open questions: e.g. improving and completing the upper and lower bounds in Theorems 2.1, 2.2 and 3.3. The most interesting seems to be the problem of drawing of $K_n, K_{m,n}$ and Q_n in the surface of orientable genus g. We can find drawings that have $O(\sqrt{g})$-times more crossings than the proved lower bounds.

Acknowledgment. The authors thank to Professor Kurt Mehlhorn, Max Planck Institut für Informatik, Saarbrücken for all his support. Many thanks also go to Hristo Djidjev for helpful discussions.

References

[1] Boshier, D. O., Enlarging properties of graphs, Ph.D. Thesis, Royal Holloway and Bedford New Colledge, University of London, 1987.

[2] Cheeger, J., A lower bound for the smallest eigenvalue of the Laplacian, In: Problems in Analysis, Princeton University Press, 1970, 159-199.

[3] De Millo, R.A., Eisenstat, S. C., Lipton, R. J., Preserving average proximity in arrays, Communications of the ACM, 21, 1978, 228-230.

[4] Diks, K., Djidjev, H. N., Sýkora, O., Vrťo, I., Edge separators for planar graphs and their applications, In: Proc. of the 13-th Symposium on Mathematical Foundations of Computer Science, LNCS 324, Springer Verlag, 1988, 280-290.

[5] Diks, K., Djidjev, H. N,. Sýkora, O., Vrťo, I., Edge separators for planar graphs with applications, Journal of Algorithms, to appear.

[6] Djidjev, H. N., A liner algorithms for partitioning graphs of fixed genus, SERDICA Bulgaricae Mathematicae Publications, 11, 1985, 369-387.

[7] Djidjev, H. N., A separator theorem, Comptes Rendus de l'Academie Bulgare des Sciences, 34, 1981, 643-645.

[8] Gazit, H., Miller, G. L., Planar separators and Euclidean norm, In: Proc. Algorithms, International Symposium SIGAL'90, 1990, 338-347.

[9] Gilbert, J. R., Hutchinson, J., Tarjan, R. E., A separator theorem for graphs of bounded genus, Journal of Algorithms, 5, 1984, 391-401.

[10] Guy, R. K., Jenkins, T., Schaer, J., The toroidal crossing number of the complete graph, Journal of Combinatorial Theory, 4, 1968, 376-390.

[11] Guy, R. K.,Jenkins, T., The toroidal crossing number of the $K_{m,n}$. Journal of Combinatorial Theory, 6, 1969, 235-250.

[12] Kainen, P. C., On the stable crossing number of cubes, In: Proceedings of the American Mathematical Society, 36, 1972, 55-62.

[13] Leighton, F. T., New lower bound techniques for VLSI, In: Proc. of the 22-nd Annual IEEE Symposium on Foundations of Computer Science, 1981, 1-12.

[14] Leiserson, C. E., Area-efficient graph layouts (for VLSI), In: Proc. of the 21-st Annual IEEE Symposium on Foundations of Computer Science, 1980, 270-280.

[15] Lipton, R. J., Rose, D. J., Tarjan, R. E., Generalized nested dissection, SIAM Journal on Numerical Analysis, 16, 1979, 346-358.

[16] Lipton, R. J., Tarjan, R. E., Application of a planar separator theorem, SIAM Journal on Computing, 9, 1980, 615-627.

[17] Lipton, R. J., Tarjan, R. E., A separator theorem for planar graphs, SIAM Journal on Applied Mathematics, 36, 1979, 177-189.

[18] Miller, G. L., Finding small simple cycle separators for 2-connected planar graphs, Journal of Computer and System Science, 32, 1986, 265-279.

[19] Miller, G. L., Thurston, W., Separators in two and three dimensions, Proc. of the 22-nd Annual ACM Symposium on Theory of Computing, 1990, 300-309.

[20] Mohar, B., Isoperimetric numbers of graphs, Journal of Combinatorial Theory, B, 47, 3, 1989, 274-291.

[21] Monien, B., Sudborough, I. H., Comparing interconnection networks, In: Proc. of the 13-th Symposium on Mathematical Foundations of Computer Science, LNCS 324, Springer Verlag, 1988, 138-153.

[22] Noga, A., Seymour, P., Thomas, R., A separator theorem for graphs with an excluded minor and its applications, In: Proc. of the 22-nd Annual ACM Symposium on Theory of Computing, 1990, 203-299.

[23] Yannakakis, M., Linear and book embedding of graphs, In: Proc. of the Aegean Workshop on Computing, LNCS 227, Springer Verlag, 1986, 229-240.

Line Digraph Iterations and the Spread Concept—with Application to Graph Theory, Fault Tolerance, and Routing

Ding-Zhu Du,* Yuh-Dauh Lyuu,† and D. Frank Hsu‡

Abstract

This paper is concerned with the spread concept, line digraph iterations, and their relationship. A graph has spread (m, k, l) if for any $m + 1$ distinct nodes $x, y_1, \ldots, y_m$ and m positive integers $r_1, \ldots, r_m$ such that $\sum_i r_i = k$, there exist k node-disjoint paths of length at most l from x to the y_i, where r_i of them end at y_i. This concept contains, and is related to, many important concepts used in communications and graph theory. The line digraph of a digraph $G(V, E)$ is the digraph $L(G)$ where nodes represent the edges of G and there is an edge (x, y) in $L(G)$ if and only if x represents the edge (u, v) in G and y represents the edge (v, w) in G for some $u, v, w \in V(G)$. Many useful graphs, like the de Bruijn and Kautz digraphs, can be generated by line digraph iterations. We prove an *optimal* general theorem about the spreads of digraphs generated by line digraph iterations. Then we apply it to the de Bruijn and Kautz digraphs to derive optimal bounds on their spreads, which improve, re-prove, or resolve previous results and open questions on the connectivity, diameter, k-diameter, diameter vulnerability, and some other issues related to length-bounded disjoint paths, of these two graphs.

Key words. graph, connectivity, spread, line digraph iteration, fault tolerance, diameter vulnerability, de Bruijn graph, Kautz graph, container, k-diameter.

1 Introduction

In this paper, we use graph-theoretical terms to discuss networks. A (direct) interconnection network can be represented by a directed graph (**digraph**) where each node of the digraph is a processor with its local memory, buffer, and switching element, and each

---
*Computer Science Department, Princeton University and Center for Discrete Mathematics and Computer Science at Rutgers. This research is supported by grant NSF-STC88-09648. On leave from Institute of Applied Mathematics, Chinese Academy of Sciences, Beijing.

†NEC Research Institute, Princeton.

‡Department of Computer and Information Science, Fordham University.

edge of the graph is a link of the network. Many concepts in graph theory have found their use in the design and analysis of communications networks, especially in areas like transmission delay and fault tolerance [18]. This paper brings them together under a unifying and general concept called spread and proves a general result, which bounds tightly the spreads of digraphs generated by line digraph iterations. We then apply the result to the familiar de Bruijn and Kautz digraphs.

A graph has spread (m, k, l) if for any $m + 1$ distinct nodes $x, y_1, \ldots, y_m$ and m positive integers $r_1, \ldots, r_m$ such that $\sum_i r_i = k$, there exist k node-disjoint paths of length at most l from x to the y_i, where r_i of them end at y_i. We also say a graph has spread (k, l) if it has spreads (i, k, l) for $1 \leq i \leq k$. Sometimes, we mean by spread the last component l only when the first component k is understood to be fixed. We say (k, l) is a tight spread for a graph if $(k, l - 1)$ is *not* its spread.

Spread is a good measure for fault tolerance and transmission delay due to its simultaneous bounds on both path length and number of node-disjoint paths. Surely path length lower-bounds the transmission delay for a message. The consideration of fault tolerance further requires that the number of node-disjoint paths be large *and* the paths short, the first because of less vulnerability to disconnection and the second because shorter paths reduce a message's probability of encountering a faulty component. In fact, the spread concept contains or is related to many important concepts used in communications and graph theory. They include connectivity, diameter, k-diameter, diameter vulnerability, and many others related to length-bounded node-disjoint paths (see Remark 2.7 and Theorem 2.8); hence, bounds on spread immediately imply bounds on those related concepts. The spread concept is also general enough to address many other concerns, including one-to-many communication cases such as (1) send k messages to a node, (2) send one message to each of k other nodes, (3) send r_i messages to node y_i ($1 \leq i \leq m$), and (4) choose r_i large for more important nodes y_i and send r_i identical messages to y_i ($0 \leq i \leq m$) lest some of them get lost, all *via* node-disjoint paths of length at most l.

The line digraph of a digraph $G(V, E)$ [4] is the digraph $L(G)$ where nodes represent the edges of G and there is an edge (x, y) in $L(G)$ if and only if x represents the edge (u, v) in G and y represents the edge (v, w) in G for some $u, v, w \in V(G)$. Line digraph iterations simply refers to the process of generating digraphs by iteratively applying the L operator. Many useful classes of graphs, such as the de Bruijn and Kautz graphs, can be generated by line digraph iterations (see Section 2). These two graphs furthermore have many good properties for use in interconnection networks [1, 17]. The concept of line digraph is due to Krausz [4].

The connection between spread and line digraph iterations is through our main result, Theorem 3.3, which says if graph $G = L^i(H)$, where H has spread (k, l) and spread[*1] $(k - 1, l)$, then G has spread $(k, l + i)$. This result reduces the calculation of spread

---

[1] The definition of spread[*] is the same as that of spread except that x is allowed to be one of the y_i, paths must be of *positive* length, and two consecutive edges of H's are deleted (see Definition 3.1 for details).

to that of the ground graph H. This theorem is also **optimal** in that it achieves the *tight* spreads for some classes of graphs—the Kautz and de Bruijn graphs, to be specific (Claims 5.1 and 5.3).

In Section 4, Theorem 3.3 is applied to the de Bruijn and Kautz graphs to show that the d^D-node de Bruijn graph with diameter D has spread $(d-1, D+1)$ and the $d^D + d^{D-1}$-node Kautz graph with diameter D has spread $(d, D+2)$. These results imply previous results or resolve open questions on their connectivity, diameter, k-diameter, diameter vulnerability, and other properties related to length-bounded disjoint paths [1, 8, 9, 10, 13, 15].

Line digraph iterations has limitations in that there are classes of digraphs that cannot be thus generated, for instance the symmetric digraphs (Section 6). In Section 7, we conclude with some remarks and open questions.

2 Definitions and Related Concepts

We will restrict ourselves to *strongly connected* digraphs and use $V(G)$ to denote the nodes, $E(G)$ the edges, $\kappa(G)$ the connectivity, and $D(G)$ the **diameter** of graph G. All paths will have *positive* lengths. Given an edge e, e' will denote the node it is incident *from* and e'' the node it is incident *to*. Note that if e is a self-loop, then $e' = e''$. We will allow at most a *single* self-loop at each node and not allow parallel edges.

Remark 2.1 It is known [1, 5, 6, 15] that if G is d-regular and has diameter D, then $L(G)$ is d-regular and has diameter $D+1$ if G is not a circuit. Furthermore, the node (edge) connectivity of $L(G)$ is at least that of G. Note also that if e is a self-loop in G, then there is a self-loop at e in $L(G)$, and *vice versa*. Furthermore, there is at most one self-loop at any node in $L(G)$.

Two graphs will be of special interest in this paper. The **de Bruijn graph** $B(d, D)$ with in- and out-degree d and diameter D is the graph whose nodes are labeled with words of length D from an alphabet of d letters. There is an edge from $(x_1, \ldots, x_D)$ to the nodes $(x_2, \ldots, x_D, \alpha)$ where α is any letter. This graph has d^D nodes. In the **Kautz graph** [11] $K(d, D)$ with in- and out-degree d and diameter D, the nodes are labeled with words $(x_1, \ldots, x_D)$, where x_i belongs to an alphabet of $d+1$ letters and $x_i \neq x_{i+1}$ for $1 \leq i \leq D-1$. The node $(x_1, \ldots, x_D)$ is incident to the d nodes $(x_2, \ldots, x_D, \alpha)$ where α is any letter different from x_D. This graph has $d^D + d^{D-1}$ nodes. Both the de Bruijn and the Kautz graphs can be constructed by line digraph iterations:

Fact 2.2 *([6, 15]) Let K'_d be the complete graph on d nodes with a loop at each node, then $L^{D-1}(K'_d)$ is $B(d, D)$. Let K_{d+1} be the complete graph on $d+1$ nodes (without loops), then $L^{D-1}(K_{d+1})$ is $K(d, D)$.*

Before embarking on our new concept, spread, we give two notions that lead naturally to it. Given a graph G, let a k-**container**, defined by Meyer and Pradhan [12], between nodes u and v be a set of k node-disjoint paths between u and v, its **length** defined by the length of the longest path in it. A k-**best container** between two nodes u and v is a k-container with the shortest length. We define the k-**distance** [8] between u and v to be the length of a k-best container between them. The following definition is given by Hsu and Lyuu:

Definition 2.3 *([8]) For any graph G, the k-diameter $d_k(G)$ is the maximum k-distance over all pairs of distinct nodes in G.*

It was shown by Imase, Soneoka, and Okada [10] that $B(d, D)$ has $(d - 1)$-diameter equal to $D+1$, just one greater than its diameter. They and others [9, 10, 15] also showed that $K(d, D)$ has d-diameter equal to $D + 2$, only greater than its diameter by two. We note that the connectivity of $B(d, D)$ and $K(d, D)$ is precisely $d-1$ and d, respectively (see [1, p. 115] for the references). Results on k-diameter for loop networks, hypercubes, and some other graphs are also known [8, 12, 16].

We now turn to the second notion formalized by Hsu and Lyuu [8] with motivation from [14]. Given positive integers k and l, let $\mathcal{R}(k, l)$ be the class of graphs G so that for any $k + 1$ distinct nodes $x, y_1, \ldots, y_k \in V(G)$, there exist node-disjoint paths (except at x) of length at most l from x to $y_1, \ldots, y_k$. $\mathcal{R}_k(l)$ will denote the subclass of graphs in $\mathcal{R}(k, l)$ such that k is exactly their connectivity. The next theorem relates connectivity and membership in $\mathcal{R}(k, l)$:

Theorem 2.4 *([3, Theorem 2.6]) G is k-connected iff $G \in \mathcal{R}(k, l)$ for some l.*

Unfortunately, the above theorem has little to say about l for general graphs. Furthermore, it is NP-hard to decide, given k and l, if a graph is in $\mathcal{R}(k, l)$ [8]. For specific graphs, Rabin [14] showed that the n-dimensional hypercube is in $\mathcal{R}_n(n + 1)$, which is the best possible. Hsu and Lyuu [8] showed that the loop network with d^n nodes is in $\mathcal{R}_n((d - 1)n + 1)$, generalizing Rabin's result. We now define the spread concept:

Definition 2.5 *Given positive integers m, k, l, where $m \leq k$, let $\mathcal{L}^*(m, k, l)$ be the class of graphs G so that for any $m + 1$ distinct nodes $x, y_1, \ldots, y_m \in V(G)$ and m positive integers $r_1, \ldots, r_m$ such that $\sum_i r_i = k$, there exist k node-disjoint paths (except at x) each of length at most l from x to $y_1, \ldots, y_m$, where r_i of them go to y_i for $1 \leq i \leq m$. A graph's spread is (m, k, l) if it is in $\mathcal{L}^*(m, k, l)$.*

Definition 2.6 *Define $\mathcal{L}^*(k, l)$ to be $\displaystyle\bigcap_{1 \leq m \leq k} \mathcal{L}^*(m, k, l)$. $\mathcal{L}^*_k(l)$ will denote the subclass of graphs in $\mathcal{L}^*(k, l)$ with connectivity k. We also say a graph G has spread (k, l) if $G \in \mathcal{L}^*(k, l)$. G has tight spread (k, l) if $G \in \mathcal{L}^*(k, l) - \mathcal{L}^*(k, l - 1)$.*

Remark 2.7 *If a graph is in $\mathcal{L}^*(1, k, l)$ (resp. $\mathcal{L}^*(1, 1, l)$), then its k-diameter (resp. diameter) is at most l, and vice versa. Note that $\mathcal{L}^*(k, k, l) = \mathcal{R}(k, l)$ and $\mathcal{L}^*(k, l) \subseteq \mathcal{R}(k, l)$. Define s-diameter vulnerability [10] of a graph as the maximum of the diameters of the graphs formed by removing s arbitrary nodes. Clearly, if $G \in \mathcal{L}^*(1, k + 1, l)$, then its s-diameter vulnerability is at most l for $s \leq k$.*

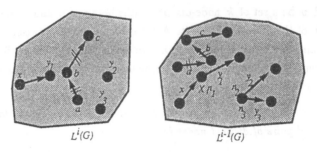

$$L^i(G) \qquad\qquad L^{i-1}(G)$$

Figure 1: Note that, in $L^{i-1}(G)$, n_2 and n_3 are identical; so are X and n_1.

As for the connection between spread and connectivity, we have:

Theorem 2.8 G *is* k-*connected if and only if* $G \in \mathcal{L}^*(k, l)$ *for some* l.

Proof: ($\Rightarrow$) Consider $m + 1$ distinct nodes, $x, y_1^0, \ldots, y_m^0$. Let $r_1, \ldots, r_m$ be m positive integers such that $\sum_i r_i = k = \kappa(G)$. Note that $m \leq k$. For each y_i^0, choose $r_i - 1$ of its neighbors, $y_i^1, \ldots, y_i^{r_i-1}$, that are incident *to* it such that the y_i^j are $\sum_i r_i$ *distinct* nodes different from x, which can be done because $\sum_i r_i = \kappa(G)$. Theorem 2.4 implies that there are node-disjoint paths from x to the y_i^j. By extending each path ending at y_i^j, where $j > 0$, to y_i^0 using the edge (y_i^j, y_i^0), we have shown that there exist $\sum_i r_i = k$ node-disjoint paths from x to $y_1^0, \ldots, y_m^0$, where r_i of them go to y_i^0 for $1 \leq i \leq m$.

($\Leftarrow$) Apply Theorem 2.4 with the fact that $\mathcal{L}^*(k, l) \subseteq \mathcal{R}(k, l)$.

$\square$

3 Spread and Line Digraph Iterations

Our main focus here is the relationship between line digraph iterations and $\mathcal{L}^*(k, l)$. We need an intermediate definition for the proof.

Definition 3.1 *Given positive integers* k, l, *let* $\mathcal{L}^\bullet(k, l)$ *be the class of graphs* G *so that for any* $m \leq k$ *distinct nodes* $y_1, \ldots, y_m \in V(G)$, *a node* $x \in V(G)$, 2 *edges* e_0 *and* e_1 *where* $e_0'' = e_1'$, *and* m *positive integers* $r_1, \ldots, r_m$ *such that* $\sum_i r_i = k$, *there exist* k *node-disjoint paths (except at* x *) each of positive length at most* l *from* x *to* $y_1, \ldots, y_m$, *where* r_i *of them go to* y_i *for* $1 \leq i \leq m$, *and none involves* e_0 *and* e_1. *Note:* x *may be one of the* y_i. *(The* INTRODUCTION *calls this notion* spread*.*)*

Lemma 3.2 $L^h(G) \in \mathcal{L}^\bullet(k, l+h)$ *if* $G \in \mathcal{L}^\bullet(k, l)$, *for* $h \geq 0$.

Proof: The proof is by induction on i, the depth of line digraph iterations. This lemma surely holds for $i = 0$. Assume it is true for i up to $h - 1$ and consider $i = h > 0$.

Consider a node x and m distinct nodes, $y_1, \ldots, y_m$, in $L^i(G)$. These nodes are edges in $L^{i-1}(G)$. Let $r_1, \ldots, r_m$ be positive integers such that $\sum_j r_j = k$. Define $n_j = y'_j$ and $X = x''$, in $L^{i-1}(G)$. Assume, in $L^i(G)$, two edges (a, b) and (b, c) are to be avoided. Figure 1 illustrates the arrangements. Since the n_j may *not* be distinct (some of them may even be X), let $U = \{n_j \mid 1 \le j \le m\} = \{n_1^\#, n_2^\#, \ldots\}$, where the $n_j^\#$ are distinct, and associate each $n_j^\# \in U$ with a positive number $r_j^\# = \sum_{l \text{ where } n_l = n_j^\#} r_l$. Clearly,

$\sum_j r_j^\# = k$.

By the induction hypothesis, there are k node-disjoint paths, in $L^{i-1}(G)$, from X to the $n_j^\#$ such that $r_j^\#$ of them go to $n_j^\#$, each path has length at most $l + i - 1$, and none involves edges a and b. Or, equivalently, these node-disjoint paths are from X to the n_j, where r_j of them go to n_j (here we treat n_a and n_b with $a \ne b$ as distinct even if $n_a = n_b$). A typical path looks like

$$X \xrightarrow{\hspace{1.5cm}\alpha\hspace{1.5cm}} \cdots \to n_j .$$

Note that none of the edges $y_1, \ldots, y_m$ can appear in α. Extend α to

$$x' \xrightarrow{x} X \overbrace{\to \cdots \to n_j}^{\alpha} \xrightarrow{y_j} y''_j$$

with β being its induced path in $L^i(G)$. Surely, (a, b) and (b, c) cannot appear in β (since none of the α's contains edges a and b and $|\alpha| > 0$ by assumption, avoiding cases like $x = a$, $y_j = b$, and $|\alpha| = 0$, which produces (a, b) in β), and the β's are node-disjoint (since the y_j are distinct and in none of the α's). $\qquad\square$

Theorem 3.3 $L^h(G) \in \mathcal{L}^*(k, l + h)$ *if* $G \in \mathcal{L}^\bullet(k - 1, l) \cap \mathcal{L}^*(k, l)$, *for* $h \ge 0$.

Proof: The proof is by induction on i, the depth of line digraph iterations. This theorem surely holds for $i = 0$. Assume it is true for i up to $h - 1$ and consider $i = h > 0$.

Consider $m + 1$ distinct nodes, $x, y_1, \ldots, y_m$, in $L^i(G)$, where $m \le k$. These nodes are edges in $L^{i-1}(G)$. Let $r_1, \ldots, r_m$ be positive integers such that $\sum_j r_j = k$. Assume $x'' = y'_1 = y'_2 = \cdots = y'_s$, in $L^{i-1}(G)$ for some $0 \le s \le m$, with $s = 0$ meaning no (x, y_i) is an edge in $L^i(G)$. Hence, in $L^{i-1}(G)$, the edge x is incident to edges y_i, $1 \le i \le s$. See Figure 2 for illustration.

There are two cases to consider: $s = 0$ and $s > 0$. Suppose $s = 0$. By the induction hypothesis, there are k node-disjoint paths from x'' to $y'_1, \ldots, y'_m$, r_j of them ending at y'_j and each of length at most $l + i - 1$, in $L^{i-1}(G)$ (here we treat y'_a and y'_b with $a \ne b$ as distinct even if $y'_a = y'_b$). Typical paths look like

$$x'' \xrightarrow{\hspace{1.5cm}\alpha\hspace{1.5cm}} \cdots \to y'_j .$$

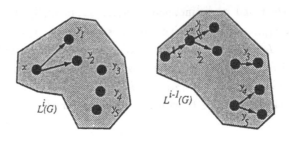

Figure 2: Here, $s = 2$ and $m = 5$.

The paths extended to

$$x' \xrightarrow{x} x'' \to \cdots \to y'_j \xrightarrow{\overset{\alpha}{\frown}} y''_j,$$

they induce, in $L^i(G)$, the desired node-disjoint paths from x to the y_i.

Suppose, on the other hand, $s > 0$. Define $y_{m+j} = y_j$ for $1 \le j \le s$ with $r_{m+1} = r_1 - 1$ and $r_{m+j} = r_j$ for $2 \le j \le s$ (if $r_{m+1} = 0$, just drop node y_{m+1} from the induction step). Consider the graph obtained from $L^{i-1}(G)$ by deleting edges x and y_1. By Lemma 3.2 and considering $y'_{s+1}, \ldots, y'_{m+1} (= y'_{m+2} = \cdots = y'_{m+s} = x'')$, we conclude that there are $\sum_{j=s+1}^{m+s} r_j = k - 1$ node-disjoint paths in $L^{i-1}(G)$ from x'' to $y'_{s+1}, \ldots, y'_{m+s}$, r_j of them ending at y'_j (here we treat y'_a and y'_b with $a \ne b$ as distinct even if $y'_a = y'_b$), each of (positive) length at most $l + i - 1$, and none involving x and y_1. Typical paths look like

$$x'' \to \cdots \to y'_j \quad \text{for } s+1 \le j \le m+s$$

The paths extended to

$$x' \xrightarrow{x} x'' \to \cdots \to y'_j \xrightarrow{y_j} y''_j, \quad \text{for } s+1 \le j \le m+s,$$

they induce, in $L^i(G)$, $k-1$ node-disjoint paths from x to the y_i, each of length at most $l + i$ and none involving the edge (x, y_1). Add (x, y_1) to get the kth paths in $L^i(G)$ and the proof is complete.

□

Remark 3.4 The above theorem starts from the ground case: $G \in \mathcal{L}^*(k-1, l) \cap \mathcal{L}^*(k, l)$. As we shall see later, this seems to be the only source that introduces "non-optimality" when applied to specific graphs (see Section 5 and CONCLUSIONS).

4 Application to de Bruijn and Kautz Graphs

Theorem 3.3 is powerful in that, to know if $L^i(G) \in \mathcal{L}^*(k, l)$, we only have to check if the ground graph $G \in \mathcal{L}^*(k-1, l-i) \cap \mathcal{L}^*(k, l-i)$, which is extremely easy for the de

Bruijn and Kautz graphs because both start with a complete graph (Fact 2.2). It is easy to show that $B(d,1) \in \mathcal{L}^\bullet(d-2,2) \cap \mathcal{L}^*(d-1,2)$ and $K(d,1) \in \mathcal{L}^\bullet(d-1,3) \cap \mathcal{L}^*(d,2)$ for $d \geq 3$. Applying Theorem 3.3, we have

Theorem 4.1 *For $d \geq 3$, $B(d,D) \in \mathcal{L}^*_{d-1}(D+1)$ and $K(d,D) \in \mathcal{L}^*_d(D+2)$.*

The following corollary, which solves the open problem posed in a manuscript to [8], follows from Remark 2.7 and Theorem 4.1.

Corollary 4.2 *For $d \geq 3$, $B(d,D) \in \mathcal{R}_{d-1}(D+1)$ and $K(d,D) \in \mathcal{R}_d(D+2)$.*

Hence, we have re-proved, improved, or resolved previous results or conjectures on connectivity, diameter vulnerability, k-diameter, and membership in $\mathcal{R}(k,l)$ of the de Bruijn and Kautz graphs; in fact, more is proved here since our results are about the much stronger spread concept.

5 Optimality of Theorem 3.3

We show that Theorem 3.3 is optimal in the sense that there exists a class of graphs, generated by line digraph iterations, whose tight spreads are matched by the theorem. Specifically, we show that the $(d-1)$-diameter vulnerability of $K(d,D)$ is exactly $D+2$ and that of $B(d,D)$ is $D+1$, which, in conjunction with Remark 2.7 and Theorem 4.1, proves our contention.

Claim 5.1 *For $d \geq 3$ and $D \geq 2$, the $(d-1)$-diameter vulnerability of $K(d,D)$ is $D+2$.*

Proof: Let $a = (\ldots,0,1,0,1,0,1,0)$ and $b = (0,2,0,\ldots)$ and suppose the following $d-1$ nodes are to be avoided: $(\ldots,0,1,0,1,0,1,0,i)$ for $2 \leq i \leq d$. Clearly, any path from a to b must use the edge from a to $c = (\ldots,0,1,0,1,0,1)$, and on any shortest path from c to b, the next node from c must be some $d = (\ldots,0,1,0,1,0,1,i)$ with $i \neq 0$ in order not to return to a. The proof is complete by observing that the distance from d to b is D.

$\square$

Remark 5.2 Here we have a case where $G \in \mathcal{L}^*(k,l)$ but $L(G) \notin \mathcal{L}^*(k,l+1)$; i.e., $K(d,1) \in \mathcal{L}^*(d,2)$ but $L(K(d,1)) = K(d,2) \in \mathcal{L}^*(d,4) - \mathcal{L}^*(d,3)$. Note that $K(d,1) \in \mathcal{L}^\bullet(d-1,3) - \mathcal{L}^\bullet(d-1,2)$.

Similar arguments applied to the distance between $(1,1,\ldots,1)$ and $(d,d,\ldots,d)$ with nodes $(1,1,\ldots,i)$ for $3 \leq i \leq d$ deleted can prove:

Claim 5.3 *For $d \geq 3$ and $D \geq 2$, the $(d-2)$-diameter vulnerability of $B(d,D)$ is $D+1$.*

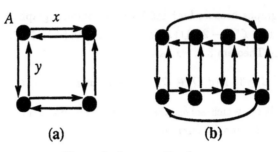

Figure 3: A non-optimal case.

That Theorem 3.3 gives very tight bounds is "intuitively" obvious from the fact that each line digraph iteration increases the diameter by one—except when the graph is a circuit (Remark 2.1)—exactly the same *increment* by which the theorem bounds the more general spread. That seems to leave the ground case (Remark 3.4) as the only source of non-optimality for Theorem 3.3 when applied to specific graphs.[2] We give one more example of non-optimality at the ground case, besides Remark 5.2. Figure 3a shows a graph that is in $\mathcal{L}^{\bullet}(1,4) \cap \mathcal{L}^{\star}(2,3) \subseteq \mathcal{L}^{\bullet}(1,4) \cap \mathcal{L}^{\star}(2,4)$ (by considering a path from node A back to itself with edges x and y deleted for the $\mathcal{L}^{\bullet}(1,4)$ part). Theorem 3.3 would imply that its line digraph, shown in Figure 3b, is in $\mathcal{L}^{\star}(2,5)$, but the sharper bound is actually $\mathcal{L}^{\star}(2,4)$.

6 Limitation of Line Digraph Iterations

Equipped with the powerful Theorem 3.3, we have derived sharp bounds for two important classes of graphs. However, limitations exist in this approach to generate graphs: there are classes of graphs which are not derivable from line digraph iterations. One such example is symmetric digraphs, where (a,b) is an edge if (b,a) is. To see that a symmetric digraph G with at least three nodes, a, b, and c, cannot be a line digraph of any graph H, consider, without loss of generality, edge (a,b). Now, (b,a) is also an edge, since G is symmetric, and, without loss of generality, (a,c) and (c,a) are both edges, since G is strongly connected. But they together mean that, for G to be a line digraph of H, between two nodes a' and a'' in H there are at least three edges: a, b, and c, contradicting the assumption that H has no parallel edges. In general, sufficient and necessary conditions for a digraph to be a line digraph can be found in, for instance, [7, Theorem 2] and [2, Theorem 5].

---

[2]To be more specific, it may be the case that $G \in \mathcal{L}^{\star}(k,l)$ but $G \notin \mathcal{L}^{\bullet}(k-1,l)$, forcing Theorem 3.3 to conclude $G \in \mathcal{L}^{\star}(k,l+2)$ instead of $G \in \mathcal{L}^{\star}(k,l+1)$, which may actually hold as well, as the next example in text shows.

7 Conclusions

We have given a general graph-theoretical notion, spread, that might be helpful in measuring fault tolerance and transmission delay for interconnection networks. An optimal theorem is also proved, linking spread with line digraph iterations. We now comment on some possible future directions. The method of line digraph iterations is a good way of generating graphs; however, its also has severe limitations. Can line digraph iteration be generalized to generate graphs without losing its basic properties? It is not true that if $G \in \mathcal{L}^*(k, l) - \mathcal{L}^*(k, l-1)$, then $L(G) \in \mathcal{L}^*(k, l+1) - \mathcal{L}^*(k, l)$, one counter-example being the Kautz graph (see Remark 5.2), at least in the *first* iteration. It would be interesting to know if it holds for G, itself generated by a line digraph iteration. Finally, it would be interesting to see if, as in the case of diameter, a tight spread of (k, l) for G implies $L(G) \notin \mathcal{L}^*(k, l)$ when G is not a circuit.

Acknowledgments. The second author would like to thank C.W. Gear and anonymous referees for improvements over the manuscript.

References

[1] J.-C. Bermond, N. Homobono, and C. Peyrat, *Large Fault-Tolerant Interconnection Networks*, Graphs and Combinatorics, 5, No. 2 (1989), pp. 107–123.

[2] L.W. Beineke, *On Derived Graphs and Digraphs*, Beiträge zur Graphentheorie, Teubner, Leipzig, 1968, pp. 17–23.

[3] B. Bollobás, *Extremal Graphs Theory*, Academic Press, New York, 1978.

[4] F. Buckley and F. Harary, *Distance in Graphs*, Addison-Wesley, Reading, Massachusetts, 1990.

[5] G. Chartrand and M.J. Stewart, *The Connectivity of Line-Graphs*, Mathematische Annalen 182 (1969), pp. 170–174.

[6] M.A. Fiol, J.L.A. Yebra, and I. Alegre, *Line Digraph Iterations and the (d, k) Digraph Problem*, IEEE Trans. on Computers, C-33, No. 5 (May 1984), pp. 400–403.

[7] F. Harary and R.Z. Norman, *Some Properties of Line Digraphs*, Rendiconti del Circolo Matematico di Palermo, 9 (1960), pp. 161–168.

[8] D.F. Hsu and Y.-D. Lyuu, *A Graph-Theoretical Study of Transmission Delay and Fault Tolerance*, to appear in Proc. Fourth ISMM International Conference on Parallel and Distributed Computing and Systems, 1991.

[9] M. Imase, T. Soneoka, and K. Okada, *Connectivity of Regular Directed Graphs with Small Diameters*, IEEE Trans. on Computers, C-34, No. 3 (March 1985), pp. 267–273.

[10] M. Imase, T. Soneoka, and K. Okada, *Fault-Tolerant Processor Interconnection Networks*, Systems and Computers in Japan, 17, No. 8 (August 1986), pp. 21–30. Translated from Denshi Tsushin Gakkai Ronbunshi, 68-D, No. 8 (August 1985), pp. 1449–1456.

[11] W.H. Kautz, *Bounds on Directed (d, k) Graphs*, Theory of Cellular Logic Networks and Machines, AFCKL-68-0668 Final Report, 1968, pp. 20–28.

[12] F.J. Meyer and D.K. Pradhan, *Flip-Trees: Fault-Tolerant Graphs with Wide Containers*, IEEE Trans. on Computers, C-37, No. 4 (April 1988), pp. 472-478.

[13] D.K. Pradhan and S.M. Reddy, *A Fault-Tolerant Communication Architecture for Distributed Systems*, IEEE Trans. on Computers, C-31, No. 9 (September 1982), pp. 863–870.

[14] M.O. Rabin, *Efficient Dispersal of Information for Security, Load Balancing, and Fault Tolerance*, J. ACM, 36, No. 2 (April 1989), pp. 335–348.

[15] S.M. Reddy, J.G. Kuhl, S.H. Hosseini, and H. Lee, *On Digraphs with Minimum Diameter and Maximum Connectivity*, Proc. 20th Annual Allerton Conference on Communication, Control, and Computing, 1982, pp. 1018–1026.

[16] Y. Saad and M.H. Schultz, *Topological Properties of Hypercubes*, IEEE Trans. on Computers, C-37, No. 7 (July 1988), pp. 867-872.

[17] M.R. Samatham and D.K. Pradhan, *The De Bruijn Multiprocessor Network: A Versatile Parallel Processing and Sorting Network for VLSI*, IEEE Trans. on Computers, C-38, No. 4 (April 1989), pp. 567-581.

[18] A.S. Tanenbaum, *Computer Networks*, Englewood Cliffs, Prentice-Hall, New Jersey, 1981.

A Generalized Encryption Scheme
Based on Random Graphs

Luděk Kučera

Charles University

Prague, Czechoslovakia

Abstract

A probabilistic encryption scheme is generalized in such a way that an encryption is sometimes ambiguous (the same object might be an encryption of both 0 and 1), but a probability of this event is very low. Such a generalization is sufficient for a large number of applications.

An implementation of the scheme presented in the paper is based on random graphs. From the point of view of the worst-case complexity, the problem of decrypting is provable NP-complete, while classical encryption schemes use always problems from $NPP \cap coNPP$ (and most graph problems of this class are either easily solvable or related to linear programming, which makes it possible to use the ellipsoidal method to break such a scheme). However, it is not known whether there are polynomial time deterministic or probabilistic decoding algorithms with probability of success bounded away form 1/2. It is shown that some obvious polynomial time attacks are too weak to break the scheme with sufficiently large probability.

1 Introduction

An encryption scheme is a sequence of functions

$$F_n : \{0,1\} \times \{0,1\}^m \to \mathcal{R}, \tag{1}$$

where $\mathcal{R}$ is some class of mathematical objects and $m = m(n)$ is a polynomial time computable function of n, such that

$$F_n(0, w_0) \neq F_n(1, w_1) \qquad \text{for any } w_0, w_1 \in \{0,1\}^m, \tag{2}$$

and there is a polynomial time algorithm computing functions F_n.

An encryption scheme can be used to encrypt one bit of information as follows:

Suppose that n, the security parameter, is given (and publicly known). Given a bit b, compute $m = m(n)$, generate a sequence w of m (private) random bits and publish $F_n(b, w)$.

There is no obvious way to decrypt easily using some (secret) additional information, but it is possible to make the information about the bit b public so that it is hopefully difficult to find the actual value of b, but later the encryptor can prove to everyone what the value of b was. Among different applications, let us mention zero knowledge proofs based on cryptographical assumptions [4].

Properties of the functions F_n imply that the encryption can be performed by an algorithm working in polynomial (in n) time and is unambiguous, i.e. the decryption is unique, see (2).

The main problem of any implementation is the security of the scheme. Even if we suppose that $P \neq NP$, there are two principal sources of potential attacks:

(i) First of all, problems that are difficult from the worst-case point of view are quite often solvable almost surely by very simple and fast algorithms. An algorithms giving a correct decryption with large probability would make an encryption scheme useless.

(ii) Moreover, if we are given $D \in \mathcal{R}$, it is not only easy to prove that D encrypts 1 (i.e. $D = F_n(1, w)$ for some $w \in \{0, 1\}^m$), but also to *disprove* that D encrypts 1 (by proving that it encrypts 0). This implies that the worst case complexity of the decryption might be lower than that of NP-complete problems, because it shows that the problem is a premise problem [EY] belonging to $NPP \cap coNPP$ with premise in NP: the premise

$$\{F_n(b, w) \mid n \in N, b \in \{0, 1\}, w \in \{0, 1\}^m\},$$

the language $\{F_n(1, w)\}$ and the intersection of its complement with the premise are all members of the class NP.

As one of the main goals of the paper is to use random graphs as a base for encryption scheme, the second obstacle is principal, because classical techniques and the ellipsoidal method appear to be strong enough to solve practically all natural graph theoretic problems that belong to $NP \cap coNP$ or $NPP \cap coNPP$.

Therefore we first introduce a generalization of the above definition, which can be used in most cases when (1) is usually applied, and which admit implementations such that, from the worst-case point of view, the decryption problem is NP-complete.

A generalized encryption scheme (GES) is a sequence of polynomial time computable functions

$$G_n : \{0, 1\} \times \{0, 1\}^k \times \{0, 1\}^m \to \mathcal{R} \tag{3},$$

where $k = k(n)$ and $m = m(n)$ are polynomial time computable function of n.

A string $w \in \{0, 1\}^k$ is *fair* (with respect to a given GES) if

$$G_n(0, w, w_0) \neq G_n(1, w, w_1) \qquad \text{for any } w_0, w_1 \in \{0, 1\}^m, \tag{4}$$

and GES is *fair* if a random string $w \in \{0, 1\}^k$ is almost surely fair (i.e. the relative frequency of fair strings in $\{0, 1\}^k$ tends to 1 as $n \to \infty$).

Given a security parameter n, a fair GES can be used to encrypt a bit b as follows:

Determine $k = k(nk)$ and $m = m(n)$, generate a sequence u of k *public* random bits, a sequence v of m *private* random bits, and publish $G_n(b, u, v)$.

It is clear that we can encrypt in polynomial time and if GES is fair then it can be publicly verified that the probability that the encryption is not ambiguous is close to 1. Such a scheme can successfuly replace the classical one in most applications (e.g. in cryptographical zero knowledge proofs).

Our implementation, given in the paragraph 2, guarantees unambiguity with probability $1 - O(exp(-n^c))$ for some constant $c > 0$. The main problem is a potential existence of fast algorithms that would be able to give a correct decryption with probability bounded away from 1/2. We analyze some possible attacks in the paragraph 3, showing that they are unable to have high probability of success in polynomial time.

2 Graph Encryption Scheme (based on the Largest Independent Set Problem)

Throughout the rest of the paper, we suppose that natural numbers n, s and real numbers p, $P \in [0, 1]$ are given (s, p, P are supposed to be polynomial time computable functions of n). To simplify proof, we suppose that there are constants α, β, κ such that

$$0 < \beta < \alpha < \kappa < 1/2, \quad 3\alpha < 2\kappa \qquad \text{and}$$

$$p = n^{-\alpha}, \quad P = n^{-\beta}, \quad s = \lfloor n^{\kappa} \rfloor. \tag{5}$$

Though these values represent the preferred values of p, P, s, the scheme seems to have good properties for many other choices of parameters. (However, if e.g. p and P are constant and s is substantially greater than the square root of n, then there is an easy way to find the decryption in polynomial expected time).

A generalized encryption scheme investigated in the present paper is based on the next algorithm defining functions G_n:

Graph Encryption Scheme

1. $X := \{1, \ldots, n\}$;

2. $Y := \{1, \ldots, s + b\}$;

3. using a *public* source of random bits, build a relation E_{publ} on X as follows: elements i, j are not connected if both belong to Y, otherwise they are connected with probability p (and the probabilities of an existence of edge are independent for different pairs of vertices);

4. using a *private* source of random bits, build a relation E_{priv} on X as follows: elements i, j are not connected if both belong to Y, otherwise they are connected with probability P (and the probabilities of an existence of edge are independent for different pairs of vertices);

5. $E := E_{publ} \cup E_{priv}$;

6. using a *private* source of random bits, build a permutation τ of X;

7. $E_{\tau} := \{ \{\tau(x), \tau(y)\} \mid \{x, y\} \in E \}$;

8. return the graph (X, E_{τ}).

We will call this scheme Graph Generalized Encryption Scheme and denote by GGES. It is obvious that an encryption of 1 is always a graph H with property

$$H \text{ contains an independent set } U \text{ of the size } s + 1, \tag{6}$$

namely the set $\{\tau(x) \mid x \in Y\}$.

Let us say that an independent set Z of a graph H with vertex set X is t-bounded if any vertex $z \in X - Z$ is connected with at least t vertices of Z.

Theorem 1 *With probability at least $1 - n\exp(-n^{\kappa-\alpha}/12)$, a public random string is such that for any possible choice of a private random string the resulting graph H satisfies*

$$H \text{ contains a } sp/2 \text{-bounded independent set } V \text{ of the size } s. \tag{7}$$

Proof: It is clear that it is sufficient to show that any vertex $x \in X - Y$ is likely to be connected by edges of E_{publ} with at least $sp/2$ elements of Y.

Given a fixed $x \in X - Y$, the distribution of the number of vertices of Y such that $\{x, y\} \in E_{publ}$ is the binomial distribution $B(s, p)$ and therefore it follows from the well known bounds to the tail of the binomial distribution (see e.g. [2], I.3,thm.7]) that the probability that this number is smaller then $sp/2$ is at most $\exp(-sp/12)$. Consequently, the probability that there is such a vertex x is at most $(n - s) \exp(-sp/12)$. ♣

Now we prove that the graph generalized encryption scheme is fair. The proof is based on the fact that the output of the algorithm is very unlikely to satisfy both (6) and (7).

Theorem 2 *GGES is fair.*

Proof: Suppose that a graph (X, E_τ), obtained as an encryption of a bit b by the above algorithm is also an encryption of $1 - b$. In view of Theorem 1, (X, E_τ) is likely to satisfy both (6) and (7), which implies that (X, E) also verifies both properties.

Since $|U| > |V|$, there is $x \in U - V$, which is connected with at least $sp/2$ elements of V. Therefore $|U \cup V| \geq s + 1 + sp/2$, $|(U \cup V) - Y| \geq sp/2$ (where Y is the set from the step 2), which implies that there is independent set Z of (X, E) such that $Z - Y$ has at least $sp/4$ elements (either $Z = U - Y$ or $Z = V - Y$).

No matter how E_{priv} is build, no edge of E_{publ} might connect two elements of Z, and the last event occurs with probability at most

$$(1 - p)^{\binom{|Z-Y|}{2}} \leq (1 - p)^{(sp)^2/33} \leq \exp(-p(sp)^2/33) = \exp(-n^{2\kappa - 3\alpha}/33)$$

for large n.♣

3   Analysis of GGES

Any graph can be an output of the algorithm defining GGES and therefore the problem of decrypting is NP-complete. However, there are algorithms solving many NP-complete problems in polynomial expected time with respect to certain input distributions. It would therefore be desirable to prove that the decryption is difficult "on average".

The only theory that allows such proofs is that of Levin [5]. There are two reasons why we are not able to prove that the decrypting of GGES is difficult in his sense. First of all, a list of problems known to be complete on average is still very limited and it does not contain problems with natural input distributions. Second reason seems to be more serious: the average time complexity of known algorithms solving problems that are complete on average is exponential [6], but there exists an $O(n^2 2^{(1+\varepsilon)n^\alpha \ln n})$ time algorithm decrypting GGES with probability $1 - o(1)$:

GGES Decryption Algorithm
Parameter: a constant $\varepsilon > 0$.
if for some set $Z \subset X$ such that $|Z| = (1 + \varepsilon)n^\alpha \ln n$ the set

$$\{y \mid \{y, z\} \text{ is an edge for no } z \in Z\}$$

is an independent set with $s + 1$ elements
then return 1
else return 0.

The time complexity of the algorithm is clearly $O(n^2 2^{(1+\varepsilon)n^\alpha \ln n})$.

Theorem 3 *The GGES Decryption Algorithm is almost surely correct.*

Proof: Since GGES is almost surely fair, an encryption of 0 is unlikely to contain an independent set of the size $s + 1$.

Suppose that $b = 1$ and $Z \subset Y$ is a set of size $(1 + \varepsilon)n^\alpha \ln n$ (for Y, see the step 2 of GGES). Of course, no element of Y is connected with any element of Z by an edge of E. On the other hand, the probability that a fixed element of $X - Y$ is connected to no element of Z is $(1 - p)^{|Z|}$ and therefore the probability that

$$Y \neq \{y \mid \{y, z\} \text{ is an edge of } E \text{ for no } z \in Z\}$$

is not greater than

$$n(1 - p)^{|Z|} \leq n\exp(-p|Z|) \leq n\exp(-(1 + \varepsilon)\ln n) = n^{-\varepsilon}. \qquad \clubsuit$$

Other difficulties might arise from the fact the outputs of the GGES algorithm belong almost surely to a smaller class which could have lower complexity level than NP-complete problems.

First of all, it can be proved that the solution of the largest independent set problem, which gives the decryption, is almost surely unique. However, the class UP of problems with unique solutions has been proved to be polynomial time solvable only if no one-way function exist, which means that the complexity level of UP is sufficiently high for cryptographical purposes.

Since the properties (6) and (7) almost surely excludes each other, it is possible to indicate that D is not likely to be an encryption of a bit b by showing that it encrypts $1 - b$. But this method does not give a valid proof in all cases and therefore we can hope that the complexity level of decrypting GGES is higher than that of the class $NPP \cap coNPP$. However a further complexity theoretic study of GGES seems to be necessary.

The second source of potential breaks of GGES are polynomial time heuristics, that are quite often successful from the point of view of average behavior.

We will investigate two approximation heuristics for the largest independent set problem. Both of them are variant of the next algorithmic scheme:

Largest Independent Set Problem Heuristics
Input: a graph $H = (X, E)$.
begin
$Z := X$; $IS := \emptyset$;
while Z is not empty **do begin**
  choose $z \in Z$;
  $IS := IS \cup \{z\}$;
  $Z := Z - \{z\} - \{x \in Z \mid \{z, x\}$ is an edge of $E\}$;
  end;
end;

The heuristic algorithm that we call RANDOM selects $z \in Z$ in the first statement of the body of the while-loop randomly among all elements of Z. A more sophisticated algorithm SMALLEST chooses always an element of $z \in Z$ which minimize the degree of z in the graph H (if there are more such vertices, one of them is chosen randomly). This strategy is based on the fact that elements of the set Y have slightly smaller expected degree in the graph (X, E) built by the algorithm GGES. (There is even better heuristic, which is similar to SMALLEST, but selects vertices by their degrees with respect to the set Z. It can be shown that if the first vertex is chosen from the largest independent set, the chances that the next vertices will be also chosen from this set increase. However the analysis of the improved algorithm is more complicated and gives essentially the same results as the analysis of SMALLEST and therefore it will be contained in the full version of the paper only).

Since an encryption of 0 is unlikely to contain an independent set of the size $s + 1$ (see Theorem 2), all the above heuristics are likely to decrypt correctly an encryption of 0.

In order to analyze a behavior of heuristics applied to an encryption of 1, we need the next

Theorem 4 *With probability $1 - o(1)$, the independent set $\tau(Y)$ is the only independent set of the size $s + 1$ in an encryption of 1 by GGES*

Proof: Similar to the proof of Theorems 1,2. $Z = \tau(Y)$ is an independent set of the size $s + 1$. If there is another independent set T of the same size, there is $x \in T - Z$. With large probability, x has at least $sp/2$ neighbours in Z and therefore $|T - Z| \geq sp/2$. However, the probability that $T - Z$ is independent is at most

$$(1 - p)^{\binom{sp/2}{2}}.$$

♣

If applied to an encryption of 1, the above heuristics must give an exact solution of the problem, because if they find a smaller solution, they answer 0. Since we have proved that the optimal solution is almost surely unique, the first vertex z chosen in the while loop of the algorithm must belong to the optimal solution, because otherwise it has almost surely at least $sp/2$ neighbours in $\tau(Y)$, which means that the optimal solution is not found.

Theorem 5 *With probability* $1 - o(1)$, *RANDOM does not decrypt correctly an encryption of 1.*

Proof: The probability that the first vertex chosen by RANDOM is in the set $\tau(Y)$ is $(s+1)/n = n^{-(1-\kappa)}$. ♦

The SMALLEST heuristic is much better tool to search for the optimal solution in the output of GGES. It follows from the proof of the next theorem that it would almost surely give the correct decryption if $\kappa > (1 + \beta)/2$. However, if $\kappa < 1/2$ (and it would be sufficient to suppose that $\kappa < (1 + \beta)/2$), we get

Theorem 6 *With probability* $1 - o(1)$, *SMALLEST does not decrypt correctly an encryption of 1.*

Proof: Let $\varepsilon > 0$ be a constant, put $\mathcal{P} = p + P - pP$ (the edge probability of E). We prove later that, with probability $1 - o(1)$,

$$|\deg(y, E) - (n - s - 1)\mathcal{P}| < \sqrt{2(1 + \varepsilon)(n - s - 1)\mathcal{P}(1 - \mathcal{P})\ln(s + 1)} \qquad (8)$$

for any $y \in \tau(Y)$,

$$|\deg(x, E) - (n - 1)\mathcal{P}| < \sqrt{2(1 + \varepsilon)(n - 1)\mathcal{P}(1 - \mathcal{P})\ln(n - s - 1)} \qquad (9)$$

for any $x \in X - \tau(Y)$, and there exists $x \in X - \tau(Y)$ such that

$$\deg(x, E) < (n - 1)\mathcal{P} - \sqrt{2(1 - \varepsilon)(n - 1)\mathcal{P}(1 - \mathcal{P})\ln(n - s)} \qquad (10)$$

As a consequence, if

$$s > 2\sqrt{2(1 + \varepsilon)n(\mathcal{P})^{-1}(1 - \mathcal{P})\ln(n)}$$

then vertices from $\tau(Y)$ are likely to represent the first $s + 1$ vertices with smallest degrees, which would immediately imply that SMALLEST finds the largest independent set. However, if $\kappa < 1/2$ and if we put, say, $\varepsilon = 1/4$ then

$$s\mathcal{P} + \sqrt{2(1 + \varepsilon)(n - s - 1)\mathcal{P}(1 - \mathcal{P})\ln(s + 1)} -$$
$$\sqrt{2(1 - \varepsilon)(n - 1)\mathcal{P}(1 - \mathcal{P})\ln(n - s)} =$$
$$= s\mathcal{P} + (1 + o(1))\sqrt{2(1 + \varepsilon)n\mathcal{P}\kappa\ln n} - (1 + o(1))\sqrt{2(1 - \varepsilon)n\mathcal{P}\ln n} =$$
$$= s\mathcal{P} - (1 + o(1))(\sqrt{1 - \varepsilon} - \sqrt{(1 + \varepsilon)\kappa})\sqrt{2n\mathcal{P}\ln n} <$$
$$< \Theta(n^{\kappa - \beta}) - \Theta(n^{(1 - \beta)/2}) < 0$$

for large n, which implies that the smallest degree vertex is almost surely in $X - \tau(Y)$, which means that the algorithm SMALLEST would not find the set $\tau(Y)$.

Let us now prove (8)-(10). The proof is a generalization of methods used to determine the expected value of the largest degree of a vertex of a random graph. Let Z be a subset of X. Let z be a fixed element of Z and put $W = X - \{z\}$ if $z \in X - \tau(Y)$, $W = X - \tau(Y)$ if $z \in \tau(Y)$ (W is the set of potential neighbours of z). It follows from DeMoivre-Laplace theorem that the probability that the random variable

$$\mathcal{D} = \frac{\deg(z, E) - |W|\mathcal{P}}{\sqrt{|W|\mathcal{P}(1 - \mathcal{P})}} \qquad (11)$$

satisfies

$$\text{Prob}(|\mathcal{D}| > \xi) \sim \frac{2}{\xi\sqrt{2\pi}}e^{-\xi^2/2} < e^{-\xi^2/2} = |Z|^{-(1+\varepsilon)}, \qquad (12)$$

where $\xi = \sqrt{2(1+\varepsilon)\ln|Z|}$.

Since $|\mathcal{D}| > \xi$ means violating (8) or (9) by z in the cases $Z = X$, $Z = \tau(Y)$), resp., the probability that there is such a z is at most $|Z||Z|^{-(1+\varepsilon)} = |Z|^{-\varepsilon} \to 0$.

Let k be the number of vertices satisfying (10). In the same way as above we can prove that the probability that a vertex $x \in X$ satisfies (10) is almost $|X - \pi(Y)|^{-1+\varepsilon} \sim n^{-1+\varepsilon}$, which means that k is roughly n^{ε}. This does not mean immediately that such vertices are likely to exist. However, we can show easily (in the way very similar to [2],chapter III,1) that $(E(k^2) = (1 + o(1))E(k)^2$, and use the Chebychev inequality to show that the probability that $k = 0$ is $o(1)$. ♣

Proofs of theorems 5 and 6 give weak bounds to the probabilities of the failure of investigated heuristics. Sometimes we need sharper bounds, e.g. if we want to prove that prove that GGES can not be broken using the next modification of the above heuristics:

let r, t be a fixed natural number. For each r-element subset A of X apply SMALLEST (RANDOM,resp.) n^t times using A as the initial value of Z (instead of $\emptyset$).

More detailed analysis shows that the probability that even these improved algorithms succeed to find the optimal solution tends to 0 faster than any rational function.

Finally let us mention another potencial way to compromize the GGES, which is based on the fact that a part of the encryption is made public. If the relation E_{priv} were not used, the resulting graph would be an isomorphic copy of (X, E_{publ}), which is one from two publicly known graphs $((X, E_{publ})$ for $b = 0$ and $b = 1$ and the known public random string) and the decryption would reduce to solving the isomorphism problem. Though the position of isomorphism testing in the complexity hierarchy within NP seems to be high, there are fast algorithms solving the problem in most cases(see e.g. [1]). However, the relation E_{priv}, which overlays E_{publ}, contains the majority of all edges of the resulting graph and therefore we conjecture that the knowledge of E_{publ} is of very limited use.

Acknowledgement: The author thanks Sylvio Michali for pointing out a possible application of random graphs in cryptography and for fruitful discussions.

References

[1] Babai,L.,Kučera,L.,Canonical labelling of graphs in linear average time, 20th Annual Symposium on Foundations of Computer Science, Puerto Rico 1979, 39-46

[2] Bollobás B., Random Graphs, Academic Press, London 1985.

[3] Even,S. and Yakobi,Y., Cryptocomplexity and NP-completeness, In Proc. 8th Colloq. on Automata, Languages, and Programming, Lecture Notes in Computer Science,195-207, Springer Verlag, Berlin, 1980.

[4] Goldwasser,S., Micali,S., Rackoff,C., The knowledge complexity of interactive proofs, 17th Annual Symposium on Foundations of Computer Science, Providence, RI, 1985, 291-305.

[5] Levin,L., Average case complete problems, SIAM J. Computing, 15 (1986), 285-286.

[6] Levin,L., personal communication.

Dynamic Algorithms for Shortest Paths in Planar Graphs [*]

Esteban Feuerstein [†]   Alberto Marchetti-Spaccamela [‡]

Abstract

We propose data structures for maintaining shortest path in planar graphs in which the weight of an edge is modified. Our data structures allow us to compute after an update the shortest-path tree rooted at an arbitrary query node in time $O(n\sqrt{\log\log n})$ and to perform an update in $O((\log n)^3)$. Our data structure can be also applied to the problem of maintaining the maximum flow problem in an $s-t$ planar network.

As far as the all pairs shortest path problem is concerned, we are interested in computing the shortest distances between q pairs of nodes after the weight of an edge has been modified. We obtain different bounds depending on q. We also obtain an algorithm that compares favourably with the best off-line algorithm for computing the all pairs shortest path if we are interested in computing only a subset of the $O(n^2)$ possible pairs. Namely, we show how to obtain an $o(n^2)$ algorithm for computing the shortest path between q pairs of nodes whenever $q = o(n^2)$.

1   Introduction

In the last years there has been considerable research interest in the area of dynamic algorithms for graph problems. The aim is to design data structures that support operations that allow one to modify the graph in addition to answer several types of queries. The goal is to compute the new solution in the modified graph without having to recompute it from scratch. Dynamic algorithms are both of theoretical and of practical interest in several applications areas such as high level languages for incremental computations, incremental data flow analysis, interactive network design as well as many other interactive applications ([19], [18], [17], [2]).

Finding shortest paths has been a fundamental and well studied problem in computer science for a long time. In the particular case of planar graphs, Frederickson [10], using the separator theorem for planar graphs of Lipton and Tarjan [14] and a new data structure, the topology-based heap, achieves the best algorithm known up-to-date for the single-source version of the problem with a running time of $O(n\sqrt{\log n})$. As for the all-pairs version of the problem is concerned, in the same paper an $O(n^2)$ algorithm is

---

[*]Work supported by the ESPRIT II Basic Research Action Program of the European Community under contract No.3075 (Project ALCOM) and by MURST project Algoritmi e Strutture di calcolo.

[†]Dipartimento di Informatica e Sistemistica, Università di Roma "La Sapienza", Roma, Italy. On leave from ESLAI (Escuela Superior Latinoamericana de Informática), partially supported by a grant from Fundación Antorchas, Argentina.

[‡]Dipartimento di Matematica, Università dell'Aquila, L'Aquila, Italy

proposed. If we allow a succinct encoding of the shortest path information, this result has been improved to $O(np)$ where p is the minimum cardinality of a subset of the faces that cover all vertices of the graph [11]. Moreover, if we are not interested in knowing the distance between all pairs of nodes, but only q of them, then it is possible to achieve a running time of $O(n \log n + p^2 + q \log n)$ [6]. Note that in the worst case p is $O(n)$.

There have been several papers that consider shortest path problems in dynamic graphs under insertions and/or deletions of edges ([16],[8]). We observe that the running times of the proposed algorithms are not satisfactory: in fact the deletion of an edge has the same complexity as running the best off-line algorithm from scratch; in the case of insertion of new edges the running time for the insertion of m edges is $O(mn^2)$. In the particular case of insertions of new edges with integer edge cost Ausiello et al. improve this bound by an order of magnitude in the amortized sense [1]. If we restrict ourselves to planar graphs we observe that data structures have been proposed for incremental planarity test [3],[4] minimum spanning tree [7], and connectivity [4], but no dynamic data structures have been proposed for the shortest path problem.

In this work, we propose data structures for maintaining shortest path in planar graphs in which the weight of an edge is modified. Particularly, our data structures and algorithms allow to compute after an update the shortest-path tree rooted at an arbitrary query node in time $O(n\sqrt{\log \log n})$ and to perform an update in $O((\log n)^3)$. As far as the all pairs shortest path problem is concerned, we are interested in computing the shortest distances between q pairs of nodes after the weight of an edge has been modified. We obtain different bounds depending on q.

As an interesting application of our data structure, we obtain an algorithm that, in the worst case, compares favourably with the best off-line algorithm for computing the all pairs shortest path if we are interested in computing only a subset of the $O(n^2)$ possible pairs. Namely, we show how to obtain an $o(n^2)$ algorithm for computing the shortest path between q pairs of nodes whenever $q = o(n^2)$. Also in this case the running time depends on q.

As a final application of our results we observe that our data structure can be also applied to the problem of maintaining the maximum flow problem in an $s - t$ planar network.

2 Preliminaries

We refer to [15] for basic definitions and algorithms concerning shortest path problems and planar graphs.

In [10], fast algorithms are proposed for solving single-source and all-pairs shortest paths problems in planar graphs. In this section we will briefly describe the techniques used in that work. The algorithms are based on a *topological division* of the planar graph into regions. A *region* contains two types of vertices, boundary vertices and interior vertices. An *interior vertex* is contained in exactly one region, and is adjacent only to vertices contained in that region, while a *boundary vertex* is shared among at least two (and at most three) regions. To achieve this, the initial planar graph is transformed into a planar graph with no vertex having degree greater than 3. A well known transformation in graph theory [12] that consists in replacing each node with degree greater than 3 by a cycle with cost 0 may be used to do this.

Lipton and Tarjan [14], showed how to find in linear time a set of $O(\sqrt{n})$ nodes that separates the remaining nodes of the graph in two parts that are separated (i.e. not connected).

Definition 1 *Given a planar graph with n vertices and an integer r, a suitable r-division of the graph is a division in $\theta(n/r)$ regions, each with $O(r)$ vertices and $O(\sqrt{r})$ boundary vertices such that i) each boundary vertex belongs to at most 3 regions and ii) each region that is not connected consists of connected components all which share boundary vertices with exactly the same set of either one or two connected regions.*

Clearly the total number of boundary vertices in a suitable r-division is $\theta(n/r)$; moreover, if we define a boundary set to be a maximal set of boundary vertices, then the number of boundary sets that are obtained in a suitable r-division is $O(n/r)$.

Frederickson showed how repeated applications of the separator procedure proposed by Lipton and Tarjan to the subgraphs obtained allows to obtain a suitable r-division.

Fact 1 *[10]. There exists a $O(n\sqrt{\log n})$ algorithm that divides the graph in a suitable r-division.*

The single-source algorithm works as follows: first shortest paths between all pairs of boundary vertices within each region are found. Using Dijkstra's algorithm this costs $O(n\sqrt{r}\log r)$. Then the single-source shortest path tree between all boundary vertices is computed *(main thrust)*. Finally a *mop-up* phase is performed within each region, that consists in extending the shortest paths to every interior vertex. Dijkstra's algorithm is also used for the second and third steps, with the addition of a special data structure, a *topology based heap* that allows to perform the second step in $O(n + (n/\sqrt{r})\log n)$. Letting $r = \log n/\log\log n$ the total complexity of the algorithm is $O(n\sqrt{\log n}\sqrt{\log\log n})$ time.

The topology based heap is a heap with the boundary vertices in the leaves such that vertices from any boundary set being in consecutive leaves. A topology based heap allows to perform the following operations: initialize, deletemin and batched update. A batched update is an operation that updates the values associated with all vertices of a boundary set.

Fact 2 *[10] Given a n vertices planar graph divided in a suitable r-division. Using the associated topology-based heap, the running time of the main thrust of the algorithm is $O(n + n/\sqrt{r}\log n)$.*

A slightly more complicated algorithm based on recursively dividing each region with a faster division algorithm is proposed, that achieves the same results in $O(n\sqrt{\log n})$ time.

In the same paper Frederickson proposes another algorithm that achieves a bound of $O(n^2)$ for the all-pairs shortest path problem. The result is based on the following fact.

Fact 3 *[10] Given a n vertices planar graph it is possible to perform a $O(n\log n)$ preprocessing in such a way to answer each single source shortest path tree in $O(n)$.*

3   Dynamic single-source shortest path

We are interested in implementing the following two operations on a weighted planar digraph:

- *update(i, j, w)*: modify the weight of the edge between nodes i and j, assigning to it the value w, $w \geq 0$.

- *shortest − path(s)*: return the shortest distances from s to all other nodes.

Later in this section we will show how to modify *shortest − path(s)* to obtain the shortest path tree rooted at s.

The data structure we propose for the problem maintains a suitable r-division of the graph (the value of r will be determined later). Besides we maintain the all-pairs shortest distances between boundary vertices within each region. Since the total number of boundary vertices is $O(n\sqrt{r})$ then the space requirement of this data structure is $O(n)$. The shortest-path query can be answered with the following algorithm:

Algorithm for shortest-path (s)

1. If s is an interior vertex, compute the shortest distances from s to the boundary vertices of its region;

2. compute the shortest distances from s to all boundary vertices *(main thrust)*;

3. compute the shortest distances from boundary to interior vertices in every region *(mop-up)*.

The correctness of the algorithm is trivial. As far as the running time is concerned, step 1. will take $O(r\sqrt{\log r})$, as we will run Frederickson's algorithm within the region containing s. For step 2. we will use Frederickson's topology based heap, starting the search from the set of boundary vertices of the region containing s in the case that s is an interior vertex, or in s itself otherwise. By Fact 2, this step will take $O(n + (n/\sqrt{r})\log n)$. For step 3. we will run Frederickson's algorithm in all regions, with the heap initialized with the distances from s to each boundary vertex of the region. This will cost $O(r\sqrt{\log r})$ for each of the $O(n/r)$ regions, $O(n\sqrt{\log r})$ in total. Letting $r = (\log n)^2$ we get an algorithm that allows us to query for the shortest distances from an arbitrary node in $O(n\sqrt{\log \log n})$ time.

As for the *update* operation, it will consist in modifying the values of the all-pairs shortest distances among boundary vertices for the (at most 3) regions in which the edge is included. We could do this running Frederickson's single source shortest-path algorithm for each of the $O(\sqrt{r})$ boundary vertices of these regions, with a running time of $O((\log n)^3 \sqrt{\log \log n})$.

It is possible to improve this bound by observing that we have to re-compute a set of shortest path trees after a modification of an edge's weight, within a single region. To this aim we can use the algorithm presented above, that allows to answer each query in $O(n\sqrt{\log \log n})$ time instead of Frederickson's $O(n\sqrt{\log n})$ time, provided that an appropriate division of the graph in regions and a certain preprocessing has been previously done. Consequently, we will divide each region (of size $O(r)$) in *subregions* of size $O(r')$

with the same characteristics as before, with the additional fact that all "level-one" boundary vertices will automatically be considered also as "level-two" boundary vertices. Note that the total number of boundary vertices in each region is $O(\sqrt{r} + r/\sqrt{r'})$. In order to implement an *update* operation, we have to re-compute all-pairs shortest distances between all boundary vertices within the regions in which the arc is included. As we have these regions divided in sub-regions, the *update* operation can be performed as follows:

Algorithm for update(i,j,w)

1. compute all-pairs shortest distances between level-two boundary vertices of the sub-region that includes the modified arc;

2. compute the all-pairs shortest distances between all level-one boundary vertices of the regions that include the modified arc.

Note that in step 2. no mop-up is needed since we are interested only in boundary vertices. Step 1. can be done using $O(\sqrt{r'})$ times Frederickson's algorithm, or even Dijkstra's, with a total cost of $O(\sqrt{r'}r'\sqrt{\log r'})$ or $O(\sqrt{r'}r'\log r')$, depending on which algorithm is chosen. As for step 2., it can be done running $O(\sqrt{r})$ times the main thrust phase within the region; by Fact 2 this costs $O(\sqrt{r}(r + (r/\sqrt{r'})\log r))$. If we let $r' = (\log r)^2$ then the time complexity of *update* is $O((\log n)^3)$.

A preprocessing phase needs to be done, that consists, besides of the division of the graph in regions and sub-regions, in computing all pairs shortest distances between the boundary vertices of *every* sub-region and *every* region. By Fact 1 the division can be performed in $O(n \log n)$ time. As for the second part of the preprocessing, for each subregion of size r' we need to compute single source shortest path within the region. Applying Frederickson algorithm this costs $O(r'\sqrt{\log r'})$ for each node; by observing that the total number of level-two boundary vertices is $O(n/\sqrt{r'})$ and that each one of them is part of at most a constant number of sub-regions we have a cost of $O(n\sqrt{r'}\sqrt{\log r'})$. The total number of level-one boundary vertices is $O(n/\sqrt{r})$. At most three single-source calculations in regions of size $O(r)$ must be done for each, which, by Fact 2, will require $O((n/\sqrt{r})(r + (r/\sqrt{r'})\log r))$ time. This term clearly dominates the other one, and hence, with the values chosen for r and r' this yields a total preprocessing time of $O(n \log n)$.

We have proved the following

Theorem 1 *There exists a data structure that allows one to mantain shortest distances from arbitrary sources on an n-vertex planar graph with nonnegative edge costs under the modification of an edge's weight with costs $O((\log n)^3)$ for update and $O(n\sqrt{\log \log n})$ for the query. A preprocessing phase is needed with cost $O(n \log n)$. The space requirement is $O(n)$.*

It is easy to modify the data structure so as to be able to answer not only the shortest distances but also the shortest paths tree without additional cost in the running times of our algorithms. However, the space requirement will be of $O(n \log n)$, as it is sufficient to maintain, for each boundary vertex, the shortest paths tree in the region. Since each tree requires $O(r)$ space, and there are $O(n/\sqrt{r})$ boundary vertices in total, the total space needed is of $O(n\sqrt{r})$. Since $r = (\log n)^2$ the claim follows.

Corollary 1 *There exists a data structure that allows one to maintain single-source shortest path trees rooted at arbitrary nodes on an n-vertex planar graph with nonnegative edge costs under the modification of an edge's weight with costs $O((\log n)^3)$ for update and $O(n\sqrt{\log\log n})$ for the query. A preprocessing phase is needed with cost $O(n\log n)$. The space requirement is $O(n\log n)$.*

In some applications, we only want to maintain the distance between a source s and a query node t; that is, we consider the query $dist(s,t)$. In this case, the mop-up need not be done in all the regions but only in that containing t. This fact is not crucial in the off-line algorithm, but it is so in the on-line version, since the mop-up phase is not dominant anymore. More precisely, in the third step of the algorithm for *shortest-path(s)* described earlier, the mop up needs only be done for t's region, and hence, letting $r = (\log n)^2$ we have that the second step becomes dominating, yielding a linear algorithm.

Theorem 2 *There exists a data structure that allows one to maintain one-pair shortest distance for arbitrary nodes on an n-vertex planar graph with nonnegative edge costs under the modification of an edge weight with costs $O((\log n)^3)$ for the modification and $O(n)$ for the query. A preprocessing phase is needed with cost $O(n\log n)$. The space requirement is $O(n)$.*

Analogous to the single-source case, it is possible to have not only the shortest distance between two nodes but also the shortest path between them without any additional cost in what refers to running time, but with an increment in the space complexity, which is in this case of $O(n\log n)$.

Corollary 2 *There exists a data structure that allows one to maintain one-pair shortest path for arbitrary nodes on an n-vertex planar graph with nonnegative edge costs under the modification of an edge weight with costs $O((\log n)^3)$ for the modification and $O(n)$ for the query. A preprocessing phase is needed with cost $O(n\log n)$. The space requirement is $O(n\log n)$.*

Our data structure can also be used to maintain dynamically a maximum flow in an $s - t$ planar network, or just it's value, which correspond respectively to the shortest distances from every node to the root s' and the shortest $s' - t'$ distance in the dual network, as described in [13] and [10].

Corollary 3 *i) A maximum flow in an $s - t$ planar network can be maintained dynamically under the modification of an edge's weight in $O(n\sqrt{\log\log n})$ time.*

ii) The value of a maximum flow in an $s - t$ planar network can be maintained dynamically under the modification of an edge's weight in $O(n)$ time.

In both cases a preprocessing phase is needed, with cost $O(n\log n)$. The space requirement is $O(n)$ for both cases.

4 Dynamic all-pairs shortest paths

In this section we consider the all-pairs shortest path problem in a planar graph in which the weights of the edges can be modified by calling the procedure $update(i,j,w)$. When we perform an all-pairs shortest paths computation, after a graph modification, we are not necessarily interested in all the $(n^2 - n)/2$ shortest distances but only in, let's say, $q = q(n)$ queries of the form $dist(i,j)$, asking for the distance between two nodes i and j.

As an example, suppose $q = O(n)$. In this case, running the best off-line algorithm known for the all-pairs shortest paths problem will take $O(n^2)$ to compute the shortest paths between all possible pairs of nodes, and thus yielding an $O(1)$ query time. The total time for one modification and $O(n)$ queries, will be $O(n^2)$. If, instead, we use the best off-line algorithm known for the single source problem, it will not do better, because, in the worst case, it will be necessary to compute from scratch one single-source problem for every query. Thus, the time required for one modification and $O(n)$ queries will be $O(n^2\sqrt{\log n})$ (or $O(n^2\sqrt{\log\log n})$ if we use the data structure of the previous section), although the modification itself will take $O(1)$.

For this problem we propose different strategies to follow depending on the value of q. Namely, for small values of q (up to $n^{1/2}$), the algorithm performs only the above preprocessing; therefore the cost of a query is $O(n)$ and the cost of an update is $O(n \log n)$. If q is greater than $n^{1/2}$ we decrease the cost of each query, while increasing the cost of updating. In particular we distinguish two cases depending on whether the value of q is greater than a certain value q_1 (to be determined later) or not.

Case a: $q \le q_1$.

In this case we add to the preprocessing the division of the graph in a suitable r-division with $\theta(n/r)$ regions of size $O(r)$ with $O(\sqrt{r})$ boundary vertices each (for a suitable value of r depending on q). We also perform the $O(n \log n)$ preprocessing that allows to answer each single source computation in $O(n)$ and in $O(r)$ within each region. Moreover we compute the shortest path tree rooted at each boundary vertex. Since there are $O(n/\sqrt{r})$ boundary vertices the total cost of the preprocessing is $O(n \log n + n^2/\sqrt{r})$. This preprocessing allows to answer each query as follows ($d(x,y)$ denotes the shortest distance between x and y already computed):

Algorithm for dist(i,j)

1. If either i or j is a boundary vertex the distance between them is already known, and hence no computation is needed;

2. Compute the shortest path tree rooted at i limited to the region containing i;

3. If both i and j are in the same region, the distance between them is already known;

4. If i and j are in different regions, $dist(i,j) = \min\{d(i,b) + d(b,j)\}$, where b is a boundary vertex of i's region.

The time complexity of $dist(i,j)$ is $O(r)$. In fact, step 2. costs $O(r)$ and the complexity of step 4. is $O(\sqrt{r})$, as i belongs to at most 3 regions, and each of these regions contains at most $O(\sqrt{r})$ boundary vertices.

To maintain this data structure, when the weight of an arc is modified, we use the following algorithm:

Algorithm for update(i,j,w)

1. process the graph so as each following single-source computation will take $O(n)$;

2. process the (at most 3) regions in which the arc is included so as each following single-source computation within the region will take $O(r)$;

3. compute single-source shortest paths trees rooted at each boundary vertex.

By Fact 4 step 1. will take $O(n \log n)$ and step 2. $O(r \log r)$. Step 3. will consist in a single-source computation for each boundary vertex of the graph; since there are $O(n\sqrt{r})$ boundary vertices and each single-source computation will cost $O(n)$ the execution time of step 3. is $O(n^2/\sqrt{r})$.

Hence, the total time to perform one update and q queries is $O(n \log n + n^2/\sqrt{r} + qr)$.

<u>Case b</u>: $q > q_1$.

In this case we can further reduce the cost of query while increasing the cost of each *update*. Namely, besides the preprocessing performed in the previous case, we compute all-pairs shortest distances for each region of the division. Since there are $\theta(n/r)$ regions and each all-pairs computation will cost $O(r^2)$, the preprocessing can be performed with a total cost of $O(n \log n + nr + n^2/\sqrt{r})$.

With this modification the algorithm for *dist(i,j)* can be simplified by eliminating step 2.; therefore the cost of each query becomes $O(\sqrt{r})$. In order to maintain the data structure, the algorithm for *update* needs also to be modified. This can be done by adding the following step 4.:

4. compute the all-pairs shortest path distances for the (at most 3) regions in which the modified arc is included.

Since the cost of step 4. is $O(r^2)$, the cost of an update becomes $O(n \log n + r^2 + n^2/\sqrt{r})$ and the total time to perform one update and q queries is $O(n \log n + r^2 + n^2/\sqrt{r} + q\sqrt{r})$.

If we are interested in minimizing the cost of answering q queries and performing one update then it is possible to show that the value of q_1 that determines the optimal choice between the two strategies is $q_1 = n^{4/5}$. With this value we can prove the following theorem.

Theorem 3 *There exists an algorithm that allows one to compute shortest distances between q pairs of vertices on an n-vertex planar graph, after the modification of an edge's weight with the following time complexities (complessive):*

1. *if $q < n^{1/2}$ the running time is $O(qn)$ and the preprocessing cost is $O(n \log n)$;*

2. *if $n^{1/2} \le q \le n^{4/5}$ the running time is $O(n^{4/3}q^{1/3})$ and the preprocessing cost is $O(n^{4/3}q^{1/3})$;*

3. *if $n^{4/5} < q \le n^{6/5}$ the running time is $O(n^{8/5})$ and the preprocessing cost is $O(n^{9/5})$;*

4. *if $n^{6/5} < q \le n^2$ the running time is $O(\sqrt{q}n)$ and the preprocessing cost is $O(n^3/q + n\sqrt{q})$.*

Proof. In case 1. the preprocessing costs $O(n \log n)$, and then each query takes $O(n)$. In case 2., we use procedures *dist* and *update* with $r = n^{4/3}/q^{2/3}$. In cases 3. and 4. we use the modified versions of *dist* and *update*, with $r = n^{4/5}$ (case 3.) and $r = n^2/q$ (case 4.). It is easy to see that the values of r chosen are optimal for each case. □

In our example, to minimize the time for a modification and $O(n)$ queries, it suffices to let $r = n^{4/5}$ to get a total time of $O(n^{8/5})$.

The above analysis can also be used to minimize the running time of an algorithm that computes off-line the shortest paths between q pairs of nodes arbitrarily chosen. In this case we choose the values of q_1 and r in order to minimize the sum of the preprocessing cost and that of answering q queries. This will lead to different values of q_1 and r with respect to the previous theorem as shown in the following.

Theorem 4 *There exists an algorithm that allows one to compute the shortest path between q pairs of vertices on an n-vertex planar graph, with the following time complexities:*

1. *if $q \leq n^{1/2}$ the running time is $O(nq + n \log n)$;*

2. *if $n^{1/2} \leq q \leq n$ the running time is $O(n^{4/3}q^{1/3})$;*

3. *if $n < q \leq n^{4/3}$ the running time is $O(n^{5/3})$;*

4. *if $n^{4/3} < q \leq n^2$ the running time is $O(n\sqrt{q})$.*

Proof. In cases 1. and 2. the analysis is the same as in the previous theorem. In cases 3. and 4. we use the modified versions of *dist* and *update* with $r = n^{2/3}$ (case 3.) and $r = n^2/q$ (case 4.). □

Theorems 3 and 4 depict different algorithms to use depending on the value of q. Note that q must be known in advance, as it determines not only the strategy to follow but also the optimal number of regions in which the graph has to be divided. In fact the analysis of the algorithm must take into account the preprocessing costs that depend on the value of q.

However, if we do not know in advance the value of q, it is possible to modify the algorithm while mantaining the same time bounds described in theorem 4. This can be achieved by an algorithm that process the queries in phases. During a phase the algorithm does not change neither the strategy nor the division of the graph. Each phase starts with a preprocessing based on the optimal strategy and the optimal value of r for the total number of queries processed until that moment. In order to minimize the total processing cost we limit the number of phases as follows. We start a new phase each time q reaches a treshold value at which the strategy of the algorithm must change or the value of q doubles. Therefore, if we process q queries, there are $O(\log q)$ phases.

In this way, for each query we always use the optimal strategy with a division of the graph in a number of regions that is at least one half of the optimal number. Note that if the algorithm uses a division of the graph in a number of regions that is at least one half the optimal one then the running time for answering a query differs from the optimal one by a constant factor. If we consider the preprocessing required to answer q queries, then it is possible to show that, for any given value of q, the total cost incurred for the $O(\log q)$ preprocessing phases is only a constant factor more than the preprocessing cost required by theorem 4 (when we know the value of q in advance).

This implies that the running time of the modified algorithm is a constant factor away from the total cost required to answer the previous queries. Therefore the result stated in theorem 4 holds even though we do not know in advance the value of q.

Acknowledgment: We acknowledge useful discussions with G. Di Battista, G.F. Italiano and U. Nanni.

References

[1] G. Ausiello, G. F. Italiano, A. Marchetti-Spaccamela, U. Nanni, Incremental algorithms for for minimal length paths, *Proc. 1 ACM-SIAM Symp. on Discrete Algorithms*, S.Francisco, 1990.

[2] M. D. Carroll and B. C. Ryder, Incremental data flow analysis via dominator and attribute grammars, *Proc. 15th Annual ACM SIGACT-SIGPLAN Symp. on Principles of Programming Languages*, 1988.

[3] G. Di Battista and R.Tamassia, Incremental planarity testing, *Proc. 30th annual Symp. on Foundations of Computer Science*, 1989.

[4] G. Di Battista and R.Tamassia, On-line graph algorithms with SPQR-trees, *Proc. 17th Int. Coll. on Automata, Languages and Programming*, Lect. Not. in Comp. Sci., Springer-Verlag, 1990.

[5] E. W. Dijkstra, A note on two problems in connection with graphs, *Numer. Math.,1* (1959), pp. 269-271

[6] H.N. Djidjev, G.E. Pantziou, C.D. Zaroliagis, Computing Shortest Paths and Distances in Planar Graphs, *Proc. ICALP 1991*, Madrid, to appear in Lect. Not. in Comp. Sci. Springer-Verlag.

[7] D. Eppstein, G. F. Italiano, R. Tamassia, R. E. Tarjan, J. Westbrook, M. Young, Maintenance of a minimum spanning forest in a dynamic planar graph, *Proc. 1 ACM-SIAM Symp. on Discrete Algorithms*, S.Francisco, 1990.

[8] S. Even and H. Gazit, Updating distances in dynamic graphs, *Methods of Operations Research 49*, 1985.

[9] M. L. Fredman and R. E. Tarjan, Fibonacci Heaps and their uses in improved Network optimization algorithms, *Proc. 25th. IEEE Symp. on Foundations of Computer Science*, Singer Island, Oct. 1984, pp. 338-346

[10] G. N. Frederickson, Fast algorithms for shortest paths in planar graphs, with applications, *SIAM Journal Computing, 16* (1987), pp. 1004-1022

[11] G. N. Frederickson, A new approach to all pairs shortes path in planar graphs, *Proc. 19th ACM STOC*, New York City, (1987), pp. 19-28.

[12] F. Harary, *Graph Theory*, Addison-Wesley, Reading, MA, 1969

[13] R. Hassin, Maximum flow in (s,t) planar networks, *Inform. Process. Letters, 13* (1981), p. 107.

[14] R. J. Lipton and R. E. Tarjan, A separator theorem for planar graphs, *SIAM J. Appl. Math.,36* (1979), pp. 177-189.

[15] C. H. Papadimitriou, K. Steiglitz, *Combinatorial Optimization, Algorithms and Complexity*, Prentice Hall, 1982.

[16] H. Rohnert, A dynamization of the all-pairs least cost path problem, *Proc. of the 2nd Symp. on Theoretical aspects of Computer Science*, Lect. Not. in Comp. Sci., vol. 182, Springer-Verlag, 1990.

[17] J. Westbrook, Algorithms and data structures for dynamic graph problems, Ph.D. Dissertation, Tech. Rep. CS-TR-229-89, Dept. of Computer Science, Princeton University, 1989.

[18] M. Yannakakis, Graph Theoretic methods in database theory, *Proc. ACM Conf. on Principles of database Systems*, 1990.

[19] D. M. Yellin and R. Strom, INC: a language for incremental computations, *Proc. ACM SIGPLAN Conf. on Programming Language Design and Implementation*, 1988.

COMPLETE PROBLEMS FOR LOGSPACE INVOLVING LEXICOGRAPHIC FIRST PATHS IN GRAPHS

Iain A. Stewart,
Computing Laboratory, University of Newcastle upon Tyne,
Claremont Tower, Claremont Road,
Newcastle upon Tyne, NE1 7RU, England.

Abstract: It is shown that the problem of deciding whether a given vertex is on the lexicographic first path of some digraph, starting at some other specified vertex, is complete for deterministic logspace via projection translations: such translations are extremely weak forms of reductions. Other related problems involving constrained versions of the lexicographically first path problem in both digraphs and graphs are also shown to be similarly complete. The methods used to prove completeness involve the consideration of decision problems as sets of finite structures satisfying certain logical formulae.

1. INTRODUCTION

The complexity class NP (that is, nondeterministic polynomial time) has been extensively studied and many easily-stated problems have been shown to be complete for this class via logspace and polynomial time reductions (see [GJ79]). The usefulness of such a characterization is borne out by the fact that if some problem Ω is complete for NP via logspace (resp. polynomial time) reductions and we can show that Ω can be solved in logspace (resp. polynomial time), then NP = L (resp. NP = P), where the complexity class L is deterministic logspace (resp. P is deterministic polynomial time). Whilst logspace and polynomial time reductions are the usual reductions involved in NP-completeness, even weaker reductions than logspace have been considered; that is, problems have been shown to be complete for NP via projection translations (c.f. [Ste91a], [Ste91b]). Thus, showing that a problem complete for NP via projection translations is in some complexity class contained in L that is closed under projection translations, such as NC^1, would enable us to deduce that $NP = NC^1$, as well as NP = P = L (see [Coo85] for a definition of the complexity class NC^1).

Most well-known complexity classes have been shown to have complete problems via some sort of reduction. For example, many easily-stated problems have been shown to be complete for P via logspace reductions: see [GHR91] for a compendium. Some of these problems are concerned with a search for a certain substructure with respect to a lexicographic ordering on these substructures. For instance, the lexicographic first maximal independent set problem ([Coo85]), the lexicographic minimal maximal path problem ([AM87]), and the lexicographic first maximal subgraph problem ([Miy89]) have all been shown to be complete for P via logspace reductions.

The complexity class L has also been considered in a similar manner, although comparatively few complete problems have been discovered so far. Among the reductions considered in this case are NC^1-reductions and projection translations, with projection translations being weaker than NC^1-reductions (c.f. [Coo85], [CM87], [Imm87a], [IL89]). Most of the known complete problems for L have been shown to be complete via NC^1-reductions but not via projection translations.

In this paper, we show that the problems concerning the computation of certain lexicographic first paths, with respect to some simple constraints, in graphs and digraphs are complete for L via projection translations. We use a particularly powerful though often neglected technique first introduced in [Imm87a] (and also used in [Ste91a] and [Ste91b] to show that certain problems are complete for NP via projection translations). We believe that the results of this paper further emphasise the importance of adopting a logical approach to complexity theory, an approach which has recently yielded many new results (see the references mentioned above as well as [Imm87b] and [Imm88]).

2. BASIC LOGICAL DEFINITIONS

In this section we describe how we extend first-order logic using an operator corresponding to some problem and we consider the notion of a logical translation between problems. We also mention some relevant existing results. The reader is referred to [Imm87a], [Ste91a], and [Ste91b] for extensive details of the concepts mentioned here.

A *vocabulary* $\tau = <\underline{R}_1, \underline{R}_2, ..., \underline{R}_k, \underline{C}_1, \underline{C}_2, ..., \underline{C}_m>$ is a tuple of *relation symbols* $\{\underline{R}_i : i = 1, 2, ..., k\}$, with $\underline{R}_i$ of arity a_i, and *constant symbols* $\{\underline{C}_i : i = 1, 2, ..., m\}$. A *(finite) structure of size* n over τ is a tuple $S = <\{0, 1, ..., n-1\}, R_1, R_2, ..., R_k, C_1, C_2, ..., C_m>$ consisting of a *universe* $|S| = \{0, 1, ..., n-1\}$, *relations* $R_1, R_2, ..., R_k$ on the universe $|S|$ of arities $a_1, a_2, ..., a_k$, respectively, and *constants* $C_1, C_2, ..., C_m$ from the universe $|S|$. The size of some structure S is also denoted by $|S|$. We denote the set of all structures over τ by $STRUCT(\tau)$ (henceforth, we do not distinguish between relations (resp. constants), and relation (resp. constant) symbols, and we assume that all structures are of size at least 2). A *problem of arity* $t (\geq 0)$ over τ is a subset of $STRUCT_t(\tau) = \{(S, u) : S \in STRUCT(\tau), u \in |S|^t\}$ (throughout, tuples are denoted by bold type). If Ω is some problem then $\tau(\Omega)$ denotes its vocabulary.

The language of the *first-order logic* $FO_{\leq}(\tau)$ has as its (well-formed) formulae those formulae built, in the usual way, from the relation and constant symbols of τ, the binary relation symbols $=$ and s, and the constant symbols 0 and max, using the logical connectives $\vee$, $\wedge$, and $\neg$, the variables $\{x, y, z_3, ... \text{etc.}\}$, and the quantifiers $\exists$ and $\forall$. Any formula ϕ of $FO_{\leq}(\tau)$, with free variables those of the t-tuple x, is interpreted in the set $STRUCT_t(\tau)$, and for each $S \in STRUCT(\tau)$ of size n and $u \in |S|^t$, we have that $(S, u) \models \phi(x)$ if and only if $\phi^S(u)$ holds, where $\phi^S(u)$ denotes the obvious interpretation of ϕ in S, except that the binary relation symbol $=$ is always interpreted in S as equality, the binary relation symbol s is interpreted as the successor relation on $|S|$, the constant symbol 0 is interpreted as $0 \in |S|$, the constant symbol max is interpreted as $n-1 \in |S|$, and each variable of x is given the corresponding value from u. (We usually write s(x,y) as $y = x+1$, and $\neg(x = y)$ as $x \neq y$.) If we forbid the use of the successor relation in the logic $FO_{\leq}(\tau)$ then we denote the resulting logic by $FO(\tau)$: also, $FO_{\leq} = \cup \{FO_{\leq}(\tau) : \tau$

some vocabulary} (with FO defined similarly). The formula ϕ *describes* (or *specifies* or *represents*) the problem:

$\{(S,u) : (S,u) \in STRUCT_t(\tau), (S,u) \models \phi(x)\}$ of arity t.

Having detailed how we use first-order logic to describe problems, we now illustrate how we extend first-order logic with new operators to attain greater expressibility. Let τ_2 be the vocabulary consisting of the binary relation symbol E: so, we may clearly consider structures S over τ_2 as digraphs or graphs (for $i \neq j$, there is an edge (i,j) in the digraph S if and only if $E^S(i,j)$ holds, and there is an edge $\{i,j\}$ in the graph S if and only if $E^S(i,j)$ or $E^S(j,i)$ holds: we assume throughout that there are no edges from a vertex to itself in any graph or digraph). Consider the problem DTC of arity 2:

DTC = $\{(S,u,v) \in STRUCT_2(\tau_2)$: there is a path in the digraph S from vertex u to vertex v such that each vertex on the path, except perhaps v, has out-degree 1, i.e. the path is *deterministic*}.

We write $(\pm DTC)^*[FO_\leq]$ to denote the logic formed by allowing an unlimited number of nested applications of the operator DTC, where $DTC[\lambda xy \psi^S(x,y)]$, for some formula $\psi \in (\pm DTC)^*[FO_\leq]$, some k-tuples of distinct variables x and y, and some relevant structure S, denotes the digraph with vertices indexed by the tuples of $|S|^k$, and where there is an edge from u to v if and only if there is a deterministic path in the digraph described by $\psi^S(x,y)$ from u to v (this is the logic (FO + DTC) of [Imm87a]). We write $(\pm DTC)^k[FO_\leq]$ (resp. $DTC^*[FO_\leq]$) to denote the sub-logic of $(\pm DTC)^*[FO_\leq]$ where all formulae have at most k nested applications of the operator DTC (resp. where no operator appears within a negation sign): the sub-logic, $DTC^k[FO_\leq]$, of $DTC^*[FO_\leq]$ is defined similarly. Needless to say, first-order logic can be extended by other operators as we shall soon see.

In order to compare logical descriptions of decision problems, we use the notion of a logical translation (these translations play an analogous role to complexity-theoretic reductions between sets of strings). Let $\tau' = <R_1,R_2,...,R_k,C_1,C_2,...,C_m>$ be some vocabulary, where each R_i is a relation symbol of arity a_i and each C_j is a constant symbol, and let $L(\tau)$ be some logic over some vocabulary τ. Then the formulae of $\Sigma = \{\phi_i(x_i), \psi_j(y_j) : i = 1,2,...,k; j = 1,2,...,m\} \subseteq L(\tau)$, where:

(i) each formula ϕ_i (resp. ψ_j) is over the qa_i (resp. q) distinct variables x_i (resp. y_j), for some positive integer q;

(ii) for each $j = 1, 2, ..., m$ and for each structure $S \in STRUCT(\tau)$:

$S \models (\exists x_1)(\exists x_2)...(\exists x_q)[\psi_j(x_1,x_2,...,x_q) \wedge$

$(\forall y_1)(\forall y_2)...(\forall y_q)[\psi_j(y_1,y_2,...,y_q) \Leftrightarrow (x_1 = y_1 \wedge x_2 = y_2 \wedge ... \wedge x_q = y_q)]$,

are called τ'-*descriptive*. For each $S \in STRUCT(\tau)$, the τ'-*translation of S with respect to* Σ is the structure $S' \in STRUCT(\tau')$ with universe $|S|^q$, defined as follows: for all $i = 1, 2, ..., k$ and for any tuples $\{u_1,u_2,...,u_{a_i}\} \subseteq |S'| = |S|^q$:

$R_i^{S'}(u_1,u_2,...,u_{a_i})$ holds if and only if $(S,(u_1,u_2,...,u_{a_i})) \models \phi_i(x_i)$,

and, for all $j = 1, 2, ..., m$ and for any tuple $u \in |S'| = |S|^q$:

$C_j^{S'} = u$ if and only if $(S,u) \models \psi_j(y_j)$

(tuples are ordered lexicographically, with $(0,0,...,0) < (0,0,...,1) < (0,0,...,2) < ...,$ and so on). Let Ω and Ω' be problems over the vocabularies τ and τ', respectively. Let Σ be a set of τ'-descriptive formulae from some logic $L(\tau)$, and for each $S \in STRUCT(\tau)$, let $\sigma(S) \in STRUCT(\tau')$ denote the τ'-translation of S with respect to Σ. Then Ω' is an L-*translation* of Ω if and only if for each $S \in STRUCT(\tau)$, $S \in \Omega$ if and only if $\sigma(S) \in \Omega'$.

Let $\phi \in FO_\le(\tau)$, for some vocabulary τ, be of the form:

$\phi \equiv \bigvee \{\alpha_i \wedge \beta_i : i \in I\}$, for some index set I, where:

(i) each α_i is a conjunction of the logical atomic relations, s, =, and their negations;

(ii) each β_i is atomic or negated atomic;

(iii) if $i \ne j$, then α_i and α_j are mutually exclusive.

Then ϕ is a *projective formula*. If the successor relation symbol s does not appear in ϕ (that is, $\phi \in FO(\tau)$), then ϕ is a *projective formula without successor*, and if each of the β_i (above) is atomic then ϕ is a *monotone projective formula*. Consequently, we clearly have the notions of one problem being a *first-order translation*, a *quantifier-free translation*, a *projection translation*, and a *monotone projection translation* of another, as well as all of these *without successor*.

For the problem DTC of arity 2, defined earlier, let:

$DTC(0,max) = \{(S,0,n\text{-}1) \in DTC \text{ with } |S| = n\}$.

Henceforth, we equate a logic with those problems represented by some sentence of the logic.

THEOREM 2.1. ([Imm87a]) $L = (\pm DTC)^*[FO_\le] = DTC^1[FO_\le]$ and $DTC(0,max)$ is complete for L via projection translations. $\square$

DTC(0,max) and iterated multiplication in the symmetric group on n elements (c.f. [CM87], [IL89]) are (essentially) the only problems known to be complete for L via projection translations, although there are other decision problems known to be complete for L via NC^1-reductions (c.f. [Coo85], [CM87]: NC^1-reductions are defined using uniform families of Boolean circuits of O(log n) depth with oracle gates). As $FO_\le \subseteq AC^0 \subset ACC \subseteq Th^0 \subseteq NC^1 \subseteq L$ (c.f. [IL89], [BIS90] for more details and definitions of these complexity classes), then the notion of one problem being a projection translation of another is weaker than that of one problem being an NC^1-reduction of another. It is unknown whether any problem is complete for L via monotone projection translations or projection translations without successor.

3. THE PROBLEMS UNDER CONSIDERATION

In this section, we detail the problems to be considered in Section 4. These problems involve the computation of lexicographic first paths, with respect to some simple constraints, in graphs and digraphs using greedy algorithms.

Consider the following algorithm:

ALGORITHM 3.1;
```
input(G : graph)
last-v := the lowest numbered vertex;
output(last-v);
WHILE there is a neighbour of the vertex last-v not yet visited DO
    last-v := the lowest numbered neighbour of last-v not yet visited;
    output(last-v);
OD
```

The output from Algorithm 3.1 is a path of vertices in G; the *lexmin maximal path*. In [AM87] the *Lexicographic Minimal Maximal Path Problem* (LEXMMP) was considered:

   Instance of size n   : a graph G with n vertices {0,1,...,n-1};
   Yes-Instance of size n : n-1 is on the lexmin maximal path of G.
It was shown in [AM87] that LEXMMP is complete for P via logspace reductions.

   Suppose we consider digraphs and tamper with the above algorithm to obtain the following:

ALGORITHM 3.2;
   input(*G : digraph*)
   last-v : = the lowest numbered vertex;
   output(*last-v*);
   WHILE *there is a neighbour of the vertex last-v* **DO**
      last-v : = the lowest numbered neighbour of last-v;
      output(*last-v*);
      OD

Notice that all we have done (apart from moving from graphs to digraphs) is to remove the ability of Algorithm 3.1 to remember previously visited vertices. Notice also that the output from Algorithm 3.2 is a path, the *lexfirst path*, which may be infinite. Define the problem LEXFP as follows:

   Instance of size n   : a digraph G with n vertices {0,1,...,n-1};
   Yes-Instance of size n : n-1 is on the lexfirst path of G.
Even though Algorithm 3.2 might produce an infinite path as output, it should be clear that the problem LEXFP can be solved using logarithmic space, as if n-1 is on the lexfirst path of some digraph G then it appears in the initial subpath of length n.

   The reason we considered digraphs as opposed to graphs, above, is that if Algorithm 3.2 is applied to graphs then we simply traverse the edge {0,v}, in both directions, indefinitely, where v is the lowest numbered neighbour of 0. In order to similarly consider graphs, we need to alter Algorithm 3.2 slightly:

ALGORITHM 3.3;
   input(*G : graph or digraph*)
   last-v : = the lowest numbered vertex;
   previous-v : = the lowest numbered vertex;
   output(*last-v*);
   WHILE *there is a neighbour of the vertex last-v different from previous-v* **DO**
      temp-v : = last-v;
      last-v : = the lowest numbered neighbour of last-v different from previous-v;
      previous-v : = temp-v;
      output(*last-v*);
      OD

Notice that all we have done here is to modify Algorithm 3.2 so that if we move from vertex u to vertex v in a lexfirst path, then the next move cannot be back to vertex u: call the path given by Algorithm 3.3 the *lexfirst path with no U-turns*, and define the problem ULEXFPNUT (resp. LEXFPNUT) as follows:

   Instance of size n   : a graph G (resp. digraph) with n vertices {0,1,...,n-1};

Yes-instance of size n : n-1 is on the lexfirst path with no U-turns of G.
Again, it should be clear that the problems ULEXFPNUT and LEXFPNUT can be solved using logarithmic space, as if n-1 is on the lexfirst path with no U-turns of some graph or digraph, then no edge is used more than once in this path. Hence, we can restrict ourselves to looking at an initial subpath of length n^2. Notice that all of the algorithms above are simple greedy algorithms.

Just as we extended $FO_{\leq}$ using the operator DTC, in the previous section, so we can extend $FO_{\leq}$ using the operator LEXFP (or ULEXFPNUT or LEXFPNUT) in exactly the same way: the corresponding problems are over the vocabulary τ_2. For example, a structure S over τ_2, of size n, satisfies the sentence LEXFP$[\lambda xyE(x,y)](0,max)$ if and only if vertex n-1 is on the lexfirst path of S starting at vertex 0.

4. EXTENDING FIRST-ORDER LOGIC USING THE OPERATOR LEXFP

We begin by proving a normal form theorem for the logic $(\pm LEXFP)^*[FO_{\leq}]$, as was done for the logic $DTC^*[FO_{\leq}]$ in Theorem 3.3 of [Imm87a].

THEOREM 4.1. *Any problem Ω in $(\pm LEXFP)^*[FO_{\leq}]$ can be represented by a sentence of the form:*
$$LEXFP[\lambda xy\psi(x,y)](0,max),$$
where ψ is projective, x and y are k-tuples of distinct variables, and 0 (resp. max) is the constant symbol 0 (resp. max) repeated k times.

PROOF. We proceed by induction on the symbolic complexity of the sentence representing Ω.
Case (i) Ω is represented by an atomic or negated atomic sentence Φ.
Let u and v be new variables. Then for any $S \in STRUCT(\tau(\Omega))$:
  $S \models \Phi$ if and only if $S \models LEXFP[\lambda uv(u = 0 \wedge v = max \wedge \Phi)](0,max)$,
and $u = 0 \wedge v = max \wedge \Phi$ is projective.
Case (ii) Ω is represented by a sentence of the form:
  $\Phi \equiv LEXFP[\lambda xy\psi(x,y)](q,r)$,
where ψ is projective, x and y are k-tuples of variables, q and r are k-tuples of constant symbols, and all variables are distinct.
Let u_1, u_2, v_1, and v_2 be new variables. Define the formula $\phi(u_1,u_2,x,v_1,v_2,y)$ as follows:
  $(u_1 = 0 \wedge u_2 = 0 \wedge x = 0 \wedge v_1 = 0 \wedge v_2 = max \wedge y = q)$
    $\vee (u_1 = 0 \wedge u_2 = max \wedge x = r \wedge v_1 = max \wedge v_2 = max \wedge y = max)$
      $\vee (u_1 = 0 \wedge u_2 = max \wedge v_1 = 0 \wedge v_2 = max \wedge x \neq r \wedge \psi(x,y))$.
Clearly, for any $S \in STRUCT(\tau(\Omega))$:
  $S \models \Phi$ if and only if $S \models LEXFP[\lambda(u_1,u_2,x)(v_1,v_2,y)\phi]((0,0,0),(max,max,max))$,
and ϕ is projective.
Case (iii) Ω is represented by a sentence of the form:
  $\Phi \equiv \neg LEXFP[\lambda xy\psi(x,y)](0,max)$,
where ψ is projective, x and y are k-tuples of variables, and all variables are distinct.
Let u, v, w, and z be k-tuples of new variables. Define the formula $\phi(u,x,v,w,y,z)$ as follows:
  $(u = 0 \wedge w = 0 \wedge z = v+1 \wedge x \neq y \wedge x \neq max \wedge \psi(x,y))$

204

$$\bigvee (u = 0 \wedge x \neq max \wedge w = max \wedge y = max \wedge z = max)$$

("$z = v+1$" is shorthand for incrementation by 1 with respect to the lexicographic ordering). Let $S \in STRUCT(\tau(\Omega))$ be of size n. Then the digraph represented by the formula $\phi^S((u,x,v),(w,y,z))$ can be pictured as in Figure 1 (here, n-1 denotes the

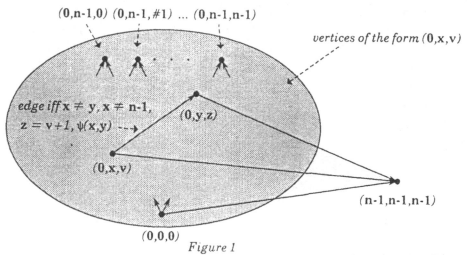

vertices of the form $(0,x,v)$

$(0,n-1,0)$ $(0,n-1,\#1)$ ... $(0,n-1,n-1)$

edge iff $x \neq y, x \neq$ n-1, $z = v+1, \psi(x,y)$

$(0,y,z)$

$(0,x,v)$

$(n-1,n-1,n-1)$

$(0,0,0)$

Figure 1

k-tuple (n-1,n-1,...,n-1) and #1 denotes 0+1, #2 denotes 1+1, etc.). If $S \models \phi$ then the lexfirst path in this digraph starting at $(0,0,0)$ reaches $(0,n-1,v)$, for some v, and goes no further. If $S \models \neg\phi$ then the lexfirst path starting at $(0,0,0)$ reaches a vertex of out-degree 1 where the only possibility is to move to the vertex (n-1,n-1,n-1) (because the lexfirst path in the digraph represented by the formula $\psi^S(x,y)$ starting at 0 either halts at $y \neq$ n-1 or loops indefinitely). Hence:

$S \models \phi$ if and only if $S \models LEXFP[\lambda(u,x,v)(w,y,z)\phi]((0,0,0),(max,max,max))$, and ϕ is projective.

<u>Case (iv)</u> Ω is represented by a sentence of the form:

$\phi \equiv LEXFP[\lambda xy\psi(x,y)](0,max) \wedge LEXFP[\lambda x'y'\psi'(x',y')](0,max)$,

where ψ and ψ' are projective, x and y (resp. x' and y') are k-tuples (resp. k'-tuples) of variables, and all variables are distinct. Proof omitted (easy).

<u>Case (v)</u> Ω is represented by a sentence of the form:

$\phi \equiv \forall z LEXFP[\lambda xy\psi(x,y;z)](0,max)$,

where ψ is projective, x and y are k-tuples of variables, z occurs free in ψ, and all variables are distinct. Proof omitted (easy).

<u>Case (vi)</u> Ω is represented by a sentence of the form:

$\phi \equiv LEXFP[\lambda xyLEXFP[\lambda uv\psi(x,y,u,v)](0,max)](0,max)$,

where ψ is projective, x and y (resp. u and v) are k-tuples (resp. k'-tuples) of variables, and all variables are distinct.

We assume, for simplicity, that $k = k' = 1$. Define the formula $\phi(x,y,u,v,x',y',u',v')$ as follows:

$(x \neq max \wedge x \neq y \wedge u \neq max \wedge x = x' \wedge y = y' \wedge u \neq u' \wedge v' = v+1$

$\wedge \psi(x,y,u,u'))$

$\bigvee (x \neq max \wedge u \neq max \wedge x = x' \wedge y' = y+1 \wedge u' = 0 \wedge v' = 0)$

$\bigvee (x \neq max \wedge y \neq max \wedge u = max \wedge x' = y \wedge y' = u' = v' = 0)$

$\bigvee (x \neq max \wedge y = max \wedge u = max \wedge x' = y' = u' = v' = max)$.

Then for $S \in STRUCT(\tau(\Omega))$ of size n, the digraph G represented by the formula $\phi^S((x,y,u,v),(x',y',u',v'))$ can be depicted as in Figure 2: here, only the edges whose

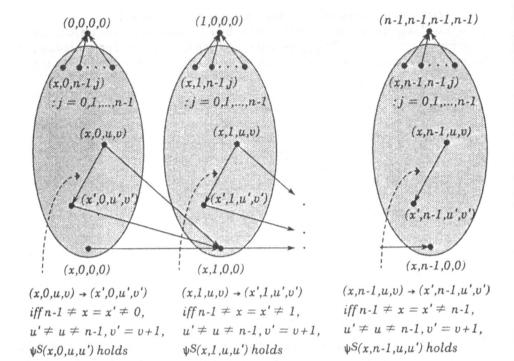

$(x,0,u,v) \rightarrow (x',0,u',v')$

$iff\ n\text{-}1 \neq x = x' \neq 0,$

$u' \neq u \neq n\text{-}1, v' = v+1,$

$\psi^S(x,0,u,u')\ holds$

$(x,1,u,v) \rightarrow (x',1,u',v')$

$iff\ n\text{-}1 \neq x = x' \neq 1,$

$u' \neq u \neq n\text{-}1, v' = v+1,$

$\psi^S(x,1,u,u')\ holds$

$(x,n\text{-}1,u,v) \rightarrow (x',n\text{-}1,u',v')$

$iff\ n\text{-}1 \neq x = x' \neq n\text{-}1,$

$u' \neq u \neq n\text{-}1, v' = v+1,$

$\psi^S(x,n\text{-}1,u,u')\ holds$

Figure 2

out-vertex is such that the first component is x are shown (and so the whole digraph consists of n similar digraphs).

LEMMA 4.1. *Let $S \in STRUCT(\tau(\Omega))$ be of size n and let $x,y \in |S|$ with $x \neq y$ and $x \neq n\text{-}1$. Let $G_{x,y}$ be the subdigraph of G induced by the vertices:*

$\{(x,y,u,v) : u,v \in |S|\} \cup \{(n\text{-}1,n\text{-}1,n\text{-}1,n\text{-}1)\}$ *if $y = n\text{-}1$;*

$\{(x,y,u,v) : u,v \in |S|\} \cup \{(y,0,0,0),(x,y+1,0,0)\}$ *if $y \neq n\text{-}1$.*

Then:

(i) if $LEXFP[\lambda uv\psi^S(x,y,u,v)](0,max)$ holds, the lexfirst path of $G_{x,y}$ starting at $(x,y,0,0)$ ends at $(n\text{-}1,n\text{-}1,n\text{-}1,n\text{-}1)$ if $y = n\text{-}1$ and at $(y,0,0,0)$ if $y \neq n\text{-}1$;

(ii) if $LEXFP[\lambda uv\psi^S(x,y,u,v)](0,max)$ does not hold, the lexfirst path of $G_{x,y}$ starting at $(x,y,0,0)$ ends at (x,y,u,v), for some u and v with $u \neq n\text{-}1$, if $y = n\text{-}1$, and at $(x,y+1,0,0)$ if $y \neq n\text{-}1$.

PROOF. Suppose that $LEXFP[\lambda uv\psi^S(x,y,u,v)](0,max)$ holds. Then there exists k, $n > k > 0$, such that:

the first v such that $\psi^S(x,y,0,v)$ holds is u_1;

the first v such that $\psi^S(x,y,u_1,v)$ holds is u_2;

...

the first v such that $\psi^S(x,y,u_{k-1},v)$ holds is u_k,

and:

$u_k = n\text{-}1$ and $0 \neq u_i \neq n\text{-}1$, for $i = 1, 2, ..., k\text{-}1$

(notice that all the u_i's are distinct). By scrutinizing ϕ, it is clear that the lexfirst path in $G_{x,y}$ starting at $(x,y,0,0)$ is:

$(x,y,0,0), (x,y,u_1,1), (x,y,u_2,2), ..., (x,y,u_k,k), (n\text{-}1,n\text{-}1,n\text{-}1,n\text{-}1)$ if $y = n\text{-}1$;

$(x,y,0,0), (x,y,u_1,1), (x,y,u_2,2), ..., (x,y,u_k,k), (y,0,0,0)$ if $y \neq$ n-1.

On the other hand, suppose that $LEXFP[\lambda uv\psi^S(x,y,u,v)](0,max)$ does not hold. So, in the digraph represented by $\psi^S(x,y,u,v)$ (where x and y are fixed) the lexfirst path starting at 0 either halts at some vertex with no place to go or loops indefinitely (without encountering n-1). In $G_{x,y}$, the lexfirst path starting at $(x,y,0,0)$ never loops indefinitely as all paths are truncated after n-1 moves. If y = n-1 then the lexfirst path in $G_{x,y}$ starting at $(x,y,0,0)$ clearly halts at some vertex with no place to go, and if $y \neq$ n-1 it reaches a vertex $(x,y,u,n-1)$, where $u \neq$ n-1, before ending at $(x,y+1,0,0)$. $\square$

LEMMA 4.2. *Let $S \in STRUCT(\tau(\Omega))$. Then:*
$$S \vDash \phi$$
if and only if:
$$S \vDash LEXFP[\lambda(x,y,u,v)(x',y',u',v')\phi]((0,0,0,0),(max,max,max,max)).$$

PROOF. Suppose that S is of size n and $S \vDash \phi$; that is, there exists k, n > k > 0, such that:

the first y such that $LEXFP[\lambda uv\psi^S(0,y,u,v)](0,max)$ holds is x_1;
the first y such that $LEXFP[\lambda uv\psi^S(x_1,y,u,v)](0,max)$ holds is x_2;
...
the first y such that $LEXFP[\lambda uv\psi^S(x_{k-1},y,u,v)](0,max)$ holds is x_k,
and:

x_k = n-1 and $0 \neq x_i \neq$ n-1, for i = 1, 2, ..., k-1
(notice that all the x_i's are distinct). In particular:

$LEXFP[\lambda uv\psi^S(0,y,u,v)](0,max)$ does not hold for all y = 0, 1, ..., x_1-1,
and so by Lemma 4.1, the lexfirst path in the digraph G starting at $(0,0,0,0)$ is:

$(0,0,0,0), ..., (0,1,0,0), ..., (0,2,0,0), ..., (0,x_1-1,0,0), ..., (0,x_1,0,0), ..., (x_1,0,0,0)$
(notice that if we are at $(x,x,0,0)$ then we are constrained to move to $(x,x+1,0,0)$, whenever $x \neq$ n-1). Clearly, applying the same reasoning and Lemma 4.1, the lexfirst path in the digraph G starting at $(x_1,0,0,0)$ leads to $(x_2,0,0,0)$, then on to $(x_3,0,0,0), ...,$ and on to $(x_{k-1},0,0,0)$, before halting at $(n-1,n-1,n-1,n-1)$.

Conversely, suppose that:
$$S \vDash LEXFP[\lambda(x,y,u,v)(x',y',u',v')\phi]((0,0,0,0),(max,max,max,max)).$$
Remembering that a vertex appears at most once on the lexfirst path between $(0,0,0,0)$ and $(n-1,n-1,n-1,n-1)$, and reversing the above argument, it is easy to see that $S \vDash \phi$. $\square$

As ϕ is projective, Case (vi) clearly follows. The other cases, such as when Ω is represented by a sentence of the form:
$$\phi \equiv LEXFP[\lambda xy\psi(x,y)](0,max) \lor LEXFP[\lambda x'y'\psi'(x',y')](0,max),$$
or:
$$\phi \equiv \exists z LEXFP[\lambda xy\psi(x,y;z)](0,max),$$
with the notation as above, follow from Cases (i) - (vi), and so the result clearly holds. $\square$

PROPOSITION 4.1. $L \subseteq (\pm LEXFP)^*[FO_{\leq}]$.

PROOF. By [Imm87a], $L = DTC^*[FO_{\leq}]$ and $DTC(0,max)$ is complete for L via projection translations. So, any problem Ω in L can be represented by a sentence of the form:

DTC$[\lambda xy\psi(x,y)](0,max)$,

where ψ is projective. Define the formula $\phi(x,y)$ as follows:

$\phi(x,y) \equiv \psi(x,y) \wedge \forall z[\neg\psi(x,z) \vee z = y]$.

Then for any structure $S \in$ STRUCT$(\tau(\Omega))$:

$S \models$ DTC$[\lambda xy\psi(x,y)](0,max)$ if and only if $S \models$ LEXFP$[\lambda xy\phi(x,y)](0,max)$,

and so the result follows. $\square$

COROLLARY 4.1. $L = (\pm LEXFP)^*[FO_\le] = LEXFP^l[FO_\le]$ and $LEXFP(0,max)$ *is complete for* L *via projection translations.*

PROOF. Immediate from Theorem 4.1 and Proposition 4.1. $\square$

By exhibiting monotone projection translations from LEXFP to the problems mentioned in·Section 3, we can easily prove the following result (see the full version of this paper for details).

THEOREM 4.2. *LEXFPNUT, ULEXFPNUT, LEXFPNKT, and ULEXFPNKT are complete for* L *via projection translations.* $\square$

REFERENCES

[AM87]   R.ANDERSON AND E.W.MAYR, Parallelism and the maximal path problem, *Inform. Process. Lett.* 24 (1987), 121-126.

[BIS90]   D.A.M.BARRINGTON, N.IMMERMAN AND H.STRAUBING, On uniformity within NC^1, *J. Comput. System Sci.* 41 (1990), 274-306.

[CM87]   S.A.COOK AND P.MCKENZIE, Problems complete for deterministic logarithmic space, *J. Algorithms* 8 (1987), 385-394.

[Coo85]   S.A.COOK, A taxonomy of problems with fast parallel algorithms, *Inform. and Control* 64 (1985), 2-22.

[GHR91]   R.GREENLAW, H.J.HOOVER AND W.L.RUZZO, A compendium of problems complete for P, Part II: P-complete problems, Tech. Rep., University of Alberta, to appear.

[GJ79]   M.R.GAREY AND D.S.JOHNSON, "Computers and Intractability: A Guide to the Theory of NP-completeness", Freeman, San Francisco, 1979.

[IL89]   N.IMMERMAN AND S.LANDAU, The complexity of iterated multiplication, *in* "Proc. 4th Symp. on Structure in Complexity Theory, 1989", 104-111.

[Imm87a]   N.IMMERMAN, Languages which capture complexity classes, *SIAM J. Comput.* 16, No.4 (1987), 760-778.

[Imm87b]   N.IMMERMAN, Expressibility as a complexity measure: Results and directions, *in* "Proc. 2nd Symp. on Structure in Complexity Theory, 1987", 194-202.

[Imm88]   N.IMMERMAN, Nondeterministic space is closed under complementation, *SIAM J. Comput.* 17, No.5 (1988), 935-938.

[Miy89]   S.MIYANO, The lexicographically first maximal subgraph problems: P-completeness and NC algorithms, *Math. Systems Theory* 22, No.1 (1989), 47-73.

[Ste91a]   I.A.STEWART, Using the Hamiltonian path operator to capture NP, extended abstract in "Proc. 2nd International Conference on Computing and Information, 1990", Lecture Notes in Computer Science Vol. 468, Springer-Verlag, 134-143: to appear, *J. Comput. System Sci.*

[Ste91b]   I.A.STEWART, Comparing the expressibility of languages formed using NP-complete operators, extended abstract in "Proc. 16th International Workshop on Graph Theoretic concepts in Computer Science, 1990": *J. Logic and Computation* 1, No. 3 (1991), 305-330.

A New Upper Bound on the Complexity of the All Pairs Shortest Path Problem[*]

Tadao Takaoka

Department of Computer Science, Ibaraki University
Hitachi, Ibaraki, JAPAN

1 Introduction

The all pairs shortest path (APSP) problem is to compute shortest paths between all pairs of vertices of a directed graph with non-negative edge costs. We present an algorithm that computes shortest distances between all pairs of vertices, since shortest paths can be computed easily as by-products as in most other algorithms. It is well-known that the time complexity of (n, n)-distance matrix multiplication (DMM) is asymptotically equal to that of the APSP problem for a graph with n vertices. See Aho, Hopcroft and Ullman [1] for example. Based on this fact, Fredman [5] invented an algorithm for DMM of $O(n^3(\log \log n / \log n)^{1/3})$ time, which is $o(n^3)$, meaning that the APSP problem can be solved with the same time complexity. In the average case, Moffat and Takaoka [6] solved this problem with $O(n^2 \log n)$ expected time.

Our algorithm in this paper solves DMM in $O(n^3(\log \log n / \log n)^{1/2})$ time, meaning that the APSP problem can be solved with the same time complexity. This is an asymptotic improvement of Fredman's algorithm by the factor of $(\log n / \log \log n)^{1/6}$. Another merit of our algorithm is that it is simple and easy to implement, whereas Fredman's algorithm is complicated and difficult to implement. Also a possible parallel implementation is mentioned. The base of logarithm is assumed to be two in this paper and fractions are rounded up if necessary.

2 Distance matrix multiplication by divide-and-conquer

Let A and B be (n, n) matrices whose components are non-negative real numbers. The product $C = A \cdot B$ is defined by

$$c_{ij} = \min_{1 \le k \le n} \{a_{ik} + b_{kj}\} \quad (i, j = 1, \dots, n). \tag{1}$$

The operation in the righthand-side of (1) is called distance matrix multiplication and A and B are called distance matrices in this context. This is also viewed as computing n^2 inner products defined for $a = (a_1, \dots, a_n)$ and $b = (b_1, \dots, b_n)$ by

$$a \cdot b = \min_{1 \le k \le n} \{a_k + b_k\}.$$

Now we divide A, B and C into (m, m)-submatrices for $N = n/m$ as follows:

$$A = \begin{bmatrix} A_{11} & \cdots & A_{1N} \\ & \cdots & \\ A_{N1} & \cdots & A_{NN} \end{bmatrix} \quad B = \begin{bmatrix} B_{11} & \cdots & B_{1N} \\ & \cdots & \\ B_{N1} & \cdots & B_{NN} \end{bmatrix} \quad C = \begin{bmatrix} C_{11} & \cdots & C_{1N} \\ & \cdots & \\ C_{N1} & \cdots & C_{NN} \end{bmatrix}.$$

[*]This work was partially supported by a research grant from Hitachi Engineering Co., Ltd.

Then C can be computed by

$$C_{ij} = \min_{1 \leq k \leq N} \{A_{ik} B_{kj}\} \quad (i, j = 1, \ldots, N), \tag{2}$$

where the product of submatrices is defined similarly to (1) and the "min" operation is defined on matrices by taking the "min" operation componentwise. Since comparisons and additions of distances are performed in a pair, we omit counting the number of additions for measurement of complexity. We have N^3 multiplications of distance matrices in (2). Let us assume that each multiplication of (m, m) matrices can be done in $T(m)$ computing time, assuming that a precomputed table is available with no cost. The time for constructing the table is reasonable when m is small and will be mentioned in Section 6. Then the total computing time is given by

$$O(n^3/m + (n/m)^3 T(m)). \tag{3}$$

In the next section we show that $T(m) = O(m^{2.5})$. Thus the time given in (3) becomes $O(n^3/\sqrt{m})$.

3 Distance matrix multiplication by table-lookup

In this section we divide (n, n)-matrices into strips for $M = n/m$ where $\sqrt{n} \leq m \leq n$ as follows:

$$A = \begin{bmatrix} \boxed{A_1} & \cdots & \boxed{A_M} \end{bmatrix} \qquad B = \begin{bmatrix} \boxed{\dfrac{B_1}{\cdots}} \\ \boxed{B_M} \end{bmatrix},$$

where A_i are (n, m) matrices and B_j are (m, n) matrices. We regard later A and B as (m, m)-submatrices in (2). Now the product $C = A \cdot B$ is given by

$$C = \min_{1 \leq k \leq M} \{A_k B_k\}. \tag{4}$$

We show later that $A_k B_k$ can be computed in $O(m^2 n)$ time, assuming that a precomputed table is available. Then the righthand-side of (4) can be computed in time $O(n^3/m + mn^2)$. Setting m to $\sqrt{n}$, this time becomes $O(n^{2.5})$. Now we show that for (n, m) matrix A and (m, n) matrix B, $A \cdot B$ can be computed in $O(m^2 n)$ time. We assume that the lists

$$(a_{1r} - a_{1s}, \ldots, a_{nr} - a_{ns}) \qquad (1 \leq r < s \leq m)$$

and

$$(b_{s1} - b_{r1}, \ldots, b_{sn} - b_{rn}) \qquad (1 \leq r < s \leq m)$$

are already sorted for all r and s such that $1 \leq r < s \leq m$. Let E_{rs} and F_{rs} are the sorted lists respectively. For each r and s, we merge lists E_{rs} and F_{rs} to form list G_{rs}. This takes $O(m^2 n)$ time. The time for sorting will be mentioned in Section 6. Let H_{rs} be the list of ranks of $a_{ir} - a_{is}$ $(i = 1, \ldots, n)$ in G_{rs} and L_{rs} be the list of ranks of $b_{sj} - b_{rj}$ $(j = 1, \ldots, n)$ in G_{rs}. Let $H_{rs}[i]$ and $L_{rs}[j]$ be the i-th and j-th components of H_{rs} and L_{rs} respectively. Then we have

$$\begin{aligned} G_{rs}[H_{rs}[i]] &= a_{ir} - a_{is} \\ G_{rs}[L_{rs}[j]] &= b_{sj} - b_{rj}. \end{aligned}$$

It is obvious that it takes $O(m^2 n)$ time to make lists H_{rs} and L_{rs} for all r and s. As observed by Fredman, we have

$$a_{ir} + b_{rj} \leq a_{is} + b_{sj} \iff a_{ir} - a_{is} \leq b_{sj} - b_{rj}$$

He observed that the information of ordering for all i, j, r and s in the righthand-side of the above formula is sufficient to determine the product $A \cdot B$ by a precomputed decision tree. We observe that for computing each (i, j) component of C it is enough to know the above ordering for all r and s by

$$a_{ir} - a_{is} \leq b_{sj} - b_{rj} \iff H_{rs}[i] \leq L_{rs}[j]. \tag{5}$$

Hereafter we use a short cut notation $a_{12}a_{13} \ldots a_{m-1\,m}$ to express the sequence $a_{12}a_{13}$ $\ldots a_{1m}a_{23}a_{24} \ldots a_{2m} \ldots a_{m-1\,1}a_{m-1\,2} \ldots a_{m-1\,m}$. The list $H[i] = H_{12}[i]H_{13}[i] \ldots H_{m-1\,m}[i]$ is encoded into a single integer in lexicographic order as shown in the next section and $h[i]$ is assumed to represent that integer. Similarly $\ell[j]$ represents the positive integer for the list $L[j] = L_{12}[j]L_{13}[j] \ldots L_{m-1\,m}[j]$. The time for this encoding for all $H[i]$ and $L[j]$ is $O(m^2 n)$. Then the precomputed table T gives the desired index for the inner product, that is,

$$T[h[i], \ell[j]] = k \iff k \text{ is the index for } \min_{1 \leq k \leq m} \{a_{ik} + b_{kj}\}. \tag{6}$$

After the encoding we can compute n^2 inner products in $O(n^2)$ time, assuming that the precomputed table T is available. Hence $A_k \cdot B_k$ can be computed in $O(m^2 n)$ time.

4 Encoding lists in lexicographic order

The sequence of m integers $x_{m-1} \ldots x_1 x_0$, where $1 \leq x_i \leq n$, is encoded into x such that

$$x = (x_{m-1} - 1)n^{m-1} + \cdots + (x_1 - 1)n + x_0 - 1.$$

We denote this encoding by the mapping f, that is, $f(x_{m-1} \ldots x_1 x_0) = x$. It is obvious that this mapping f is one-to-one and onto. The mapping f can be computed in $O(m)$ time using Horner's method, assuming that arithmetic operations on integers up to n^m can be preformed in $O(1)$ time. We assume later that m and n are small enough so that these operations can be done in $O(1)$ time. We note that the inverse f^{-1} of f, that is, decoding can be done in $O(m)$ time by successive devision. The encoding/decoding method in this section is used in this paper with n and m replaced by $2n$ and $m(m-1)/2$ respectively.

5 How to compute table T

Let $H = H_{12}H_{13} \ldots H_{m-1\,m}$ and $L = L_{12}L_{13} \ldots L_{m-1\,m}$ be any sequences of $m(m-1)/2$ integers whose values are between 1 and $2n$. Let h and ℓ be positive integers representing H and L in lexicographic order. Then T is defined by

$$T[h, \ell] = \begin{cases} k, & \text{if there is } k \text{ such that } H_{ks} < L_{ks} \text{ for all } s > k \\ & \text{and } H_{rk} > L_{rk} \text{ for all } r < k \\ \text{undefined}, & \text{otherwise.} \end{cases}$$

This table can be used for table T in (6). The time for computing table T is given with a constant $c > 0$ by

$$O(m^2 n^{2\binom{m}{2}}) = O(c^{m^2 \log n}) = O(c^{n \log n}).$$

Example: Let H and L be given in two dimensional forms as follows, where $m = 3$ and $n = 5$:

$$H = \begin{bmatrix} - & 5 & 6 \\ - & - & 7 \\ - & - & - \end{bmatrix}, \quad L = \begin{bmatrix} - & 4 & 3 \\ - & - & 10 \\ - & - & - \end{bmatrix}$$

$$k = 2 \text{ since } H_{12} > L_{12} \text{ and } H_{23} < L_{23}$$
$$h = 456, \ell = 329$$

6 How to determine the size of submatrices

It is time to determine the size of submatrices in Section 1. Do not be confused between the usages of m and n in Section 2 and Section 3. Let m in Section 2 be given by

$$m = \log n / (\log c \log \log n),$$

then the time for making the table T of this size is easily shown to be $O(n)$ which can be absorbed in the main computing time. Substituting this value of m for m in $O(n^3/\sqrt{m})$ we have the overall computing time for distance matrix multiplication

$$O(n^3 (\log \log n / \log n)^{1/2})$$

as claimed, and consequently we have proven that the APSP can be solved in the same complexity. We note that the time for sorting to obtain the lists E_{rs} and F_{rs} in Section 3 is $O(n^{2.5} \log n)$. This task of sorting, which is called presort, is done for all A_{ij} and B_{ij} in Section 2 in advance, taking

$$O((n/m)^2 m^{2.5} \log m) = O(n^2 (\log n \log \log n)^{1/2})$$

time where $m = O(\log n / \log \log n)$, which is absorbed into the above main complexity.

7 Parallelization

The simple structure of our algorithm makes a parallelization possible. Here we use a PRAM with EREW memory (See Akl [2] for parallel computational models). First we separate the computation in (2) into two parts:

(I) Computation of n^2 minima of N numbers.

(II) Computation of N^3 products $A_{ik}B_{kj}$ $(i, j, k = 1, \ldots, N)$.

We design a parallel algorithm for (I) first. According to Dekel and Sahni [3], there is a PRAM-EREW algorithm for finding the minimum of N numbers with $O(N/\log N)$ processors and $O(\log N)$ time. Applying this, we have a parallel algorithm for (I) with $O(n^3 \log \log n / \log^2 n)$ processors and $O(\log n)$ time, since the n^2 minima can be computed in parallel.

Next we design a parallel algorithm for (II). In the computation of $A_k B_k$ in (4), there are m^2 independent mergings of lists of length n. Such computation is done for $k = 1, \ldots, n/m$. There are also $M - 1$ "min" operations on (n, n)-matrices. It is obvious that there is a parallel algorithm for these operations with $O(n^{1.5}/\log n)$ processors and $O(n \log n)$ time, where $m = \sqrt{n}$. It is also obvious that the tasks of encoding and table-lookup in Section 3 can be done within the above complexities of processors and time.

Since N^3 products in (II) can be computed in parallel, we have a parallel algorithm for (II) with P processors and T time, where $m = \Theta(\log n / \log \log n)$ and

$$P = O((n/m)^3 m^{1.5} / \log m) = O(n^3 (\log \log n)^{1/2} / (\log n)^{3/2})$$
$$T = O(m \log m) = O(\log n).$$

The lists E_{r_s}'s and F_{r_s}'s have to be broadcast to the computation of N^3 products of $A_{ik} B_{kj}$. This is done with $O(n^3 \log \log n / \log n)$ processors and $O(\log n)$ time only once at the beginning. The task of table construction and presort described in Sections 5 and 6 can be done within the above comlexities. Note that the complexities of (II) dominates that of (I).

To solve the APSP problem, we use the so-called repeated squaring (see [3] for example) instead of recursion for sequential algorithms in [1]. Since $\log n$ squarings of distance matrices solve the problem, we can say that we can solve the APSP with a parallel algorithm on a PRAM-EREW with P processors and T time where

$$P = O(n^3 (\log \log n)^{1/2} / (\log n)^{3/2}), \ T = O(\log^2 n).$$

This is compared with the algorithm by the combination of Dekel, Nassimi and Sahni [3] and Dekel and Sahni [4], with $O(n^3 / \log n)$ processors and $O(\log^2 n)$ time. In the average case, there is a more efficient parallel algorithm by Takaoka [7] with $P \doteqdot O(n^{2.5})$ and $T \doteqdot O(\log \log n)$.

Acknowoledgement

The author thanks Ernst W. Mayr who pointed out that the PRAM-EREW model is sufficient to solve the present problem with the claimed complexities while the author was using PRAW-CREW model.

References

[1] Aho, A.V., Hopcroft, J.E. and Ullman, J.D., The design and analysis of computer algorithms, Addison-Wesley (1974).

[2] Akl, S.G., The design and analysis of parallel algorithms, Prentice-Hall (1989).

[3] Dekel, E., Nassimi, D. and Sahni, S., "Parallel matrix and graph algorithms," SIAM Jour. on Computing, Vol. 10, No. 4, pp. 657–675 (1981).

[4] Dekel, E. and Sahni, S., "Binary trees and parallel scheduling algorithms," IEEE Trans. on Computers, Vol. C-32, No. 3, pp. 307–315 (1983)

[5] Fredman, M.L., "New bounds on the complexity of the shortest path problem," SIAM Jour. on Computing, Vol. 5, No. 1, pp. 83–85 (1976).

[6] Moffat, A. and Takaoka, T., "An all pairs shortest path algorithm with expected running time $O(n^2 \log n)$," SIAM Jour. on Computing, Vol. 16, No. 6, pp. 1023–1031 (1987).

[7] Takaoka, T., "An efficient parallel algorithm for the all pairs shortest path problem," WG 88, Lecture Notes in Computer Science 344, Springer-Verlag, pp. 276–287 (1988).

ON THE CROSSING NUMBER OF THE HYPERCUBE AND THE CUBE CONNECTED CYCLES

Ondrej Sýkora and Imrich Vrťo [*]
Computing Centre, Slovak Academy of Sciences
Dúbravská 9, 84235 Bratislava, Czecho-Slovakia

Abstract

We prove tight bounds for the crossing number of the n-dimensional hypercube and cube connected cycles (CCC) graphs.

1 Introduction

The n-dimensional hypercube graph Q_n is defined as the 1-skeleton of the n-dimensional cube. Hypercubes have been much studied in the graph theory. Interest in hypercubes has been increased by the recent advent of massively parallel computers [5]. The cube connected cycles graph (CCC_n) is obtained from Q_n by replacing each vertex by the n-vertex cycle. Computers with CCC_n architecture have a similar computing power as those with Q_n structure [10].

The crossing number $cr(G)$ of a graph G is defined as the least number of crossings of its edges when G is drawn in a plane. All that is known on the exact value of $cr(Q_n)$ is $cr(Q_3) = 0, cr(Q_4) = 8$ and $cr(Q_5) \leq 56$ (see the survey paper [4]). Although, Eggleton and Guy [2] announced that

$$cr(Q_n) \leq \frac{5}{32}4^n - \left\lfloor \frac{n^2+1}{2} \right\rfloor 2^{n-1} \tag{1}$$

a gap has been found in the description of the construction (see again [4]). Anyway, Erdös and Guy [3] conjectured equality in 1.

In this paper we prove the following bounds for $cr(Q_n)$ and $cr(CCC_n)$:

$$\frac{4^n}{20} - (n+1)2^{n-2} < cr(Q_n) \leq \frac{4^n}{6} - \left(\frac{n^2}{2} + \frac{2}{3} \right) 2^{n-2}$$

$$\frac{4^n}{20} - 3(n+1)2^{n-2} < cr(CCC_n) \leq \frac{4^n}{6} + \left(\frac{3}{2}n^2 - 6n - \frac{4}{3} \right) 2^{n-2}.$$

In practice, crossing numbers appear in the fabrication of VLSI circuits. The crossing number of the graph corresponding to the VLSI circuit has strong influence on the

---

[*]Both authors were supported by a research grant from Humboldt Foundation, Bonn, Germany

area of the layout as well as on the number of wire - contact cuts that should be as small as possible. Leighton [7] was the first who pointed out the importance of crossing numbers in VLSI complexity theory. He showed that the crossing number serves as a good area lower bound argument. Thus our lower bounds for $cr(Q_n)$ and $cr(CCC_n)$ give immediately an alternative proof that the area complexity of hypercube and CCC computers realized on VLSI circuits is $A = \Omega(4^n)$. Another proofs are in [1,8]. Optimal layouts are proposed in [1,10].

2 Upper bounds

First we give a simple recursive drawing of Q_n in a plane. Consider the real axis x in the 2-dimensional Euclidean plane. Let D_{n-1} be a drawing of Q_{n-1} in the plane such that the vertices of Q_{n-1} are the points $0, 1, 2, ..., 2^{n-1} - 1$ on x. Produce a symmetrical drawing to D_{n-1} arround the line normal to x in the point $2^{n-1} - 0.5$. If n is even (odd) then join the points i and $2^n - 1 - i$, $i = 0, 1, 2, ..., 2^{n-1} - 1$ by circular arcs above (below) x.

Theorem 2.1 *Let* $cr_0(Q_n)$ *denote the number of crossings in the above construction. Then*

$$cr_0(Q_n) \leq \frac{4^n}{6} - \left(\frac{n^2}{2} + \frac{2}{3}\right) 2^{n-2}.$$

Proof: It is easy to show that $cr_0(Q_n)$ satisfies the following recurrent relation

$$cr_0(Q_n) = 2cr_0(Q_{n-1}) + \sum_{i=1}^{\lfloor \frac{n}{2} \rfloor - 1} 4^i \sum_{j=1}^{n-2i} (2^{n-2i} - 2).$$

The direct solution of the relation gives the claimed upper bound for $cr_0(Q_n)$. □

Corollary 2.1

$$cr(Q_n) \leq \frac{4^n}{6} - \left(\frac{n^2}{2} + \frac{2}{3}\right) 2^{n-2}.$$

Corollary 2.2

$$cr(CCC_n) \leq \frac{4^n}{6} + \left(\frac{3}{2}n^2 - 6n - \frac{4}{3}\right) 2^{n-2}.$$

Proof: Consider the above drawing D_n of Q_n in the plane. Around each vertex of Q_n we find a region containing no crossings. In each region we replace the vertex by a cycle of length n. Thus we have constructed a drawing of CCC_n in the plane while the number of crossings increased at most by $\binom{n-1}{2} 2^n$. □

3 Lower bounds

We apply the lower bound method proposed by Leighton [7]. Let $G_1 = (V_1, E_1)$ and $G_2 = (V_2, E_2)$ be graphs. An embedding of G_1 in G_2 is a couple of mappings (ϕ, ψ) satisfying

$$\phi : V_1 \rightarrow V_2 \quad \text{is an injection}$$

$$\psi : E_1 \rightarrow \{\text{set of all simple paths in } G_2\}$$

such that if $(u,v) \in E_1$ then $\psi((u,v))$ is a path between $\phi(u)$ and $\phi(v)$. For any $e \in E_2$ define

$$cg_e(\phi,\psi) = |\{f \in E_1 : e \in \psi(f)\}|$$

and

$$cg(\phi,\psi) = \max_{e \in E_2}\{cg_e(\phi,\psi)\}.$$

The value $cg(\phi,\psi)$ is called congestion.

Lemma 3.1 [7] *Let (ϕ,ψ) be an embedding of G_1 in G_2 with congestion $cg(\phi,\psi)$. Then*

$$cr(G_2) \geq \frac{cr(G_1)}{cg^2(\phi,\psi)} - \frac{|E_2|}{2} \tag{2}$$

Theorem 3.1

$$cr(Q_n) > \frac{4^n}{20} - (n+1)2^{n-2}.$$

Proof: Let $2K_m$ denote the complete multigraph of m vertices, in which every two vertices are joined by two parallel edges. Set $G_1 = 2K_{2^n}$ and $G_2 = Q_n$. In what follows, we show, that there exists an embedding (ϕ,ψ) of $2K_{2^n}$ in Q_n with

$$cg(\phi,\psi) \leq 2^n. \tag{3}$$

Kleitman's paper [6] implies

$$cr(K_{2^n}) \geq \frac{2^n(2^n-1)(2^n-2)(2^n-3)}{80}. \tag{4}$$

According to Kainen [9] it holds

$$cr(2K_{2^n}) = 4cr(K_{2^n}). \tag{5}$$

Substituting 3, 4 and 5 into 2, we obtain the desired result. Now we will show an embedding satisfying 3. Let ϕ be any bijection of $2K_{2^n}$ into Q_n. For any two vertices of Q_n, we have to design two paths between them. Consider two arbitrary vertices u and v of Q_n. Let d be their distance. Then there exists the unique path of length d starting in u, traversing dimensions in ascending order and ending in v. Let the second path be the symmetrical one starting in v and ending in u. Let $e = (x,y)$ be an arbitrary edge of Q_n lying in a dimension $i, 1 \leq i \leq n$. Now we count the number of edges of $2K_{2^n}$ whose images (paths) traverse the edge (x,y). Let A (B) be the subcube of Q_n that contains x (y) and lies in dimensions $1,2,...,i-1(i+1,i+2,...,n)$. (If $i = 1$ or n then A or B is a single vertex, i.e. Q_0.) Similarly, let C (D) be the subcube of Q_n that contains y (x) and lies in dimensions $0,1,2,...,i-1$ $(i+1,i+2,...,n)$. It is easy to show that when an above defined path contains the edge (x,y) it must start in A (or C) and end in B (or D). Thus

$$cg_e(\phi,\psi) \leq 2^{i-1}2^{n-i} + 2^{i-1}2^{n-i} = 2^n$$

and consequently

$$cg(\phi,\psi) \leq 2^n. \quad \square$$

We use the same method to prove the lower bound on $\mathrm{cr}(CCC_n)$. Formally, the graph CCC_n is defined as follows. The set of vertices consists of tuples (i,j), $i = 0,1,2,3,...,2^n - 1, j = 0,1,2,...,n-1$. Vertices (i_1,j_1) and (i_2,j_2) are adjacent if and only if $i_1 = i_2$ and $\mid j_2 - j_1 \mid \mathrm{mod}\, n = 1$ or $j_1 = j_2$ and the binary representations of i_1, i_2 differ only in the j_1-th bit.

Theorem 3.2

$$\mathrm{cr}(CCC_n) > \frac{4^n}{20} - 3(n+1)2^{n-2}.$$

Proof: Denote by CCP_n (Cube Connected Paths) the graph which is obtained from CCC_n by removing edges $((i,0),(i,n-1))$, for $i = 0,1,2,3,...,2^n - 1$. Observe that the graph CCP_n has a nice recursive structure. Clearly it holds

$$\mathrm{cr}(CCC_n) \geq \mathrm{cr}(CCP_n). \tag{6}$$

Set $G_1 = K_{2^n,2^n}, G_2 = CCP_n$. In what follows we shall construct an embedding (ϕ_n, ψ_n) of $K_{2^n,2^n}$ in CCP_n such that

$$\mathrm{cg}(\phi_n, \psi_n) = 2^n. \tag{7}$$

Once more the Kleitman's result [6] implies

$$\mathrm{cr}(K_{2^n,2^n}) \geq \frac{2^{2n-1}(2^n - 1)(2^{n-1} - 1)}{5} \tag{8}$$

Substituting 7 and 8 into 2 and noting 6 we obtain the desired result.

Assume $n \geq 2$. Let ϕ_n be an injection that maps the first (second) 2^n mutually nonadjacent vertices of $K_{2^n,2^n}$ in the set $\{(i,0) \mid i = 0,1,2,3,...,2^n - 1\}$ ($\{(i,n-1) \mid i = 0,1,2,3,...,2^n - 1\}$). We design ψ_n by induction. Let $n = 2$. The 16 paths between the vertices $\{(i,0) \mid i \leq 3\}$ and $\{(i,1) \mid i \leq 3\}$ are the following:
$(k,0)(k,1)$
$(k,0)(k+1,0)(k+1,1)$
$(k,0)(k,1)((k+2) \bmod 4, 1)$
$(k,0)(k+1,0)(k+1,1)((k+3) \bmod 4, 1)$ for $k = 0, 2$
$(k,0)((k-1),0)((k-1),1)$
$(k,0)(k,1)$
$(k,0)((k-1),0)((k-1),1)((k+1) \bmod 4, 1)$
$(k,0)(k,1)((k+2) \bmod 4, 1)$ for $k = 1, 3$.
Clearly $\mathrm{cg}(\phi_2, \psi_2) = 4$.
Assume we have constructed (ϕ_{n-1}, ψ_{n-1}) such that $\mathrm{cg}(\phi_{n-1}, \psi_{n-1}) = 2^{n-1}$. Consider vertices $(i_1, 0), (i_2, n-1)$ of CCP_n.

1. If $i_1, i_2 < 2^{n-1}$ or $i_1, i_2 \geq 2^{n-1}$ then we first create a path between $(i_1, 0)$ and $(i_2, n-2)$ using ψ_{n-1} and then prolong this path to $(i_2, n-1)$.

2. If $i_1 < 2^{n-1}$ and $i_2 \geq 2^{n-1}$ then we first create a path between $(i_1, 0)$ and $(i_2 - 2^{n-1}, n-2)$ using ψ_{n-1} and then prolong this path to $(i_2, n-1)$ through $(i_2 - 2^{n-1}, n-1)$. The case $i_1 \geq 2^{n-1}, i_2 < 2^{n-1}$ is analogical. One can easily see that

$$\mathrm{cg}(\phi_n, \psi_n) = \max(2\mathrm{cg}(\phi_{n-1}, \psi_{n-1}), 2^n) = 2^n. \quad \square$$

4 Acknowledgement

This work was done while the authors were visiting Max-Planck Institut für Informatik, Saarbrücken. The authors thank Professor Kurt Mehlhorn for all his support.

References

[1] Brebner, G., Relating routing graphs and two dimensional array grids, In: Proceedings VLSI: Algorithms and Architectures, North Holland, 1985.

[2] Eggleton, R. B., Guy, R. P., The crossing number of the n-cube, Notices of the American Mathematical Society, 17, 1970, 757.

[3] Erdös, P., Guy, R. P., Crossing number problems, American Mathematical Monthly, 80, 1, 1973, 52-58.

[4] Harary, F., Hayes, J. P., Horng-Jyh Wu, A survey of the theory of hypercube graphs, Computers and Mathematics with Applications, 15, 4, 1988, 277-289.

[5] Heath, M. I. (editor), Hypercube Multicomputers, Proceedings of the 2-nd Conference on Hypercube Multicomputers, SIAM, 1987.

[6] Kleitman, D. J., The crossing number of $K_{5,n}$, Journal of Combinatorial Theory, 9, 1971, 315-323.

[7] Leighton, F. T., New lower bound techniques for VLSI, In: Proceedings of the 22-nd Annual Symposium on Foundations of Computer Science, 1981, 1-12.

[8] Leiserson, C. E., Area efficient graph layouts (for VLSI), In: Proceedings of the 21-st Annual IEEE Symposium on Foundations of Computer Science, 1980, 270-281.

[9] Kainen, P. C., A lower bound for crossing numbers of graphs with applications to K_n, $K_{p,q}$, and $Q(d)$, Journal of Combinatorial Theory (B), 12, 1972, 287-298.

[10] Preparata, F. P., Vuillemin, J. E., The cube-connected cycles: a versatile network for parallel computation, In: Proceedings of the 20-th Annual IEEE Symposium on Foundations of Computer Science, 1979, 140-147.

Logic Arrays for Interval Indicator Functions

Peter Damaschke
Fernuniversität Hagen, Theoretische Informatik II
Postfach 940, W-5800 Hagen, Germany

0 Introduction

Programmable logic arrays (PLA) are highly regular circuits which compute suitable Boolean functions. PLAs are a popular circuit design especially in microcontrollers. PLA folding is a technique for minimizing the area of such PLA. Folding can be formulated in purely graph-theoretic terms.

In this paper, we study PLA for a very special class of Boolean functions, namely indicator functions of intervals of binary numbers. As a possible application, we can imagine PLA which shall supervise that an input value lies in a prescribed tolerance interval. From a mathematical point of view, the connection of these interval indicator functions with threshold graphs and cographs is interesting. Our study naturally leads to the folding problem in cographs which is shown to be solvable in linear time.

In sections 1-2 we introduce the basic notions. In section 3 we prove a theorem on minimal normal forms of interval indicator functions, thereby motivating the folding problem on cographs. A linear time algorithm solving this problem is derived in section 4. Finally, in section 5, we mention some related results and possible directions of further research.

1 Normal Forms of Boolean Functions

First of all, we recall some well-known facts on Boolean functions. Every Boolean function can be written in disjunctive normal form (DNF) as well as in conjunctive normal form (CNF).

A literal is a variable x (positive literal) or a negated variable $\bar{x}$ (negative literal). As usual, we define $x^0 = \bar{x}$ and $x^1 = x$. A DNF is an expression of the form $p_1 \vee \ldots \vee p_k$ where the "summands" p_i are conjunctions of literals. A CNF is an expression of the form $q_1 \wedge \ldots \wedge q_k$ where the clauses q_i are disjunctions of literals. We will simply write fg instead of $f \wedge g$. By the duality principle, there is a 1-1-correspondence between the CNF of $\bar{f}$ and the DNF of f: negate all literals and replace all symbols $\wedge$ by $\vee$, and vice versa. In the following, we consider only DNF, and we get automatically the dual definitions and statements for CNF.

For two DNF A, B of f, we define $A \subset B$ if A is obtained from B by deleting literals from summands and/or by deleting complete summands. B is called a minimal DNF (for short: MDNF) of f if B is a DNF of f, and there is no DNF A of f with $A \subset B$. Analogously we define MCNF. In general, the MDNF of a Boolean function is not uniquely determined. The reason for this is the behaviour of prime implicands. A conjunction p of clauses is called an implicand of f if $p = 1$ implies $f = 1$. Clearly, all p_i in a DNF are implicands of f. An implicand p is called a prime implicand if there is no implicand q of f with $q \subset p$. One easily sees that an MDNF consists only of prime implicands. But in general, some of the prime implicands may be absent. For Boolean n-tuples $x = (x_1, \ldots, x_n)$ and $y = (x_1, \ldots, x_n)$ we define $x \subseteq y$ if $x_i \leq y_i$ for all i. An n-ary Boolean function f is called monotone if $x \subseteq y$ implies $f(x) \leq f(y)$. f is called antimonotone if $\overline{f}$ is monotone. For monotone functions f, the following is well-known: f has a unique MDNF consisting of all prime implicands of f, and every prime implicand contains only positive literals.

Lemma 1 . Let X, Y be disjoint sets of variables, and let $g(X), h(Y)$ be non-constant Boolean functions with MDNF $p_1 \vee \ldots \vee p_k$ and $q_1 \vee \ldots \vee q_l$, respectively.
Then $p_1 \vee \ldots \vee p_k \vee q_1 \vee \ldots \vee q_l$ is an MDNF of $g \vee h$, and the disjunction of all $p_i q_j$ is an MDNF of gh. Furthermore, all MDNF of $g \vee h$ and gh can be obtained in this way by MDNF of g and h. □

Lemma 2 . Let g, h be non-constant Boolean functions not depending on z, with g monotone, h antimonotone. Let $p_1 \vee \ldots \vee p_k$ and $q_1 \vee \ldots \vee q_l$ be the (unique) MDNF of g and h, respectively. Let var denote the set of variables occuring in a summand. Then the prime implicands of $f = zg \vee \overline{z}h$ are exactly the expressions $zp_i, \overline{z}q_j$, and $p_i q_j$ with $\text{var}(p_i) \cap \text{var}(q_j) = \emptyset$. Further, the disjunction of all terms $zp_i, \overline{z}q_j$ is already an MDNF of f. □

The proofs are straightforward.

2 Logic Arrays and Folding

Programmable logic arrays (PLA) are circuits of simple, regular structure which compute Boolean functions given in DNF (or CNF). Apart from physical details, a PLA for a fixed DNF can be considered as a matrix where the rows represent the literals, and the columns correspond to the summands of the DNF. The entry m_{ij} of the matrix is 1 if the i-th literal occurs in the j-th summand, otherwise $m_{ij} = 0$.
The number of clauses can be exponential in the number of literals. Clearly, to represent a Boolean function by a PLA makes only sense if there exists a DNF having not too many summands (or a CNF having not too many clauses).
We can reduce the number of rows by the technique of folding. In a "folded" PLA, some rows take two literals of the input. Here we only describe the graph-theoretic problem of folding. For details and for the transformation of the circuit problem into the graph problem we refer to [4].

The incompatibility graph (IC-graph) G of a DNF is defined as follows: The vertices of G are the literals; two literals x, y are adjacent iff x, y occur in some common summand. A folding set F in G is a set of additional directed edges such that:

- $xy \in F$ only if x, y are not adjacent in G.

- No edges of F share a common vertex.

- There exists no alternating cycle.
  (An alternating cycle is a sequence $(v_1, v_2, \ldots, v_{2k})$ of vertices with $v_i v_{i+1} \in F$ for i odd, v_i, v_{i+1} adjacent in G for i even, and v_{2k}, v_1 adjacent in G.)

The first two conditions say that F, cosidered as a set of undirected edges, is a matching in the complement graph $\overline{G}$.

Let $\phi(G)$ be the maximum number of edges of a folding set in G. Then we can construct a PLA with $n(G) - \phi(G)$ rows, where $n(G)$ denotes the number of vertices (see [4].) Hence the area of a PLA for a DNF with c summands and IC-graph G can be reduced to $c(n(G) - \phi(G))$.

A PLA of minimum area can be obtained only from an MDNF or an MCNF.

This is obvious, since $A \subseteq B$ for two DNF A, B implies that the number of columns of the A-PLA is not larger than the number of columns in the B-PLA, and the edge set of the IC-graph of A is a subset of the edge set of the IC-graph of B, hence ϕ can only increase.

So, for finding an optimal PLA for a given function f we have, in general, to determine all MDNF and MCNF and then to fold their IC-graphs. This task is NP-hard, since it is difficult to compute the minimal normal forms, and furthermore, the problem of determining $\phi(G)$ is also NP-hard [4].

3 Indicator Functions of Intervals

In the following, let $(x_{n-1}, \ldots, x_0)$ be the binary representation of the integer $x = 2^{n-1}x_{n-1} + \ldots + 2^1 x_1 + 2^0 x_0$. For fixed s, t with $0 \le s \le t \le 2^n - 1$ we define $f_{[s,t]}$ by $f_{[s,t]}(x) = 1$ iff $s \le x \le t$, and $f_{[s,t]}(x) = 0$ otherwise. Further we set $f_{\ge s} = f_{[s,2^n-1]}$ and $f_{\le t} = f_{[0,t]}$. Obviously, the function $f_{\ge s}$ ($f_{\le t}$) is monotone (antimonotone) and has therefore both a unique MDNF and a unique MCNF.

For $0 \le s \le 2^n - 1$ with binary representation $(s_{n-1}, \ldots, s_0)$ we define the graph $T_s = (V, E)$ as follows: $V = (v_{n-1}, \ldots, v_l)$ where l is the smallest index with $s_l = 1$. For $j > i$, $v_j v_i \in E$ holds iff $s_j = 1$. This gives a 1-1-correspondence between the integers $\{0, 1, \ldots, 2^n - 1\}$ and the threshold graphs of at most n vertices. (For properties of threshold graphs see e.g. [3].)

Proposition 1 . The IC-graph of the MDNF of $f_{\ge s}$ is T_s .

Proof. We define $M = \{x_i : s_i = 1\}$ and $N = \{x_i : s_i = 0, i > l\}$ where l is the smallest index with $s_l = 1$. Verify that the prime implicands of $f_{\ge s}$ are exactly the following expressions:

- For each $x_i \in N$, the conjunction of x_i and all $x_j \in M$ with $j > i$.

- The conjunction of all $x_i \in M$.

This yields the assertion. □

For $0 \leq x \leq 2^n - 1$ we define $\bar{x} = 2^n - 1 - x$. We get the binary representation of $\bar{x}$ from that of x by replacing all symbols 1 by 0, and vice versa. Note that $f_{\leq t}(x) = f_{\geq \bar{t}}(\bar{x})$, $f_{\geq s}(x) = \bar{f}_{\leq s-1}(x) = \bar{f}_{\geq \bar{s}+1}(\bar{x})$, and $f_{\leq t}(x) = \bar{f}_{\geq t+1}(x)$.
From these identities, the duality principle, and Proposition 1 we get immediately:

Proposition 2 .

- The MDNF of $f_{\leq t}$ has the IC-graph $T_{\bar{t}}$.
- The MCNF of $f_{\geq s}$ has the IC-graph $T_{\bar{s}+1}$.
- The MCNF of $f_{\leq t}$ has the IC-graph T_{t+1}. □

We define the symbols k, l, m as follows:

$$
\begin{aligned}
k &= \max \ \{i : s_i \neq t_i\} \ . \ (\text{Clearly, } s_k = 0 \text{ and } t_k = 1 \ .) \\
l &= \min \ \{i : s_i = 1\} \ . \\
m &= \min \ \{i : t_i = 0\} \ .
\end{aligned}
$$

As we have seen above, $f_{\geq s}$ depends only on $x_{n-1}, \ldots, x_l$, and $f_{\leq t}$ depends only on $x_{n-1}, \ldots, x_m$. Let be $x' = 2^{k-1} x_{k-1} + \ldots + 2^0 x_0$. We define s', t' analogously. Then we have:

Proposition 3 . $f_{[s,t]}(x) = x_{n-1}{}^{s_{n-1}} \ldots x_{k+1}{}^{s_{k+1}} (\bar{x}_k f_{\geq s'}(x') \vee x_k f_{\leq t'}(x'))$. □

Let r be the factor before parentheses and f' the term in parentheses. Lemma 1 and the duality principle yield:

Proposition 4 . If A is a MDNF of f' then we get an MDNF of rf' by multiplying each summand of A by r. If B is an MCNF of f' then we get an MCNF of rf' by adding the clauses $x_i{}^{s_i} (i > k)$. Further, all MDNF and MCNF of rf' can be obtained from the minimal normal forms of f' in this way. □

To get the minimal normal forms of $f_{[s,t]}$ it remains to determine the MDNF and MCNF of f'.

Proposition 5 . The MDNF or the MCNF of f' is uniquely determined.

Proof. First suppose $t' < s'$. Let $f_{\geq s'} = p_1 \vee \ldots \vee p_a$ and $f_{\leq t'} = q_1 \vee \ldots \vee q_b$ be (unique) MDNF, according to Proposition 1 and 2. Applying Lemma 2 with $z = \bar{x}_k$ we find that $\bar{x}_k p_i, x_k q_j$, and $p_i q_j$ $(\text{var}(p_i) \cap \text{var}(q_j) = \emptyset)$ are the prime implicands of f'. Let $k' < k$ be the largest index where $s_{k'} \neq t_{k'}$. Since $t' < s'$, that means $t_{k'} = 0$ and $s_{k'} = 1$. Further note $f_{\leq t'}(x') = f_{\geq \bar{t'}}(\bar{x'})$. So we have $s_k = \bar{t}_k = 0$, $s_i \neq t_i$ for $k > i > k'$, and $s_{k'} = \bar{t}_{k'} = 1$. Now one can easily check that always $\text{var}(p_i) \cap \text{var}(q_j) = \emptyset$. Hence the terms $\bar{x}_k p_i, x_k q_j$ are the only prime implicands of f'. Due to Lemma 2 they build the unique MDNF of f'.

Now let be $t' \geq s'$. Then we use a simple trick to reach the first case: We replace x_k by $\overline{x}_k$, and vice versa. This manipulation does not disturb the structure of the function, but now we have:

$$f' = x_k f_{\geq s'}(x') \vee \overline{x}_k f_{\leq t'}(x') \text{, hence}$$
$$\overline{f'} = (\overline{x}_k \vee f_{\leq s'-1}(x'))(x_k \vee f_{\geq t'+1}(x')) = \overline{x}_k f_{\geq t'+1}(x') \vee x_k f_{\leq s'-1}(x'),$$

where $s' - 1 < t' + 1$. Thus $\overline{f'}$ has a unique MDNF, and f' has a unique MCNF. □

We introduce some further graph-theoretic notions: The union $G + H$ of graphs is obtained from G and H by building the disjoint unions of the vertex sets and the edge sets of G and H. For building the join $G * H$, we add to $G + H$ all possible edges between vertices of G and H. K_n is the complete graph of n vertices. I_n is the graph of n vertices without any edge.

G is a cograph if G can be obtained from $n(G)$ copies of K_1 by using only the operations $+$ and $*$. Threshold graphs are special cographs.

Altogether we have shown:

Theorem 1 . $f_{[s,t]}$ has a unique MDNF or a unique MCNF. The IC-graph G of this normal form is $((K_1 * T_{s'}) + (K_1 * T_{t'})) * K_{n-1-k}$ in the case of unique MDNF and $(K_1 * T_{t'-2^k}) + (K_1 * T_{s'+2^k}) + I_{n-1-k}$ in the case of unique MCNF. In particular, G is a cograph. □

Next we will show that cographs are easy to fold. In order to find an optimal PLA for $f_{[s,t]}$ we have only to fold G and then to study the minimal normal forms of the other type (which are, unfortunately, not necessarily unique) and their IC-graphs. So we have at least halved the necessary efforts.

4 Folding in Cographs

Lemma 3 . $\phi(G * H) = \max\{\phi(G), \phi(H)\}$.

Proof. This follows immediately from the observation that a folding set cannot have edges in both G and H. □

Lemma 4 . Suppose w.l.o.g. $n(G) \leq n(H)$. Then we have:

$$\phi(G + H) = n(G) + \min\left\{\left\lceil \frac{n(H) - n(G)}{2} \right\rceil, \phi(H)\right\}.$$

Proof. Consider a maximum folding set F in $G + H$. We rebuild F ⠂ that the number of directed edges does not decrease and F remain⠂ (In particular note carefully that the following procedure does not c⠂ cycles.)
   1. Orientate all edges of F between G and H such that the s⠂ and the end vertex belongs to H.

2. If there is a vertex z in G not covered by F then remove an edge xy within H from F and add zx to F.

3. If there is an edge $xy \in F$ and an uncovered vertex u in H then replace xy by xu in F.

4. If there is an edge $xy \in F$ within G and an edge $uv \in F$ within H then replace xy and uv by xu and yv in F.

Repeat steps 2-4 as long as possible.

Note that there cannot be vertices x in G and y in H, respectively, both uncovered by F. (In this case, we could add xy to F which contradicts the maximality of F.) Further, there cannot exist both an F-edge xy within G and two uncovered vertices u, v in H. (In this case, we could xy replace by xu and yv, a contradiction.)

So, carrying out our procedure, we reach a situation where all vertices of G are covered by F-edges between G and H. So, given a maximum folding set F' in H, we can compute a maximum folding set in $G + H$ as follows:

1. Remove as few as possible edges within H from F' such that $n(H) - n(G)$ vertices of H become uncovered.

2. Add $n(G)$ disjoint undirected edges between the vertices of G and the uncovered vertices of H.

This completes the proof. $\qquad\qquad\qquad\qquad\qquad\qquad\qquad\qquad\qquad$ □

Now Lemma 3-4 and the definition of cographs yield:

Theorem 2 . A maximum folding set in a cograph can be computed in linear time.
□

For the special case of threshold graphs, we get the following very simple algorithm for folding T_s :

$$F := \emptyset;$$
$$\text{for } i := l \text{ to } n \text{ do}$$
$$\qquad \text{if } s_i = 0 \text{ and } v_j \text{ is not covered by } F \text{ for some } j < i$$
$$\qquad\qquad \text{then } F := F \cup \{v_i v_j\} \ ;$$

5 Concluding Remarks

Threshold graphs build a small subclass of interval graphs. In [2] we prove by an ʳ reduction to MAXIMUM MATCHING that folding for interval graphs is also ɔmial-time solvable. Studying the influence of pendant vertices to $\phi(G)$ we find ꞁe algorithms e.g. for trees and P_4-reducible graphs. We further show in [2] ɔleteness of folding for split graphs and bipartite graphs. An interesting is the complexity of folding e.g. for permutation graphs and completely s.

ꞁ variant of folding, called BLOCK FOLDING or VERTEX SEPA-
ꞁmially solvable even for trapezoid graphs, as we show also in [2].

References

[1] C. Arbib. Two polynomial problems in PLA folding. In *Proc. 16th International Workshop on Graph-Theoretic Concepts in Computer Science (WG90)*, volume 484 of *Lecture Notes in Computer Science*. Springer-Verlag, 1991.

[2] P. Damaschke. PLA folding in special graph classes. In *2nd Twente Workshop on Graphs and Combinatorial Optimization*. Twente Universiteit, 1991. Memorandum 964.

[3] M. C. Golumbic. *Algorithmic Graph Theory and Perfect Graphs*. Academic Press, New York, 1980.

[4] R. H. Möhring. Graph problems related to gate matrix layout and PLA folding. Technical Report 223, TU Berlin, 1989.

On the broadcast time of the Butterfly network

Elena Stöhr

Karl-Weierstraß-Institut für Mathematik,
Mohrenstr. 39, O-1086 Berlin, Germany

Broadcasting is the process of message dissemination in a communication network in which a message originated by one processor is transmitted to all processors of the network. In this paper we consider undirected graphs as models for communication networks. Thus, a given vertex in a graph has a message, which it wishes to disseminate to all other vertices; each vertex can transmit a message to exactly one vertex to which it is adjacent during one unit of time, and each vertex can either transmit or receive the message per unit of time.

Assume the message originates at vertex u of the connected graph G. The *broadcast time of the vertex u, $b(u)$*, is the minimum number of time units required to complete broadcasting from vertex u. The *broadcast time of a graph G, $b(G)$*, is the maximum broadcast time of any vertex u in G.

Broadcasting in graphs of bounded degree has been studied in [Bermond, Peyrat, 1988], [Bermond, Hell, Liestman, Peters, 1990], [Hromkovic, Jeschke, Monien, 1990], [Liestman, Peters, 1988], [Stöhr,1990]. The maximum degree is an important parameter in the design of interconnection networks and this is one of the motivations to investigate broadcasting in graphs with fixed maximum degree, in particular in bounded-degree "approximations" of the hypercube such as cube-connected cycles, butterfly, shuffle-exchange and deBruijn networks. Lower and upper bounds on the time required to broadcast in graphs with maximum degrees 3 and 4 were given in [Liestman, Peters, 1988]. The results of Liestman and Peters were improved in [Bermond, Hell, Liestman, Peters,1990] where general lower bounds were obtained.

The trivial lower bound on the broadcast time is the diameter of a graph. For cube-connected cycles and shuffle-exchange networks broadcast time and diameter differ at most by a constant [Hromkovic, Jeschke, Monien, 1990], [Liestman, Peters, 1988]. In this paper we present non-trivial lower bound on broadcast time of the Butterfly graph. We show that $b(B(m)) > 3m/2 + t_0 m$ for some $t_0, t_0 = 0.045...$, and all sufficiently large m. Note, that the diameter of $B(m)$ is $D = \lceil 3m/2 \rceil - 1$. Thus, $b(B(m)) > cD$ for some constant $c > 1$.

An upper bound on broadcast time of the Butterfly graph was given in [Stöhr, 1990]. It was shown that $b(B(m)) \leq 2m$.

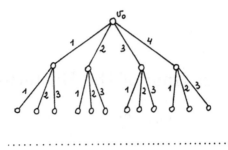

Figure1: Labeling of edges of T

Definition. The *order-m Butterfly graph $B(m)$* has vertex-set

$$V_m = \{0, 1, ..., m-1\} \times \{0,1\}^m,$$

where $\{0,1\}^m$ denotes the set of length-m binary strings. For each vertex $v = <\ell, \vec{\beta}> \in V_m$, $\ell \in \{0,1,...,m-1\}, \vec{\beta} \in \{0,1\}^m$, we call ℓ the level and $\vec{\beta}$ the position-within-level (PWL) string of v. The edges of $B(m)$ are of two types: For each $\ell \in \{0,1,...,m-1\}$ and each $\vec{\beta} = \beta_0\beta_1...\beta_{m-1} \in \{0,1\}^m$, the vertex $<\ell, \vec{\beta}>$ on level ℓ of $B(m)$ is connected by a straight-edge with vertex $<\ell+1(mod\ m), \vec{\beta}>$ on level $\ell+1(mod\ m)$ of $B(m)$, and is connected by a cross-edge with vertex $<\ell+1(mod\ m), \beta_0\beta_1...\beta_{\ell-1}(\beta_\ell \oplus 1)\beta_{\ell+1}...\beta_{m-1}>$, on level $\ell+1(mod\ m)$ of $B(m)$, where $\oplus$ denotes addition modulo 2.

Theorem. Let $B(m)$ be the order m Butterfly graph. Then $b(B(m)) > 3m/2 + t_0 m$ for some t_0, $t_0 = 0.045...$, and all sufficiently large m.

Proof. To obtain a contradiction suppose that broadcasting can be completed on $B(m)$ in time $3m/2 + tm$, $0 \le t < 1/2$.

Since the Butterfly graph is a Cayley graph [Annexstein, Baumslag, Rosenberg, 1990], and every Cayley graph is vertex symmetric [Akers, Krishnamurthy, 1986], we can assume that the message originates at the vertex $v_0 = <0, 00...0>$, and the originator learns the message at time 0.

In any connected graph G, a broadcast from a vertex u determines a rooted spanning tree of G [Farley, Hedetniemi, Mitchell, Proskurowski,1979]. Denote by T a spanning tree of $B(m)$ with the root v_0 such that broadcasting can be done in T in time at most $3m/2 + tm$. Since the Butterfly graph has maximum degree 4, the tree T has maximum degree at most 4. Clearly, the distance in T between any leaf of T and the root v_0 is at most $3m/2 + tm$.

Now we label the edges of T as follows:

The edges incident to v_0 are labeled by 1,2,3,4, so that the vertex connected with v_0 by the edge labeled by i receives the message at time i, $i = 1,2,3,4$.

Let $v \ne v_0$ be any non-leaf vertex of T and v_1, v_2, v_3 be vertices of T adjacent to v. Suppose that the vertex v receives the message at time t and the vertex v_i receives the message at time $t+i$, $i = 1,2,3$. Then we label the edge connecting v with v_i by i, $i = 1,2,3$. See Fig. 1.

Let v be any vertex of T. Denote by $P_v = \{v_0, v_1, ..., v_n\}$ the shortest path in T connecting the vertices v and v_0. The length of P_v is the distance in T between v and v_0. Denote by $\sharp_i(P_v)$ the number of edges in P_v labeled by i, $i = 2,3,4$.

Let T_n be the set of vertices of T such that the distance in T between any vertex $v \in T_n$ and v_0 is equal to n, $n \geq 3m/2 - tm$. The vertex $v \in T_n$ receives the message from v_0 at time

$$n + \natural_2(P_v) + 2\natural_3(P_v) + 3\natural_4(P_v) \leq 3m/2 + tm$$

Let $T_n(p,q,r) \subseteq T_n$ be a subset of T_n. Suppose that for the path P_v connecting any vertex $v \in T_n(p,q,r)$ with v_0 we have $\natural_2(P_v) = p$, $\natural_3(P_v) = q$, $\natural_4(P_v) = r$, $r = 0,1$.

Let $\natural_4(P_v) = 0$. Then

$$|T_n(p,q,0)| \leq \binom{n}{p}\binom{n-p}{q}.$$

Since $3m/2 - tm \leq n \leq 3m/2 + tm$, $p \leq 2tm$, $q \leq tm$, $t < 1/2$, we have

$$\binom{n}{p} \leq \binom{3m/2 + tm}{p} \leq \binom{3m/2 + tm}{2tm}$$

$$\binom{n-p}{q} \leq \binom{3m/2 + tm}{q} \leq \binom{3m/2 + tm}{2tm}.$$

Therefore,

$$|T_n(p,q,0)| \leq \left[\binom{3m/2 + tm}{2tm}\right]^2.$$

Since only one edge incident to v_0 is labeled by 4 we have

$$|T_n(p,q,1)| \leq \left[\binom{3m/2 + tm}{2tm}\right]^2.$$

Thus,

$$(1) \qquad |T_n| \leq 2\sum_{p=0}^{2tm}\sum_{q=0}^{tm}|T_n(p,q,0)| \leq cm^2\left[\binom{3m/2 + tm}{2tm}\right]^2$$

for some constant c.

Let $\check{V} = \{\bigcup T_n, \quad 3m/2 - tm \leq n \leq 3m/2 + tm\}$ be the set of vertices of T such that the distance in T between any vertex $v \in \check{V}$ and v_0 is at least $3m/2 - tm$. Using (1) we get

$$(2) \qquad |\check{V}| \leq \sum_n |T_n| \leq cm^3\left[\binom{3m/2 + tm}{2tm}\right]^2.$$

Let $L = \{v = < m/2, \vec{\delta} > | \vec{\delta} \neq \alpha 0^k \beta \text{ for some } k \geq tm/2, \quad \alpha 0^k \beta \in \{0,1\}^m\}$ be the subset of the set of level-$m/2$ vertices of $B(m)$.

It is not difficult to show that

$$(3) \qquad |L| \geq 2^m - m2^{m-tm/2}.$$

From the definition of the Butterfly graph it follows that the distance between any vertex v from L and v_0 is at least $3m/2 - tm$.

Therefore, $L \subseteq \check{V}$. Using (2) and (3) we have

$$cm^3\left[\binom{3m/2 + tm}{2tm}\right]^2 \geq 2^m - m2^{m-tm/2}.$$

For large m an approximate expression for the factorial is given by Stirling formula $m! \approx m^m e^{-m} \sqrt{2\pi m}$. Using Stirling formula we obtain

$$\binom{3m/2 + tm}{2tm} \approx \frac{(3/2 + t)^{3m/2 + tm}}{(2t)^{2tm}(3/2 - t)^{3m/2 - tm}}.$$

Thus,

(4)
$$cm^3 \left(\frac{(3/2 + t)^{3m/2 + tm}}{(2t)^{2tm}(3/2 - t)^{3m/2 - tm}} \right)^2 \geq 2^m - m2^{m - tm/2}.$$

Taking the mth root of the both side of (4) we have for large m

$$\frac{(3/2 + t)^{3 + 2t}}{(2t)^{4t}(3/2 - t)^{3 - 2t}} \geq 2.$$

The latter inequality is not true for $0 < t < t_0$, for some $t_0 = 0.045....$ This contradiction establishes the Theorem.

Acknowledgment.

The author thanks B. Monien for suggesting this problem.

References:

S.B. Akers, B. Krishnamurthy (1986), A group-theoretic model for symmetric interconnection networks parallel processing, International Conference Parallel Processing, pp.216-233.

F. Annexstein, M. Baumslag, A.L. Rosenberg (1990), Group action graphs and parallel architectures, Siam. J. Comput. 19, N 3, pp.544-569.

J.-C. Bermond, P. Hell, A.L. Liestman, J.G. Peters (1988), Broadcasting in bounded degree graphs, to app. in SIAM J. Discr. Math.

J.-C. Bermond, C. Peyrat (1988), Broadcasting in deBruijn networks, in Proc. Nineteenth SE Conf. on Combinatorics, Graph Theory and Computing, Congressus Numerantium, pp. 283-292.

A. Farley, S. Hedetniemi, S. Mitchell, A. Proskurowski (1979), Minimum broadcast graphs, Discrete Math. 25, pp. 189-193.

J. Hromkovic, C.D. Jeschke, B. Monien (1990), Optimal algorithms for dissemination of information in some interconnection networks, Proc. MFCS'90, LNCS 452, pp 337-346.

A.L. Liestman, J.G. Peters (1988), Broadcast networks of bounded degree, Siam. J. Discr. Math. 1, pp. 531-540.

E.A. Stöhr (1990), Broadcasting in the Butterfly network, IPL, to app.

ON DISJOINT CYCLES*

Hans L. Bodlaender

Department of Computer Science, University of Utrecht

P.O.Box 80.089, 3508 TB Utrecht, the Netherlands

Abstract

It is shown, that for each constant $k \geq 1$, the following problems can be solved in $\mathcal{O}(n)$ time: given a graph G, determine whether G has k vertex disjoint cycles, determine whether G has k edge disjoint cycles, determine whether G has a feedback vertex set of size $\leq k$. Also, every class $\mathcal{G}$, that is closed under minor taking, or that is closed under immersion taking, and that does not contain the graph formed by taking the disjoint union of k copies of K_3, has an $\mathcal{O}(n)$ membership test algorithm.

1   Introduction.

In this paper we consider the following problem: given a graph $G = (V, E)$, does G contain at least k vertex disjoint cycles. If k is part of the instance of the problem, then the problem is NP-complete, because then it contains PARTITION INTO TRIANGLES as a special case (take $k = |V|/3$). (See [12].) In this paper we consider the problem for fixed k.

The problem can be seen as a special case of the MINOR CONTAINMENT problem. A graph G is a minor of a graph H, if G can be obtained from H by a series of vertex deletions, edge deletions and edge contractions (an edge contraction is the operation to replace two adjacent vertices v, w by one vertex that is adjacent to all vertices that were adjacent to v or w). Robertson and Seymour [16] showed that for every fixed H there exists an $\mathcal{O}(n^3)$ algorithm to test whether a given graph G contains H as a minor. If H is planar, then they give an $\mathcal{O}(n^2)$ algorithm. Using recent results of Lagergren [13] or of Arnborg et. al. [2] it is possible to improve on the $\mathcal{O}(n^2)$ bound. In the former case, using an approximation for treewidth, one arrives at an $\mathcal{O}(n \log^2 n)$ algorithm. In the latter case, using graph rewriting, one gets an $\mathcal{O}(n \log n)$ algorithm, or an algorithm that uses $\mathcal{O}(n)$ time in the uniform cost measure and polynomial (not linear) space. In [6] (extending results in [11]) a class of graphs is given for which minor tests can be done in $\mathcal{O}(n)$ time with the help of depth first search. This class does not include the graph consisting of k disjoint copies of K_3, (here denoted as $k \cdot K_3$.) As a graph G contains $k \cdot K_3$ as a minor, if and only if G contains k vertex disjoint cycles, it follows from this paper that testing whether $k \cdot K_3$ is a minor of a given graph G can be done in $\mathcal{O}(n)$ time for fixed k.

We also consider the problem of finding k edge disjoint cycles in a graph G. This problem is related to the problem of finding immersions: a graph G has $k \cdot K_3$ as an immersion, if and only if G contains k edge disjoint cycles. (A graph G is an immersion of H, if G can be obtained from H by a series of vertex deletions, edge deletions and edge lifts. An edge lift is the operation of replacing two edges $(v, w), (w, x)$ by an edge (v, x).)

The problem in this paper is related to the (much more difficult) problem of finding vertex or edge disjoint paths between a number of fixed pairs of vertices. For an overview of recent results on this problem, see e.g. [17].

---

*This work has been partially supported by the ESPRIT II Basic Research Actions Program of the EC, under contract No. 3075 (project ALCOM).

We also consider the problem of finding a minimum size feedback vertex set is a graph. A feedback vertex set in graph $G = (V, E)$, is a set $W \subseteq V$ such that the graph $(V - W, E - \{(v, w) | v \in W \vee w \in W\})$ is cycle-free. The problem of finding a minimum feedback vertex set is NP-complete [12]. Here we show that for fixed k, the problem of finding a feedback vertex set of size $\leq k$, if it exists, is solvable in $\mathcal{O}(n)$ time.

It must be pointed out that, although we give $\mathcal{O}(n)$ algorithms, these algorithms may be far from practical, due to the large constant factor, hidden in the "$\mathcal{O}$" notation. However, we believe that with further optimizations, practical algorithms can be derived, that have an $\mathcal{O}(n)$ worst case time bound. The purpose of this paper is to show that there exist linear time algorithms for the considered problems, not to derive the best possible algorithm.

We end this introduction with some definitions and notations, used in this paper.

The subgraph of $G = (V, E)$, induced by $W \subseteq V$ is denoted by $G[W] = (W, \{(v, w) \in E | v, w \in W\})$.

A tree-decompostion of a graph $G = (V, E)$ is a pair $(\{X_i | i \in I\}, T = (I, F))$ with $\{X_i | i \in I\}$ a family of subsets of V, and T a tree, such that $\cup_{i \in I} X_i = V_j$; $\forall (v, w) \in E : \exists i \in I : v, w \in X_i$; for all $v \in V : \{i \in I | v \in X_i\}$ forms a connected subtree of T.

The treewidth of a tree-decompositon $(\{X_i | i \in I\}, T)$ is $\max_{i \in I} |X_i| - 1$. The treewidth of a graph $G = (V, E)$ is the minimum treewidth over all possible tree-decompositions of G.

All graphs considered in this paper are assumed to be finite, undirected and simple. n denotes the number of vertices of graph $G = (V, E)$.

2 Finding two vertex or edge disjoint cycles.

In this section we give a linear time algorithm to test whether a given graph contains two vertex or edge disjoint cycles. We need the following lemmas.

Lemma 2.1
Let $G = (V, E)$ be an undirected graph. Suppose G contains a cycle c and four vertex disjoint paths, that have their endpoints on c, and are edge disjoint from c. Then G contains two vertex disjoint cycles.

Proof.
Let $v_{i_1}, v_{i_2} (1 \leq i \leq 4)$ be the endpoints of the four paths. For $i = 1, \ldots, 4$, look at a path formed by going from v_{i_1} to v_{i_2} over the cycle c. If one of these paths is entirely contained in another, or two of these paths are disjoint, then G contains two vertex disjoint cycles, as shown in Figure 1.

If this case does not appear for any pair $i_1 \neq i_2$, then we are in a situation as shown in Figure 2. Again G contains two vertex disjoint cycles. □

Lemma 2.2
Let $G = (V, E)$ be an undirected graph. Suppose G contains two cycles c_1, c_2 that share exactly one vertex and suppose G contains a path p, that is vertex disjoint from c_1, has both endpoints in c_2, and is edge disjoint from c_2. Then G contains two vertex disjoint cycles.

Proof.
The result is obvious from Figure 3. □

Theorem 2.3
There exists an algorithm, that uses $\mathcal{O}(n)$ time, and that given a graph $G = (V, E)$, either finds two vertex disjoint cycles in G or outputs a feedback vertex set X with $|X| \leq 9$.

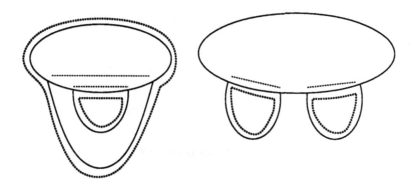

Figure 1: See Lemma 2.1

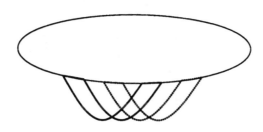

Figure 2: See Lemma 2.1

Proof.

First test whether G is cycle-free. If so, take $X = \emptyset$, and we are done. Otherwise find a cycle c in G. Let W be the set of vertices on this cycle. Consider $G[V - W]$. If $G[V - W]$ contains a cycle we have two vertex disjoint cycles and we are done. So suppose $G[V - W]$ is a forest.

The algorithm now proceeds by trying to find the four paths, mentioned in Lemma 2.1. A cycle which shares exactly one vertex with W is treated as a path with both endpoints the same vertex in W, and is not handeled separately. In the remainder of this proof, an i-path is a path, that has i endpoints in W and no other vertex in $W(i = 1, 2)$.

The algorithm uses a counter α, that denotes the number of vertex disjoint 2-paths, found so far. We also use a set $X \subseteq V$. Initially $X \subseteq \emptyset$. An i-path is X-free, if it does not contain a vertex in X.

Repeat the following procedure, until it stops.

1. If G is a cycle with possibly a number of trees attached to it (each tree sharing exactly one vertex with the cycle), then take an arbitrary vertex $v \in W$, and let $X = \{v\}$. Now stop. Clearly $G[V - X]$ is cycle-free.

2. If $\alpha = 4$, then stop. By Lemma 2.1 or Lemma 2.2 we can find two vertex disjoint cycles.

3. If $G[V - X]$ is a forest, then stop. Now $\alpha \leq 3$.

4. Otherwise, an X-free 2-path exists. We will find such an X-free 2-path p, and then put at most 3 vertices from this path in X, such that all further X-free 2-paths are disjoint from p.

Let a *junction* be a vertex $v \in V - (W \cup X)$, that is endpoint of three edge-disjoint X-free 1-paths. (It follows that these paths are vertex-disjoint, except for the endpoints.) See Figure 4 for an example.

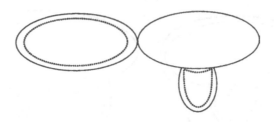

Figure 3: See Lemma 2.2

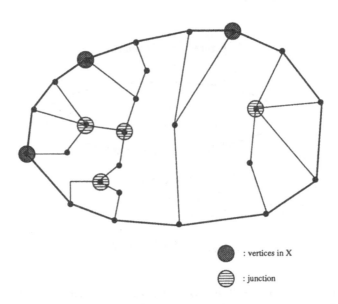

● : vertices in X

⊜ : junction

Figure 4: Example of junctions

Now one easily determines in linear time what vertices are a junction. If no junction exists, then take an X-free 2-path p, and put its endpoints in X. Increase α by 1, and continue with step 2. Note that all 2-paths, that are now X-free are vertex disjoint from p, otherwise p would contain a junction.

In case there exists at least one junction, we search for a junction, such that at least 2 of its X-free 1-paths do not contain other junctions. Such a junction can be found in the following way. Start in an arbitrary junction v_o. Consider two of its X-free 1-paths. If neither contains a junction, we are done. Otherwise, let $v_1 \neq v_o$ be a junction on one of these paths.

Repeat the process with v_1, instead of v_o, and consider the two X-free 1-paths of v_1, that do not contain v_o. As $G[V - W]$ is cycle-free, this process ends after a finite number of steps. It is not necessary to consult each edge more than a constant number of times, so the procedure has a linear time implementation.

Let w_1, w_2 be the other endpoints of the two found paths from junction v_j. Combining these two paths we have an X-free 2-path p from w_1 to w_2. Now add w_1, w_2 and v_j to X, and add 1 to α.

We claim that all 2-paths, that are X-free (for the new set X), are disjoint from p. Because if not, the first vertex v^*, that is shared by p and an X-free 2-path p', would have been a junction before the addition of w_1, w_2 and w_j to X (see Figure 5). But then $v_j \to w_1$ or $v_j \to w_2$ would have contained a junction, and hence would not have been chosen by the procedure.

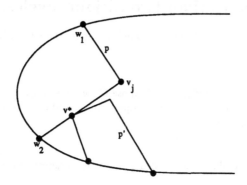

Figure 5: See Theorem 2.3

The procedure terminates, either because two vertex independent cycles are found, or because $G[V - X]$ is cycle-free. In the latter case $\alpha \leq 3$, and hence $|X| \leq 9$.

The procedure can be implemented to use time, linear in the number of vertices, even if there are more than a linear number of edges: testing acyclicity can be done in $\mathcal{O}(n)$ time, $G[V - W]$ has, when acyclic, $\mathcal{O}(n)$ edges, and one never needs to inspect more than $|X| + 3 = \mathcal{O}(1)$ edges (v, w) with $v \in V - W, w \in W$ per vertex $v \in V - W$. □

Theorem 2.4
There exists an algorithm, that uses $\mathcal{O}(n)$ time, and that given a graph $G = (V, E)$, either finds two vertex disjoint cycles in G, or outputs a tree-decomposition of G with treewidth ≤ 10.

Proof.
First use the procedure of Theorem 2.3. In case we have a feedback vertex set X with $|X| \leq 9$, then make a tree-decomposition $(\{X_i | i \in I\}, T = (I, F))$ of $G[V - X]$ with treewidth 1 (this can easily be done, see e.g. [4]).

Now $(\{X_i \cup X | i \in I\}, T = (I, F))$ is a tree-decomposition of G with treewidth ≤ 10. □

Many graph problems, including many NP-complete problems, become linear time solvable for graphs, given together with a tree-decomposition with constant bounded treewidth (see e.g. [1, 3, 7, 18]). As the question, whether the input graph contains $\geq k$ vertex (or edge) disjoint cycles can be formulated in monadic second order form (see [3]), or in the calculus of [7], the next result follows from these papers and Theorem 2.4.

Corollary 2.5
There exists an algorithm, that uses $\mathcal{O}(n)$ time, that given a graph $G = (V, E)$, decides whether G contains two vertex (or edge) disjoint cycles, and if so, outputs these.

We remark here, that the given method is not (yet) practical, due to the high constant factor. However, we believe that the procedure of Theorem 2.3 with some optimizations, followed by an extensive case analysis, can yield a practical algorithm for the problem, that has a good worst-case running time. Probably a good average case running time is obtained by a straightforward backtracking procedure. However, this approach may in some cases give exponential time, e.g., with inputs of the form $K_{3,m}$.

Also, further optimizations are possible in the case of edge disjoint cycles.

3 Finding more than two disjoint cycles.

In this section we consider the problem of finding $k \geq 3$ vertex (or edge) disjoint cycles. We first need a lemma, similar to Lemma 2.1.

Lemma 3.1
Let $G = (V, E)$ be an undirected graph. Suppose G contains two vertex disjoint cycles c_1, c_2, and 9 vertex disjoint paths, that each have one endpoint in c_1, and one endpoint in c_2. Then G contains at least three vertex disjoint cycles.

Proof.
The proof of this lemma relies on a detailed and not very interesting case analysis, which is omitted from this paper.

Instead, we give here a much shorter proof for the weaker result, where " 9 " is replaced by "26".

Without loss of generality, we may suppose that the 26 vertex disjoint paths between c_1 and c_2 do not share with c_1 or c_2 other vertices than their endpoints. Number the vertices on c_1 $v_1 \cdots v_r$ in order of a traversal of c_1, and likewise number the vertices on c_2 $w_1 \cdots w_s$. Order the paths between c_1 and c_2 with respect to their endpoints on c_1. Now consider the sequence of the 26 endpoints of the paths on c_2, in this order. This is a sequence of 26 different numbers. By a theorem of Erdös and Szekeres [9], this sequence has a subsequence of six numbers, corresponding to vertices $v_{i_1} \cdots v_{i_6}$ with $i_1 < i_2 \cdots < i_6$ or $i_1 > i_2 > \cdots > i_6$. This corresponds to a situation, shown in Figure 6. Clearly, G has three vertex disjoint cycles. □

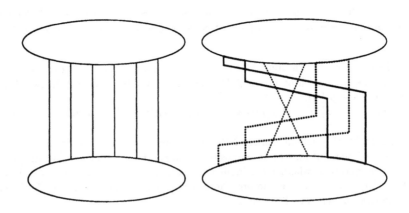

Figure 6: See Lemma 3.1

Lemma 3.2
Suppose $G = (V, E)$ contains $k-1$ vertex disjoint cycles $c_1 \cdots c_{k-1}$, and $3(k-1)+4(k-1)(k-2)+1$ disjoint paths, such that each path has both its endpoints on the cycles $c_1 \cdots c_{k-1}$, and does not share another vertex with the cycles. Then G contains k vertex disjoint cycles.

Proof.
By a pidgeonhole argument, either there are at least four paths with all endpoints on the same cycle c_i, or there are two cycles c_i, c_j with at least nine paths with one endpoint on c_i, and one endpoint on c_j. In the former case, apply Lemma 2.1. In the latter case, apply Lemma 3.1. In both cases, we get one extra, disjoint cycle. □

Theorem 3.3
For every $k \geq 2$, there exists an algorithm, that uses $\mathcal{O}(n)$ time, and that given a graph $G = (V, E)$, either finds k vertex disjoint cycles, or finds a feedback vertex set $X \subseteq V$ with $|X| \leq 12k^2 - 27k + 15$.

Proof.
If $k = 2$, then use Theorem 2.3. Otherwise, first recursively apply the algorithm for $k - 1$. Either a suitable set X is found, or we find $k - 1$ vertex disjoint cycles. In the latter case, apply a procedure, similar to the one in the proof of Theorem 2.3, but instead let W be the set of all vertices on the $k - 1$ vertex disjoint cycles, and stop when $\alpha = 4k^2 - 9k + 6(= 3(k - 1) + 4(k - 1)(k - 2) + 1)$. If we stop with $\alpha = 4k^2 - 9k + 6$, then we have enough paths to apply Lemma 3.2. Otherwise $|X| \leq 3\alpha \leq 3(4k^2 - 9k + 5)$, and $G[V - X]$ is cycle free. □

In the same way as in section 2, we can derive the following results:

Theorem 3.4
For every $k \geq 2$, there exists an algorithm, that uses $\mathcal{O}(n)$ time, and that given a graph $G = (V, E)$, either finds two vertex disjoint cycles in G, or outputs a tree-decomposition of G with treewidth $\leq 12k^2 - 27k + 16$.

Corollary 3.5
For every $k \geq 2$, there exists an algorithm, that uses $\mathcal{O}(n)$ time, that given a graph $G = (V, E)$, decides whether G contains k vertex (or edge) disjoint cycles, and if so, outputs these.

More efficient algorithms can be found with help of a more detailed case analysis. The purpose here was only to show that linear time is achievable, not to give the most efficient algorithm.

4   Some consequences.

Robertson and Seymour [15, 14] proved that for every class of graphs $\mathcal{G}$, that is closed under taking of minors (or immersions), there exists a finite set of graphs, called the obstruction set of $\mathcal{G}$, such that a graph H belongs to $\mathcal{G}$, if and only if no graph G in the obstruction set of $\mathcal{G}$ is a minor (immersion) of H. Combining this result with Theorem 3.4, the fact that minor tests can be done in $\mathcal{O}(n)$ time with a constant width tree-decomposition, and the observation that a graph G contains $k \cdot K_3$ as a minor (immersion), if and only if G contains k vertex (edge) disjoint cycles, we have the following result:

Corollary 4.1
Let $\mathcal{G}$ be a class of graphs, closed under minor (immersion) taking, that does not contain all graphs of the form $k \cdot K_3$ ($k \geq 1$). Then there exists an $\mathcal{O}(n)$ algorithm to test membership in $\mathcal{G}$.

Note that the algorithm uses also linear space, standard cost measure (in constrast with [2]), is non-constructive, and has a very large constant factor. Another consequence of our results is for the FEEDBACK VERTEX SET problem.

Corollary 4.2
For every constant k, there exists an algorithm, that uses $\mathcal{O}(n)$ time, that determines whether a given graph $G = (V, E)$ contains a feedback vertex set of size $\leq k$, and if so, outputs one.

Proof.
Use Theorem 3.4. If G contains $\geq k+1$ vertex disjoint cycles, then every feedback vertex set contains at least $k+1$ vertices. Otherwise, use the tree-decomposition, and a dynamic programming algorithm, e.g. as in [3, 5, 7] to find the optimal feedback vertex set. □

This result was also obtained, independently, by Fellows [10].

For a class of graphs $\mathcal{G}$, let *within k vertices of* $\mathcal{G}$ denote the class of graphs $\{G = (V, E) \mid$ there exists a subset $W \subseteq V$ of at most k vertices, such that $G[V - W] \in \mathcal{G}\}$, i.e., a graph G is within k vertices of $\mathcal{G}$, if we can remove $\leq k$ vertices from G and the resulting graph belongs to $\mathcal{G}$. This type of problem was considered in [8]. We have the following easy result:

Lemma 4.3
Let $\mathcal{G}$ be a class of graphs with $G \in \mathcal{G} \Rightarrow$ treewidth $(G) \leq k$. Then $G \in$ within l vertices of $\mathcal{G} \Rightarrow$ treewidth $(G) \leq k + l$.

Proof.
Use the same technique as in Theorem 2.4. □

A corollary of this result is that a graph $G = (V, E)$ with a 'partial feedback vertex set' $W \subseteq V$ such that $G[V - W]$ does not contain a cycle with length $\geq K$ has treewidth $\leq |W| + K - 2$. (Use that a graph that contains no cycle with length $\geq K$ has treewidth $\leq K - 2$ [11].) Clearly, the same results hold if we consider classes *within k edges of* $\mathcal{G}$.

References

[1] S. Arnborg. Efficient algorithms for combinatorial problems on graphs with bounded decomposability – A survey. *BIT*, 25:2–23, 1985.

[2] S. Arnborg, B. Courcelle, A. Proskurowski, and D. Seese. An algebraic theory of graph reduction. Technical Report 90-02, Laboratoire Bordelais de Recherche en Informatique, Bordeaux, 1990. To appear in Proceedings 4th Workshop on Graph Grammars and Their Applications to Computer Science.

[3] S. Arnborg, J. Lagergren, and D. Seese. Problems easy for tree-decomposable graphs (extended abstract). In *Proceedings of the 15'th International Colloquium on Automata, Languages and Programming*, pages 38–51. Springer Verlag, Lect. Notes in Comp. Sc. 317, 1988. To appear in J. of Algorithms.

[4] H. L. Bodlaender. Classes of graphs with bounded treewidth. Technical Report RUU-CS-86-22, Dept. of Computer Science, Utrecht University, Utrecht, 1986.

[5] H. L. Bodlaender. NC-algorithms for graphs with small treewidth. In J. van Leeuwen, editor, *Proc. Workshop on Graph-Theoretic Concepts in Computer Science WG'88*, pages 1–10. Springer Verlag, LNCS 344, 1988.

[6] H. L. Bodlaender. On linear time minor tests and depth first search. In *Proceedings Workshop on Algorithms and Data Structures WADS'89*, pages 577–590. Springer Verlag, Lecture Notes in Computer Science, vol. 382, 1989.

[7] R. B. Borie, R. G. Parker, and C. A. Tovey. Automatic generation of linear algorithms from predicate calculus descriptions of problems on recursive constructed graph families. Manuscript, 1988.

[8] D. J. Brown, M. R. Fellows, and M. A. Langston. Nonconstructive polynomial-time decidability and self-reducibility. *Int. J. Computer Math.*, 31:1–9, 1989.

[9] P. Erdös and G. Szekeres. A combinatorial problem in geometry. *Compos. Math.*, 2:464–470, 1935.

[10] M. R. Fellows, 1989. Personal communication.

[11] M. R. Fellows and M. A. Langston. On search, decision and the efficiency of polynomial-time algorithms. In *Proceedings of the 21th Annual Symposium on Theory of Computing*, pages 501–512, 1989.

[12] M. R. Garey and D. S. Johnson. *Computers and Intractability, A Guide to the Theory of NP-Completeness*. W.H. Freeman and Company, New York, 1979.

[13] J. Lagergren. Efficient parallel algorithms for tree-decomposition and related problems. In *Proceedings of the 31th Annual Symposium on Foundations of Computer Science*, pages 173–182, 1990.

[14] N. Robertson and P. D. Seymour. Graph minors. XV. Wagner's conjecture. To appear.

[15] N. Robertson and P. D. Seymour. Graph minors — a survey. In I. Anderson, editor, *Surveys in Combinatorics*, pages 153–171. Cambridge Univ. Press, 1985.

[16] N. Robertson and P. D. Seymour. Graph minors. XIII. The disjoint paths problem. Manuscript, 1986.

[17] J. van Leeuwen. Graph algorithms. In *Handbook of Theoretical Computer Science, A: Algorithms and Complexity Theory*, Amsterdam, 1990. North Holland Publ. Comp.

[18] T. V. Wimer. *Linear algorithms on k-terminal graphs*. PhD thesis, Dept. of Computer Science, Clemson University, 1987.

Short disjoint cycles in cubic bridgeless graphs

Andreas Brandstädt

FB 11–Mathematik FG Informatik
Universität/Gesamthochschule Duisburg
e-mail: hn 305 br @ unidui. uni-duisburg. de

Abstract

The girth $g(G)$ of a finite simple undirected graph $G = (V, E)$ is defined as the minimum length of a cycle in G. We develope a technique which shows the existence of $\Omega(n^{1/7})$ pairwise disjoint cycles of length $0(n^{6/7})$ in cubic bridgeless graphs.

As a consequence, for bridgeless graphs with $\deg v \in \{2,3\}$ for all $v \in V$ and $| \{v : \deg v = 3\} | / | \{v : \deg v = 2\} | \geq c > 0$ the girth $g(G)$ is bounded by $0(n^{6/7})$. Furthermore similarly as for cycles, the existence of many small disjoint subgraphs with k vertices and $k + 2$ edges is shown. This very technical result is useful in solving the bisection problem (configuring transputer networks) for regular graphs of degree 4 as B. Monien pointed out. Furthermore the existence of many disjoint cycles in such graphs could be also of selfstanding interest.

0. Motivation

In connection with transputer networks the bisection problem is studied where for a given graph a balanced partition with a minimal number of crossing edges has to be found (cf. for instance [HrMo 90]). This problem is well–known and well–studied and has applications not only for transputer networks but also for VLSI layout.

Transputers typically have only 4 communication links such that the bisection problem is investigated in [HrMo 90] for graphs with vertex degree 4 instead of arbitrary graphs.

A nontrivial upper bound for the bisection width of such graphs is given in [HrMo 90].

Good upper bounds are very important for automatically configuring transputer networks. As B. Monien pointed out, their upper bound could be further improved if one had the following (very technical) result (1):

let us define a (not necessarily induced) subgraph (V', E') of (V, E) to be *useful* if $| V' | +2 =| E' |$.

Furthermore let $n_i(G) =| \{v : \deg v = i\}$ and c be a constant with $c > 0$.

(1) If $G = (V, E)$ is a finite simple undirected graph and for all

$v \in V$ $\deg v \in \{2,3\}$ such that $n_3(G)/n_2(G) \geq c > 0$

then there is a useful subgraph of size $o(n)$ in G.

We show that the order of $o(n)$ can be choosen as $0(n^{13/14})$.

Furthermore we show: cubic bridgeless graphs have many short pairwise disjoint cycles.

1. Notions and helpful results

In the following all graphs $G = (V, E)$ are finite, simple (i.e. without parallel edges and selfloops) and undirected. For standard notions such as vertex degree, paths, cycles and bridges cf. a standard textbook.

The *girth* $g(G)$ is the minimum length of a cycle in G.

Let $\Delta(G)$ be the maximum degree of vertices of G.

A set $V' \subseteq V$ of vertices is *independent* iff for all $u, v \in V'$ $uv \notin E$.

Let $\alpha(G)$ be the maximum size of an independent vertex set in G.

A graph G is *regular of degree k* iff all vertices of G have degree k.

A *cubic* graph is a graph which is regular of degree 3.

A *k-factor* of G is a subgraph $G' = (V, E')$ which is regular of degree k.

A *1-factor* is thus a set of pairwise disjoint edges (a perfect matching) which cover V, and a *2-factor* is a set of pairwise disjoint cycles which cover V.

There is a well-known property of cubic graphs:

Theorem 1.1 ([Pet 1891])

Every cubic graph $G = (V, E)$ without bridges can be partitioned into a 1-factor E_1 and a 2-factor E_2, $E_1 \cap E_2 = \emptyset$, $E_1 \cup E_2 = E$.

This theorem is used in the following as well as the well-known fact that for every graph $\alpha(G) \cdot (\Delta(G) + 1) \geq n = |V|$ and two facts about circle and permutation graphs.

Circle graphs are the intersection graphs of chords in a circle (cf.[Gol 80]).

Permutation graphs are the intersection graphs of chords between two parallel lines (cf.[Gol 80]).

Let C be a circle with a set M of chords. A chord $x \in M$ is an *equator* iff x intersects all chords $y \in M \setminus \{x\}$.

Theorem 1.2 ([Gol 80])

A circle graph G is a permutation graph iff G admits a circle/chord representation with an equator.

We need only the fact that a circle graph with an equator is a permutation graph.

A subset $V' \subseteq V$ is a *clique* in G iff for all $u, v \in V'$, $u \neq v$, $uv \in E$ holds.

Let $\omega(G)$ be the maximum clique size in G.

Let $G(V')$ be the subgraph of G induced by V'.

For the notion of perfect graphs cf.[Gol 80]:

A graph $G(V, E)$ is *perfect* iff for all $V' \subseteq V$ $\alpha(G(V')) \cdot \omega(G(V')) \geq |V'|$ holds.

Permutation graphs are known to be perfect whereas circle graphs are in general not perfect.

2. The main results

Let G be a cubic bridgeless graph. Then a cycle C of the 2–factor together with the edges of the 1–factor with both endpoints in C forms a circle graph.

Theorem 2.1:

Every cubic bridgeless graph $G = (V, E)$ contains at least $\frac{1}{256} \cdot n^{1/7}$ pairwise disjoint cycles of length at most $2 \cdot n^{6/7}$.

Proof:

We first apply Petersen's Theorem. Let E_1 be the 1–factor and E_2 be the 2–factor of G.

Let $C_1, C_2, \ldots, C_k$ be the (pairwise disjoint) cycles of E_2.

Recall that E_1 is a perfect matching of $G : |E_1| = \frac{n}{2}$.

In the following we denote the edges of E_1 by m–edges. We show the theorem by a distinction of cases:

(1) $k \geq 2 \cdot n^{1/7}$.

Then the length of at most $n^{1/7}$ cycles of $C_1, \ldots, C_k$ is greater than $n^{6/7}$ i.e. the length of at least $n^{1/7}$ cycles of $C_1, \ldots, C_k$ is bounded by $n^{6/7}$.

(2) $k < 2 \cdot n^{1/7}$.

(2.1) For at least $\frac{n}{4}$ m–edges xy the endpoints x and y are in the same cycle of $C_1, \ldots, C_k$.

Since $k < 2 \cdot n^{1/7}$ there is an $i, 1 \leq i \leq k$, with at least $\frac{n}{4}/2n^{1/7} = \frac{1}{8} \cdot n^{6/7}$ m–edges within C_i.

Let $E_{1,i}$ denote the m–edges within C_i.

Then $E_{1,i}$ defines a circle graph $G_i = (E_{1,i}, E_i)$ with vertex set $E_{1,i}$ and edges $ee' \in E_i$ iff e and e' cross each other in C_i.

(2.1.1) $\alpha(G_i) \geq n^{2/7}$.

This means that there are $n^{2/7}$ m–edges within C_i which pairwise do not cross each other.

(2.1.1.1) There are at least $n^{1/7}$ m–edges e for which the cycle defined by e and one of the 2 parts of C_i given by the endpoints of e on C_i contains no other m–edges:

We call these cycles peripheral cycles.

Figure 1:

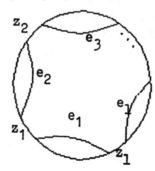

Then there are $n^{1/7}$ pairwise disjoint cycles. The length of at least one half of them is bounded by $2 \cdot n^{6/7}$.

(2.1.1.2) Less than $n^{1/7}$ m–edges form peripheral cycles. Then there are less than $n^{1/7}$ intervals z_j, $1 \leq j \leq l$ between peripheral cycles. Thus there is an interval z_j which contains endpoints of at least $n^{1/7}$ parallel m–edges in C_i.

Figure 2:

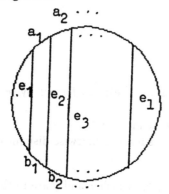

Consider the cycles formed by $\{e_1, a_1, b_1, e_2\}$, $\{e_2, a_2, b_2, e_3\}$,.... These are at least $n^{1/7}$ cycles. The length of at least one half of them is bounded by $2n^{6/7}$.

Now as a subset of pairwise disjoint cycles choose every second of these cycles – these are at least $\frac{1}{4}n^{1/7}$ cycles.

(2.1.2) $\alpha(G_i) < n^{2/7}$.

The circle graph G_i has at least $\frac{1}{8}n^{6/7}$ vertices. Let Δ_i be the maximum degree in G_i.
Then $\alpha(G_i) \cdot (\Delta_i + 1) \geq \frac{1}{8}n^{6/7}$ i.e.
$\Delta_i \geq \frac{1}{8}n^{4/7}$.
Let $e \in E_{1,i}$ be an m–edge of maximum degree in G_i.
Then the neighbourhood $N_i(e)$ of e in G_i is a set of m–edges which cross e i.e. according to Theorem 1.2 $N_i(e)$ induces a permutation graph in G_i. But permutation graphs are perfect and thus for $G_i(N_i(e))$
$\alpha(G_i(N_i(e))) \cdot \omega(G_i(N_i(e))) \geq |N_i(e)| \geq \frac{1}{8}n^{4/7}$ holds.
Since $\alpha(G_i(N_i(e))) \leq \alpha(G_i) < n^{2/7}$
we have $\omega(G_i(N_i(e))) \geq \frac{1}{8}n^{2/7}$ i.e. there are $\frac{1}{8}n^{2/7}$ chords (m–edges) in C_i which cross each other:

Figure 3:

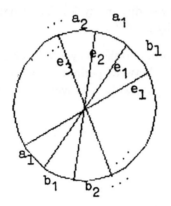

Now analogously to case (2.1.1.2) consider the cycles formed by $\{e_1, a_1, e_2, b_1\}$, $\{e_2, a_2, e_3, b_2\}, \ldots$

At least one half of them has length at most $2n^{5/7}$.

Taking every second of them yields a subset of pairwise disjoint cycles.

(2.2) There are less than $\frac{n}{4}$ m-edges $\{x, y\}$ for which x and y are in the same cycle C_i.

Thus there are at least $\frac{n}{4}$ m-edges with endpoints in different cycles C_i, C_j.

Since the number k of cycles C_i is bounded by $k < 2n^{1/7}$ there are two cycles C_i, C_j for which at least $\frac{n}{4}/(2n^{1/7} \cdot 2n^{1/7}) = \frac{1}{16}n^{5/7}$ m-edges have their endpoints in C_i and C_j.

Let e_1, e_2 be two of these edges which divide C_i into A_i, B_i and C_j into A_j, B_j:

Figure 4:

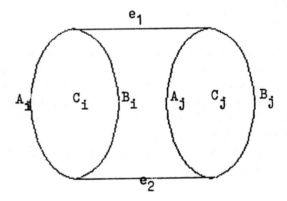

Now consider the 4 cycles $D_1 = A_i e_1 A_j e_2$, $D_2 = A_i e_1 B_j e_2$, $D_3 = B_i e_1 A_j e_2$, $D_4 = B_i e_1 B_j e_2$.

At least $\frac{1}{4}$ of the $\frac{1}{16}n^{5/7}$ m-edges with endpoints in C_i and C_j are chords of one of the cycles D_i, $i = 1, \ldots, 4$.

Thus case (2.2) reduces to (2.1) with $\frac{1}{64} \cdot n^{5/7}$ chords instead of $\frac{1}{8}n^{6/7}$ as in case (2.1).

Case (2.2.1) is as case (2.1.1) with the same estimation. Case (2.2.2) is slightly different from (2.1.2) since in this case the circle graph G_i has only at least $\frac{1}{64}n^{5/7}$ vertices.

This means that $\Delta_i \geq \frac{1}{64} \cdot n^{3/7}$ and $\omega(G_i(N_i(e))) \geq \frac{1}{64} \cdot n^{1/7}$.

This leads to at least $\frac{1}{256}n^{1/7}$ pairwise disjoint cycles of short length in case (2.2.2).

$\square$

Now to the more general case of useful subgraphs. Recall that a subgraph (V', E') is useful if $|V'| + 2 = |E'|$. Furthermore let us call cycles C *short* if if $|C| \leq n^{13/14}$ and *very short* if $|C| \leq n^{6/7}$.

Obviously the union of two vertex-disjoint subgraphs $(V_1', E_1'), (V_2', E_2')$ with $|V_i'| + 1 = |E_i'|, i = 1, 2$ yields a useful subgraph. Thus it is sufficient to estimate the number of such subgraphs:

Theorem 2.2:

Let $G = (V, E)$ be a cubic bridgeless graph. Then G contains at least $\frac{1}{16}n^{1/14}$ pairwise disjoint subgraphs (V', E') with $|V'| + 1 = |E'|$ and at most $3n^{13/14}$ vertices in V'.

Proof:

As in the proof of Theorem 2.1 let $C_1, C_2, \cdots, C_k$ be the cycles in the 2-factor E_2. We distinguish the following cases:

(1) $k \geq 2n^{1/7}$.

At least one half of these cycles have length at most $n^{6/7}$.

(1.1) There are at least $\frac{1}{2}n^{1/7}$ very short cycles C_i with at least one chord (from E_1). Then the assertion is fulfilled.

(1.2) There are less than $\frac{1}{2}n^{1/7}$ very short cycles C_i with a chord. Then at least $\frac{1}{2}n^{1/7}$ very short cycles have at least one m-edge to other cycles of E_2 since G is cubic. Denote this set of cycles by C_1. For each cycle C from C_1 choose one of the m-edges incident to C. Denote this set of m-edges by E_1':

$$|E_1'| \geq \frac{1}{2}n^{1/7}.$$

(1.2.1) At least $\frac{1}{4}n^{1/7}$ m-edges from E_1' connect the very short cycles from (1.2) with short cycles. Let E_1'' denote the set of these m-edges, $E_1'' \subseteq E_1'$, $|E_1''| \geq \frac{1}{4}n^{1/7}$.

(1.2.1.1) To every short cycle at most $n^{1/14}$ m-edges from E_1'' are incident. Choose one m-edge $e_1 \in E_1''$ and the two cycles C_1, C_2 incident to e_1:

Figure 5:

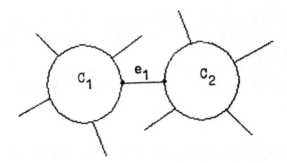

Delete C_1, C_2, e_1 and all m-edges from E_2'' incident to C_1, C_2. Since $|E_1''| \geq \frac{1}{4}n^{1/7}$, but every deletion decreases E_1'' only by $2n^{1/14} + 1$ this choice can be repeated at least $\frac{1}{8}n^{1/14}$ times.

Thus there are at least $\frac{1}{8}n^{1/14}$ vertex-disjoint-subgraphs (V', E') with $|V'| + 1 = |E'|$ with at most $2n^{13/14}$ vertices.

(1.2.1.2) There is a cycle C with $> n^{1/14}$ m-edges from E_1'' incident to C:

Figure 6:

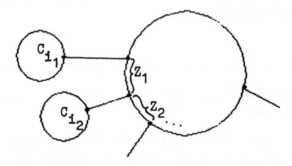

Thus there are $> n^{1/14}$ segments Z_i. At least one half of them has length $\leq n^{13/14}$. Choose every second of them. If the cycles C_{ij}, C_{ij+1} from C_1 corresponding to segment z_{ij} are different then one has a subgraph with $\leq 3 \cdot n^{13/14}$ vertices.

Figure 7:

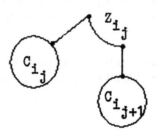

According to the choice of E_1' C_{ij}, C_{ij+1} must be different since E_1' contains only one m-edge for each cycle C from $\mathcal{C}_1$.

Thus there are $\geq \frac{1}{4}n^{1/14}$ vertex-disjoint subgraphs (V', E') with $|V'|+1 = |E'|$ and $|V'| \leq 3n^{13/14}$.

(1.2.2) Less than $\frac{1}{4}n^{1/7}$ m-edges from E_1' connect with short cycles. Then at least $\frac{1}{4}n^{1/7}$ m-edges from E_1' connect to cycles with length $> n^{13/14}$ but of this length there are only $\leq n^{1/14}$ cycles in E_2. Therefore there is a cycle C in E_2 with $\geq \frac{1}{4}n^{1/14}$ incident m-edges from E_1'.

Analogously to case (1.2.1.2) the assertion is fulfilled.

For case (2) the distinction of cases is very similar to that of the proof of Theorem 2.1. We use also the same notions as in that proof.

(2) $k < 2n^{1/7}$

(2.1) At least $\frac{n}{4}$ m-edges from E_1 have their endvertices x, y within the same cycle. Thus there is a cycle C_i of E_2 with $\geq \frac{1}{8}n^{6/7}$ m-edges within C_i.

(2.1.1) $\alpha(G_i) \geq n^{2/7}$.

(2.1.1.1) There are at least $n^{1/7}$ peripheral m-edges in C_i. At least $\frac{1}{2}n^{1/7}$ peripheral cycles have length $\leq 2n^{6/7}$. Now consider only these.

Let $Z_1', \cdots, Z_j'$, $j \geq \frac{1}{2}n^{1/7}$, denote the segments of C_i; between the peripheral cycles:

Figure 8:

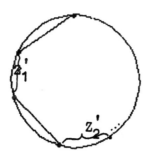

At least one half of the Z_i' $i.e.$ $\geq \frac{1}{4}n^{1/7}$ of them have length $\leq 4 \cdot n^{6/7}$.

Choose every second of them and the peripheral cycles incident to them. Thus there are $\geq \frac{1}{8}n^{1/7}$ pairs of the form

Figure 9:

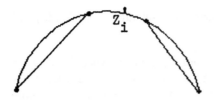

with $\leq 2n^{6/7} + 2n^{6/7} + 4n^{6/7}$ vertices.

(2.1.1.2) There are less than $n^{1/7}$ peripheral m-edges in C_i. Thus there is a segment Z_j between two peripheral m-edges with at least $n^{1/7}$ m-edges incident to Z_j i.e. parallel m-edges.

Now proceed as in the proof of Theorem 2.1 (cf. Figure 2). Let a_i' denote the segments on C_i on the a_i-side between two very short cycles $\{e_i, a_i, b_i, e_{i+1}\}$. At least one half of these segments a_i' i.e. $\geq \frac{1}{8}n^{1/7}$ of them has length $\leq 8 \cdot n^{6/7}$. Choose every second of them and the corresponding cycles.

Figure 10:

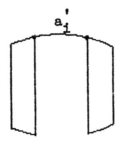

Thus the assertion is fulfilled.

(2.1.2) $\alpha(G_i) < n^{2/7}$.

Analogously to the proof of Theorem 2.1, (2.1.2) there are at least $\frac{1}{4}n^{2/7}$ pairwise crossing m-edges in C_i.

Now consider again the cycles $\{e_i, a_i, e_{i+1}, b_i\}$ (cf. Figure 3). These are at least $\frac{1}{8}n^{2/7}$ cycles. Choose every second of them for vertex-disjointness. At least one half of them i.e. at least $\frac{1}{32}n^{2/7}$ cycles have length $\leq 32n^{5/7}$.

As in case (2.1.1.2) denote the segments between two such cycles on the a_i-side of C_i by a'_i. At least one half of them i.e. $\geq \frac{1}{64}n^{2/7}$ have length $\leq 64n^{5/7}$.

Choose every second of them. Thus there are $\geq \frac{1}{128}n^{2/7}$ subgraphs of the form

Figure 11:

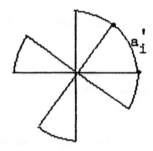

with at most $128n^{5/7}$ vertices.

(2.2) Less than $\frac{n}{4}$ m-edges $xy \in E_1$ have x, y in the cycle. Thus as in the proof of Theorem 2.1 there is a cycle $D_i, i \in \{1, 2, 3, 4\}$ with $\geq \frac{1}{64}n^{5/7}$ m-edges in D_i (cf. Figure 4 and the definition of D_i). W.l.o.g. let D_1 be this cycle.

(2.2.1) The case $\alpha(G(D_1)) \geq n^{2/7}$ can be treated as the case (2.1.1) of this proof.

(2.2.2) $\alpha(G(D_1)) < n^{2/7}$:

As in the proof of Theorem 2.1 then there is a vertex e of degree $\frac{1}{64}n^{3/7}$ in $G(D_1)$. Thus there are at least $\frac{1}{64}n^{1/7}$ pairwise crossing m-edges (instead of $\frac{1}{8}n^{2/7}$ such edges in case (2.1.2) of this proof). Thus all estimations of case (2.1.2) of this proof are valid if we replace $\frac{1}{8}n^{2/7}$ with $\frac{1}{64}n^{1/7}$. Thus there are $\geq \frac{1}{256}n^{1/7}$ vertex-disjoint subgraphs (V', E') with $|V'|+1 = |E'|$ and size $\leq 256n^{6/7}$.

$\square$

Corollary 2.1

Let $G = (V, E)$ be a bridgeless graph with $\deg(v) \in \{2, 3\}$ for all vertices $v \in V$.

Let c be a constant with $c > 0$ and
$$\frac{|\{v:\deg(v)=3\}|}{|\{v:\deg(v)=2\}|} \geq c > 0.$$
Then the following holds:

(i) $g(G) \leq c'n^{6/7}$ for a constant c'.

(ii) there is a useful subgraph in G of size $\leq c''n^{13/14}$ for a constant c''.

Proof:

Recall that $n_i = |\{v : \deg(v) = i\}|$. First contract in G all paths with vertices of degree 2 to one edge. The resulting graph G' is cubic and therefore Theorem 1 applies to G'. If $\frac{n_3(G)}{n_2(G)} \geq c$ then $\frac{n_3(G)}{n-n_3(G)} \geq c$ i.e.

$n_3(G) \geq \frac{c}{1+c} \cdot n$. Let $d = \frac{c}{1+c}$.

Then according to Theorem 1 G' has at least $\frac{1}{256} \cdot d^{1/7} n^{1/7}$ vertex–disjoint cycles of length $\leq 2 \cdot n^{6/7}$. Let $e = \frac{1}{256} \cdot d^{1/7}$.

Now we insert again the vertices of degree 2 into G'. Thus $n_2(G)$ vertices are distributed over $e \cdot n^{1/7}$ cycles.

Thus $g(G) \leq c' \cdot n^{6/7}, c' = 2 + \frac{1}{e}$.

The same argument holds for ii) with $\frac{1}{16} d^{1/7} \cdot n^{1/14}$ pairwise disjoint subgraphs instead of $\frac{1}{256} d^{1/7} \cdot n^{1/7}$ pairwise vertex-disjoint cycles.

$\square$

Acknowledgement:

I am grateful to F. Meyer auf der Heide for stimulating discussions on this topic.

References:

[Gol 80]   M.C.Golumbic, Algorithmic Graph Theory,
            Academic Press 1980

[Hr Mo 90] J. Hromkovic, B. Monien, The Bisection Problem for
            Graphs of Degree 4
            (Configuring Transputer Systems), manuscript 1990

[Pet 1891] J. Petersen, Die Theorie der regulären
            Graphen, Acta Mathematica 15 (1891), 193 - 220

List of Participants

Marc Andries
Universität Antwerpen
Universiteitsplein 1
B-2610 Antwerpen, Belgium

Dr. Luitpold Babel
Mathematisches Institut der TU München
Postfach 20 24 20
W-8000 München 2, Germany

Prof. Dr. Gunter Bär
Ernst-Moritz-Arndt-Universität
Fachrichtungen Mathematik/Informatik
Friedrich-Ludwig-Jahn-Str. 15a
O-2200 Greifswald, Germany

Dr. Ulrike Baumann
Pädagogische Hochschule Dresden
Institut für Mathematik
Wigardstr. 17
O-8060 Dresden, Germany

Dipl.-Inform. Ludwig Bayer
Fakultät für Informatik, UniBw München
Werner-Heisenberg-Weg 39
W-8014 Neubiberg, Germany

Dr. Rudolf Berghammer
Fakultät für Informatik, UniBw München
Werner-Heisenberg-Weg 39
W-8014 Neubiberg, Germany

Dr. Hans L. Bodlaender
Dept. of Computer Science
Utrecht University, Padualaan 14
3584 CH Utrecht, The Netherlands

Dr. Heidemarie Bräsel
Institut für Mathematische Optimierung
TU Magdeburg, Postfach 4120
O-3010 Magdeburg, Germany

Prof. Dr. Andreas Brandstädt
FB Mathematik / Fachgruppe Informatik
Universität / Gesamthochschule Duisburg
Postfach 10 15 03
W-4100 Duisburg 1, Germany

Dr. Klaus Brokate
Rübezahlweg 20
W-7250 Leonberg, Germany

Mathias Bull
Institut für Mathematik, PH Güstrow
Goldberger Straße 12
O-2600 Güstrow, Germany

Prof. Dr. Bruno Courcelle
Bordeaux University
Département d'Informatique
351, Cours de la Libération
F 33405 Talence Cedex, France

Dr. Peter Damaschke
Theoretische Informatik
Fernuniversität Hagen
Postfach 940
W-5800 Hagen, Germany

Ellen Farnbacher
Mathematisches Institut der TU München
Postfach 20 24 20
W-8000 München 2, Germany

Angelika Franzke
Universität Koblenz-Landau
Rheinau 3-4
W-5400 Koblenz, Germany

Dr. Rudolf Freund
Technische Universität Wien
Institut für Computersprachen
Karlsplatz 13
Wien, Austria

Dr. Kay Gürzig
Max-Planck-Ring 10 (E205)
O-6300 Ilmenau, Germany

Dr. Annegret Habel
FB Mathematik und Informatik
Universität Bremen, Postfach 33 04 40
W-2800 Bremen 33, Germany

Brigitte Haberstroh
Institut für Informationssysteme
Technische Universität Wien
Paniglgasse 16
1040 Wien, Austria

Dr. Ulrich Huckenbeck
Lehrstuhl für Informatik I
Universität Würzburg
Am Hubland
W-8700 Würzburg, Germany

Edmund Ihler
Institut für Informatik
Universität Freiburg
Rheinstr. 10-12
W-7800 Freiburg, Germany

Dr. X.Y. Jiang
Institut für Informatik
und angewandte Mathematik
Universität Bern, Länggassstr. 51
CH-3012 Bern, Switzerland

Wolfram Kahl, M. Sc.
Fakultät für Informatik, UniBw München
Werner-Heisenberg-Weg 39
W-8014 Neubiberg, Germany

Doc. dr. hab. Michał Karoński
Dept. Discrete Mathematics
Adam Mickiewicz University
Matejki 48/49
60-769 Poznan, Poland

Dipl.-Math. Andreas Kasper
Fakultät für Informatik, UniBw München
Werner-Heisenberg-Weg 39
W-8014 Neubiberg, Germany

Dipl.-Inform. Peter Kempf
Fakultät für Informatik, UniBw München
Werner-Heisenberg-Weg 39
W-8014 Neubiberg, Germany

Dr. Ralf Klasing
FB 17, Universität Paderborn
Warburger Str. 100
W-4790 Paderborn, Germany

Prof. Dr. Walter Knödel
Institut für Informatik
Universität Stuttgart, Azenbergstr. 12
W-7000 Stuttgart, Germany

Dr. Manfred Koebe
Ernst-Moritz-Arndt-Universität
FR Mathematik/Informatik
F.-L.-Jahnstr. 15a
O-2200 Greifswald, Germany

Luděk Kučera
Faculty of Mathematics and Physics
Charles University
Department of Applied Mathematics
Malostranske nam. 25,
118 00 Praha 1, Czechoslovakia

Sabine Kuske
Falkenstr. 24
W-2800 Bremen 1, Germany

Nicole Vercueil-Lafaye de Micheaux
Dépt. de Mathematiques (URA 225)
Case 901- Faculté des Sciences de Luminy
13288 Marseille Cedex 9, France

Dr. Zbigniew Lonc
Instytut Matematyki
Politechnika Warszawska
pl. Politechniki 1
00-661 Warszawa, Poland

Yuh-Dauh Lyuu
NEC Research Institute
4 Independence Way
Princeton, NJ 08540, USA

Alberto Marchetti-Spaccamela
Universita degli Studi di Roma "La Sapienza"
Dipartimento di Informatica e Sistemistica
Via Eudossiana, 18
00184 Roma, Italia

Prof. Dr. Ernst W. Mayr
Johann-Wolfgang-Goethe-Universität
FB 20/Informatik
Robert-Mayer-Str. 11-15
W-6000 Frankfurt am Main 11, Germany

Karsten Mekelburg
Karl-Liebknecht-Straße 132
O-7030 Leipzig, Germany

Satoru Miyano
Research Institute
of Fundamental Information Science
Kyushu University 33
Fukuoka 812, Japan

Prof. Dr. Rolf H. Möhring
TU Berlin, FB Mathematik
Straße des 17. Juni 136
W-1000 Berlin 12, Germany

Dr. Mohamed Mosbah
Bordeaux University
Département d'Informatique
351, Cours de la Libération
F 33405 Talence Cedex, France

Prof. Dr. Hartmut Noltemeier
Lehrstuhl für Informatik I
Universität Würzburg, Am Hubland
W-8700 Würzburg, Germany

Yasuyoshi Okada
Koyama Research Group
Information Science Research Laboratory
NTT Basic Research Laboratories
3-9-11 Midori-Cho, Musashino-shi
Tokyo 180, Japan

Christiane Rambaud
Dépt. de Mathematiques (URA 225)
Faculté des Sciences de Luminy
Case 901 - 163, Avenue de Luminy
13288 Marseille Cedex 9, France

Dr. Thomas Roos
Lehrstuhl für Informatik I
Universität Würzburg, Am Hubland
W-8700 Würzburg, Germany

Gisbert Rostock
Adenauerstr. 5
W-8039 Puchheim, Germany

Dr. Nicola Santoro
School of Computer Science
Herzberg Building, Carleton University
Ottawa K1S 5B6 1, Canada

Dipl.-Inform. Franz Schmalhofer
Fakultät für Informatik, UniBw München
Werner-Heisenberg-Weg 39
W-8014 Neubiberg, Germany

Prof. Dr. Gunther Schmidt
Fakultät für Informatik, UniBw München
Werner-Heisenberg-Weg 39
W-8014 Neubiberg, Germany

Dipl.-Math. Joachim Schreiber
Fakultät für Informatik, UniBw München
Werner-Heisenberg-Weg 39
W-8014 Neubiberg, Germany

Dipl.-Inform. Andy Schürr
Lehrstuhl für Informatik III
RWTH Aachen, Ahornstr. 55
W-5100 Aachen, Germany

Sigrid Schumacher
Institut für Kristallographie
Universität Karlsruhe, Kaiserstr. 12
W-7500 Karlsruhe, Germany

Dr. Takayoshi Shoudai
Dept. of Control Engineering and Science
Kyushu Institute
Iizuka 820, Japan

Gheorghe Ştefanescu
Math. Inst. of Romanian Academy
Bd. Pacii 220
79622 Bucharest, Romania

Iain A. Stewart
Computing Laboratory
University of Newcastle upon Tyne
Newcastle NE1 7RU, Great Britain

Dr. Elena Stöhr
Karl-Weierstraß-Institut für Mathematik
Akademie der Wissenschaften, Postfach 1304
O-1086 Berlin, Germany

Dr. Thomas Ströhlein
Institut für Informatik der TU München
Postfach 20 24 20
W-8000 München 2, Germany

Ondrej Sýkora
Max-Planck-Institut für Informatik
Im Stadtwald
W-6600 Saarbrücken, Germany

Dr. Tadao Takaoka
Dept. of Computer Science
Ibaraki University
Hitachi, Ibaraki 316, Japan

Stefan Taubenberger
Wätjenstraße 84
W-2800 Bremen 1, Germany

Prof. Dr. Gottfried Tinhofer
Mathematisches Institut der TU München
Postfach 20 24 20
W-8000 München 2, Germany

Dr. Walter Unger
FB 17, Universität Paderborn
Warburger Str. 100
W-4790 Paderborn, Germany

Imrich Vrťo
Max-Planck-Institut für Informatik
Im Stadtwald
W-6600 Saarbrücken, Germany

Dr. Frank Werner
Fakultät für Mathematik
Institut für Mathematische Optimierung
TU Magdeburg, Postfach 4120
O-3010 Magdeburg, Germany

Prof. Dr. Tilmann Würfel
Fakultät für Informatik, UniBw München
Werner-Heisenberg-Weg 39
W-8014 Neubiberg, Germany

Christos Zaroliagis
Computer Technology Institute
P.O. Box 1122
26110 Patras, Greece

Albert Zündorf
Lehrstuhl für Informatik III
RWTH Aachen, Ahornstr. 55
W-5100 Aachen, Germany

List of Authors

Lecture Notes in Computer Science

For information about Vols. 1–481
please contact your bookseller or Springer-Verlag